c.1

551.576
M445c

Matveev, L. T. (Leonid Tikhonovich)
Cloud dynamics / L.T. Matveev ; [translated from the Russian by UAT Ltd.]. -- Dordrecht ; Boston : D. Reidel Pub. Co. ; Hingham, MA : Sold and distributed in the U.S.A. and Canada by Kluwer Academic Publishers, c1984.
x, 340 p. : ill. ; 25 cm. -- (Atmospheric sciences library)
Bibliography: p. 300-333.
Includes index.
ISBN 9027717370

WMRAxc 83-24439

CLOUD DYNAMICS

ATMOSPHERIC SCIENCES LIBRARY

Cloud Dynamics

by

L. T. MATVEEV
Leningrad Hydrometeorological Institute, Leningrad, U.S.S.R.

D. Reidel Publishing Company

A MEMBER OF THE KLUWER ACADEMIC PUBLISHERS GROUP

Dordrecht / Boston / Lancaster

Library of Congress Cataloging in Publication Data

CIP

Matveev, L. T. (Leonid Tikhonovich)
Cloud dynamics.

(Atmospheric sciences library)
Bibliography: p.
Includes index.
1. Cloud physics. I. Title. II. Series.
QC921.5.M388 1984 551.57'6 83-24439
ISBN 90-277-1737-0

Published by D. Reidel Publishing Company
P.O. Box 17, 3300 AA Dordrecht, Holland

Sold and distributed in the U.S.A. and Canada
by Kluwer Academic Publishers,
190 Old Derby Street, Hingham, MA 02043, U.S.A.

In all other countries, sold and distributed
by Kluwer Academic Publishers Group,
P.O. Box 322, 3300 AH Dordrecht, Holland

Translated from the Russian by UAT Ltd.

Printed in The Netherlands

TABLE OF CONTENTS

PREFACE

During recent decades a new field of study in atmospheric science has made its appearance - the dynamics of clouds. As the name implies, the subject matter of cloud dynamics includes the causes of cloud formation and the temporal development of clouds.

At first, effort was concentrated mainly on devising models of the structure and development of convective clouds, and thus there exists considerable literature on this [9, 69, 88, 330, 411].

Although convective clouds are of great significance (thunderstorm formation and very intense turbulence are associated with these clouds), they are observed much less frequently than other cloud types. For instance, the frequency of occurrence of stratiform (frontal) clouds and wave clouds over the U.S.S.R. and Western Europe is more than 90% [2-4].

During the last 20 or 30 years there has been considerable success in studying the dynamics of stratiform clouds. Fundamental laws (equations) describing the formation, development, and dissipation of these clouds (and also of fog) have been formulated, and also laws describing the formation of humidity and temperature fields in a turbulent medium. Hydrodynamic models of clouds and fog constructed on the basis of these equations have made it possible to formulate the fundamental regularities in the formation and evolution of large-scale cloud fields, and also to ascertain the structural features of clouds of various kinds. These topics are covered in Chapters 1-4 of this monograph.

The state-of-the-art in the dynamics of stratiform, or layer clouds is such that stratiform cloudiness can not only be modeled but even predicted quite accurately with the aid of numerical methods. The methods and approaches used for this, as well as some of the results obtained, are described in Chapter 5.

During the last 10-15 years great use has been made of satellite information for studying the structure of the atmosphere. Cloud studies carried out with phototelevision, infrared, and radiometric equipment mounted aboard artificial earth satellites have provided new cloud data and have stimulated the development of a theory capable of explaining the observed features of large-scale and mesoscale cloud structure. The approaches which have been worked out in this field are discussed in Chapters 4 and 6.

Chapters 7 and 8 are devoted to an analysis of the many experimental data on clouds, fog, and precipitation which have been gathered by various technical means (aircraft, satellites, radar, and radiometry).

The experimental material utilized in this monograph serves, on the one hand, to confirm (or refute) the predictions of theory, and, on the other hand, to pose new theoretical problems. It should be noted that today there are many more unsolved problems in cloud dynamics than solved problems.

The author is most grateful to Dr. M. V. Buikov, and Professors M. E. Shvets, and N. S. Shishkin, also to the following colleagues in the Department of Cloud Physics and Active Modification at the Main Geophysical Observatory, Yu. A. Dovgalyuk, A. V. Zinchenko, V. V. Klingo, and N. V. Toropova, for the numerous useful remarks concerning the manuscript of this monograph.

Finally, the author would like to mention, too, that at various times he was fortunate enough to be able to discuss problems in the physics and dynamics of the atmosphere (including those dealt with in this monograph) with I. A. Kibel', E. N. Blinova, G. I. Marchuk, S. P. Khromov, and A. Kh. Khrgian.

INTRODUCTION

Attempts to understand the nature of clouds and precipitation have a long history. However, as an independent field of study, cloud physics, which deals with the formation and evolution of clouds, fog, and precipitation, took shape only during the last 40 or 50 years. One of the main factors stimulating the development of cloud physics was the rise of aviation. Aviation called for a deeper insight into the conditions of cloud genesis and cloud formation, and airplanes and other aircraft became powerful tools for the study of clouds. It became possible not only to observe the diversity of cloud types, but also to obtain quantitative information about clouds: the heights of their boundaries, water content, sizes and numbers of water droplets and ice crystals, electrical discharges, vertical velocities, turbulent fluctuations, etc.

The formulation of a theory of water-vapor condensation, together with laboratory and airborne studies of clouds, fog, and precipitation, made it possible to lay the foundation for this branch of cloud physics, which has come to be known as the microphysics of clouds. Cloud microphysics, then, is the study of the conditions for the origin and growth of cloud and precipitation elements: water droplets and ice crystals. The present state of cloud microphysics has been described in recent books [390, 391, 412].

In addition to the study of these microphysical processes, the scale of which is comparable to the sizes of the individual cloud and precipitation particles, it is of theoretical and practical interest to consider processes culminating in the formation and evolution of individual clouds and cloud systems, and in the formation of a precipitation field. Processes on a scale ranging from hundreds of meters to hundreds or thousands of kilometers lie within the province of *cloud dynamics*. The foundation for this branch of cloud science was laid 25 or 30 years ago with the construction of the first models of cumulus, and then also stratus, clouds. Since vertical currents and turbulent exchange play important parts in cloud formation, considerable attention is given to studying the field of motion of the atmosphere, as well as to deriving systems of equations and developing methods for their solution.

A cloud is a product of condensation of water vapor, and its makeup and phase transitions are greatly influenced by temperature. Thus, it is quite natural that one of the main tasks of cloud dynamics is to investigate the fields of moisture and temperature. With regard to this, it should be remembered that clouds in a temperature field act as a unique kind of regulator with feedback: having originated as a result of a temperature variation, clouds exert during the course of their evolution a marked inverse effect (via heat fluxes arising from condensation, radiation, and turbulence) on the temperature field, and consequently on the field of motion of the atmosphere as well.

Clouds and precipitation have considerable effect on many aspects of human activity (especially in the sphere of agriculture), as well as on the flora and fauna of the Earth. Various forms of transport are seriously affected by clouds, fog, and precipitation.

Although taking cloudiness into account involves great theoretical and technical difficulties, in recent decades more and more progress has been made in devising weather-forecasting techniques, and models of the general circulation of the atmosphere and climatogenesis, which take into account the heat of condensation and the effect of clouds [18, 28-30, 122, 169, 170, 210, 221, 227, 270, 313, 388, 389, 415, 416, 421].

The development of cloud microphysics and cloud dynamics was preceded by experimental and synoptic-aerological research, carried out in the 1930's through 1950's at the Main Geophysical Observatory, the Central Aerological Observatory, the Central Weather Forecasting Institute, the Institute of Applied Geophysics, the Mozhaiskii Air Force Engineering Academy, and the Arctic Scientific-Research Institute. This research was continued somewhat later at the Ukrainian Scientific-Research Institute, the High-Mountain Geophysical Institute, the Institute of Electromechanics, the Transcaucasian Scientific-Research Institute, and elsewhere (Goskomgidromet), at the Institute of Atmospheric Physics (USSR Academy), and also at certain Soviet universities.

Especially important for cloud dynamics has been the development of a means of measuring the water content of clouds (V. A. Zaitsev) and the organization of aircraft soundings of the atmosphere at a network of stations. The latter were, unfortunately, only continued until 1968.

The first studies of the dynamics of stratiform clouds date from about 25 years ago [177, 179, 181, 182, 197]. In [179, 197], some very simple assumptions were made concerning the vertical velocity and the turbulence coefficient (they were taken to be constant), and a solution was obtained for the unsteady-state equation of moisture transfer for the entire troposphere, the vertical water-content profiles of clouds also being calculated. The boundary conditions were specified for the Earth's surface and the tropopause, and temperature distribution with height was assumed to be known (a linear profile was adopted for the calculations). The calculations showed that for the intensities typical of the troposphere, the vertical turbulent exchange is just as influential in cloud formation as are the ordered (large-scale) vertical currents. Let us mention here one more result obtained in [179, 197]: the analysis of the solution and the calculations revealed a fundamental difference between the vertical water-content distributions of, on the one hand, the Ns-As clouds, in which the maximum water content is observed in the lower part, and, on the other hand, Sc-Ac clouds, in which the maximum water content is shifted into the upper part. Subsequent measurements of the water content (by the station network and on expeditions) served to verify this theoretical prediction.

Attempts made, for instance, by Shvets [325] to eliminate the rate of water-vapor condensation from the system of equations describing the transfer of heat and moisture in a turbulent atmosphere and to obtain differential equations for the temperature T and the specific humidity q indicated that the direct method of solving this system involves serious difficulties. Not only are the equations for T and q extremely complicated (nonlinear), but also the unknown functions have to satisfy conditions at moving cloud boundaries. Another problem is that the moment when cloud formation begins, and thus the transition from a system describing a process in unsaturated moist air to a system describing this process in saturated air, are unknown.

The author was able to overcome these, and certain other, difficulties with the aid of a method, the basic premises of which was first described in 1953-1957 [179, 181-183]. The main premise sees the cloud elements (water droplets and ice crystals) as being quite effectively entrained by those air particles (moles) which participate in the turbulent exchange. Employment of the entrainment concept enabled us to introduce two invariants, the specific moisture content and the equivalent potential temperature, and also to construct for them differential equations which have the same form prior to and during the course of cloud formation.

These equations were widely used by the author and his coworkers [136-138, 186-193, 198-202, 206, 207] to model the conditions for the formation and evolution of stratiform clouds in moving cyclones, at frontal divisions, over an inhomogeneous underlying surface, etc. In these studies, considerable attention was devoted to determining cloud boundaries, water reserves of clouds, and the moisture content of the vertical column, as well as to establishing the relationships of the unknown quantities with the temperature and humidity of the air close to the Earth's surface and with the vertical velocity and the average (over the mass) falling velocity of the cloud elements.

A new period in the study of cloud dynamics began in the 1960's, when electronic computers became available for purposes of calculation and analysis. The number of factors taken into account could then be increased considerably, the equations of motion and continuity could be added to the system describing the transfer of heat and moisture, and techniques for predicting temperature, humidity, cloudiness, and precipitation could be developed [52, 135, 160-162, 184, 193]. During recent years, a great deal of interest has been shown in modeling the processes of water-vapor condensation taking place in the near-Earth and boundary layers of the atmosphere, accompanied by the formation of fog and low clouds. The first models of fog [162, 190] were constructed on the assumption that the turbulence coefficient was a known function of the height (described, for instance, by some exponential expression) [175, 185]. In the most recent models [199, 201, 202, 206] this dependence was derived from the equation for the balance of turbulent energy, taking into account the basic results of dimensional analysis.

The main conclusion which can be drawn from an analysis of the results of calculations is that the conditions for cloud formation are extremely diverse. The number of factors influencing the horizontal and vertical cloud dimensions, the water-content profiles, the spatial locations of cloud boundaries, and so forth, is very large indeed. However, it can be said that in the construction of cloud-formation models quite a satisfactory explanation has been obtained for the observed facts and structural features of stratiform cloud systems. The main conclusions can be formulated as follows.

(1) Vertical movements, advection, and turbulent exchange are the main factors influencing cloud formation, the contributions of all these factors to the variation in the temperature, humidity, and water content of the cloud, as well as the contribution of the heat of condensation liberated during cloud formation, being about equal.

(2) The time required for clouds to form, assuming the normally observed values of vertical velocity, relative humidity, and turbulent coefficient, fluctuates from several hours to tens of hours.

(3) For regions of lower pressure, ascending vertical movements are typical, as are gradually developing frontal clouds of types Ns-As-Cs with a maximum water content in the lower part of the cloud; during the cloud formation a temperature inversion is created under the clouds, or else a zone where the temperature drops more slowly with height (frontal zone).

(4) In regions of high pressure and in mature cyclones, descending vertical movements transform the clouds from frontal stratiform (Ns-As-Cs) to stratus, stratocumulus, and altocumulus with maximum water contents in the upper part. The clouds are transported by air currents from a cyclone region to an anticyclone region, at speeds which in the middle and upper troposphere are known to exceed those of the cyclone and the front. In the lower troposphere, the clouds are transformed in the same cyclone in which they came into being as it intensified. When, however, the cyclone begins to mature, the ascending movements are replaced by descending movements. Under the influence of the latter, the frontal clouds are also transformed into St-Sc-Ac. During the transformation the clouds become stratified, and temperature inversions appear under the clouds. The transformation and stratification of the clouds require a time which is of the same order of magnitude as the time of cloud formation (several hours to some tens of

hours). St-Sc clouds can also originate from fog (of any provenance), following a rise in the temperature of the underlying surface.

The experimental data presented in Chapter 7 indicate that in more than 80% of the cases St-Sc clouds are produced by a transformation of either clouds of other types or else fog. Temperature inversions outside the near-Earth layer originate primarily under the influence of clouds (above-cloud inversions of dynamic origin).

(5) The rate of variation of the heights of the upper and lower cloud boundaries differs substantially from the vertical velocity of the air mass, right up to a difference in signs (for ascending vertical movements the lower boundary of a developing cloud, as a rule, descends).

(6) With the aid of theory, a number of more subtle features of the structure of clouds and fog can be explained; in particular, the formation of a lifted temperature inversion in a fog and the transformation of fog into low cloudiness.

(7) The following play an important role in the formation of frontal divisions in the atmosphere: large-scale ascending vertical movements produced in the baric troughs under the influence of velocity convergence in the boundary layer, and the heat condensation liberated during the formation of frontal clouds.

(8) Large-scale vertical movements play a major part in varying the thermal stratification toward instability, and, consequently, they are instrumental in producing convective movements and clouds.

The accuracy with which the meteorological quantities are measured, especially the air temperature and humidity, is very important with regard to cloud forecasting. Typical variations in air temperature and dew point over a 1- or 2-day period, as clouds form and dissipate, are 2-30°C, according to both experimental data and calculations. If, as in [194], we stipulate that the measurement error of a certain quantity has to be an order of magnitude lower than its characteristic value, then we arrive at the following demand upon the measurement accuracy: the total (instrumental and methodological) error in measuring the air temperature and dew point in the troposphere cannot exceed 0.2-0.3°C. Using the same principle, the averaging of meteorological quantities should pertain to an area 30 x 30 km in size (along the horizontal), to a layer 100-300 m thick (along the vertical), and to a time of 5 to 10 minutes.

A large series of works on the modeling of cloud-formation processes was completed by Buikov and his coworkers [43-48, 250, 315]. Using, as a basis, the studies of Belyaev [19] and Levin and Sedunov [154, 155, 271], and taking into account the effect of turbulence on the formation of the cloud-droplet spectrum, Buikov [43-45] concentrated much attention on the equation describing the droplet-size distribution function $f(r_d)$. Let us cite some of Buikov's results. In [44] it was established that, when modeling turbulent motions (eddies) in clouds, not only does the randomness of the motion have to be taken into account, but the randomness of the initial state as well, together with the repeated formation and decay of eddies of comparatively small size. An analysis of the modeled droplet spectra indicated these to be comparatively stable, so that they can be approximated by a gamma distribution with an exponent ranging from 2 to 8, depending on the height and the time. The water-condensation coefficient has a sizeable effect on the droplet spectrum. Buikov obtained an equation for the distribution function in the two extreme cases: $\omega_t \gg \omega_c$ and $\omega_t \leqslant \omega_c$ (where ω_t and ω_c are, respectively, the characteristic frequencies of the turbulent fluctuations and the rate of condensation growth of a droplet). During the cloud-formation process a transition from the first case to the second may be observed; consequently, the equations have to be checked during modeling, and their applicability must be verified by comparison with observations.

The equation for $f\ (r_d)$ was derived in a more general form by Stepanov and his coworkers [77, 213, 284, 285] and by Voloshchuk and Sedunov [62-64, 272]. However, both in its general form and the limiting cases, the equation for the distribution function is very complicated. It contains a number of terms which have not been evaluated experimentally (and which, moreover, will probably not be evaluated in the foreseeable future, since their measurement requires the use of instruments and methods of extreme accuracy). Therefore, we shall focus here on evaluating various terms and on simplifying the equations obtained for typical conditions of cloud formation.

The following general conclusion can be drawn from past studies of cloud microphysics, taking turbulence into account, to the development of which Soviet scientists have made many contributions (in addition to those cited above, the works of Mazin [163-165] and Smirnov [275] should also be cited): turbulent mixing in clouds is one of the prime factors determining the spectra of droplets and large drops. It should be noted that the idea of the entrainment of cloud elements by turbulent moles is widely used when formulating the equation for the distribution function.

The studies by Buikov and Pirnach [46, 47] with regard to modeling the properties of stratiform clouds are very pertinent to the subject of the present monograph. In these works, the equation for the droplet-size distribution function and also Maxwell's equation for the rate of water-vapor condensation on a drop, are added to the equations of heat transfer and moisture transfer.

In a number of works, it has been shown that taking microstructure into account leads to practically the same results, with regard to the water-content field, the height of the cloud boundaries, and the times of cloud formation and dissipation, as does the method which in [46, 47] is called the hydrodynamic method or method of invariants. Let us consider this question in somewhat more detail. It is posed in connection with the fact that the condition for the saturation of water vapor is used in the method of invariants. In a cloud, however, a certain amount of supersaturation is actually involved:

where q and q_m are the actual and saturated specific humidity.

In many of the studies of Buikov and others, it was found that the maximum values of Δ do not exceed 0.004 (that is, 0.4%), whereas supersaturation is, as a rule, observed only in a thin layer (30 - 50 m thick) close to the lower boundary of the developing cloud. Supersaturation is essentially absent ($\Delta \approx 0$) everywhere else in the cloud. It is obvious that even for the maximum value of Δ the error in calculating the water content of a cloud using the formula

$$\Delta = \frac{q - q_m}{q_m} \tag{1}$$

of the method of invariants will not exceed a few percent (in (1), s is the specific moisture content). Actually, if it is not assumed that the air in the cloud is saturated, then the water content of the cloud δ should be calculated using the formulas

$$\delta = s - q$$

or

$$\delta = (s - q_m) + \Delta \cdot q_m. \tag{2}$$

Here, the first term is the water content determined using formula (1). It is, as a rule, 0.2 - 0.5 g/kg. The second term is the error in calculating δ according to formula (1). Since q_m usually ranges from 0.5 to 5 g/kg (at the temperatures observed in clouds), the error in determining δ, even for the maximum values of Δ and q_m, will be $\Delta \cdot q_m$ = 0.02 g/kg, that is, 4 - 10% of the value of δ obtained using (1).

Obviously, there are other factors which cause errors in the calculated water content of a cloud which are considerably greater than those caused by neglecting supersaturation. These include the errors in calculating the vertical velocity and the turbulence coefficient, the errors in the numerical integration of the equations, and the errors in determining the initial values of the air temperature and humidity and the other parameters. Consequently, the most important cloud characteristics are the water content and the boundaries of the cloud, and the actual formation of a cloud can be determined with the aid of the equations of the large-scale physics of clouds, i.e., the equations of heat and moisture transfer in the atmosphere, with an accuracy which is quite acceptable from a theoretical point of view, and all the more so in reality. If we wish to determine simultaneously the vertical and horizontal velocity components, and also the turbulence coefficient, then we have to supplement these equations with equations of motion, continuity, radiation transfer, and turbulent-energy balance.

The need to include the equation for the droplet-size distribution function arises when the question of the evolution of the cloud-element spectrum is posed. The weighted-mean (over the mass) velocity of descent of the droplets has to be known in order to calculate the precipitation intensity. If the falling velocity is related, for instance, to the water content [150], then there is no need to include the equation for $f\cdot(r_\mathrm{d})$ when solving the precipitation problem either.

In the more general case, when the system of equations is used to determine not only the cloud characteristics, but also the velocity components, this system can (and should) be solved in three stages: (1) the components of the wind velocity are found from the equations of motion and continuity; (2) the air temperature and the cloud water content are determined from the equations of heat and moisture flux; (3) the microphysical characteristics of the cloud are found from the equation for the droplet-size distribution function (for a known water content). The acceptable error for such a divided approach to solving the system of equations must not exceed 5-10%; this was demonstrated in [148] for stratiform clouds and in [130] for cumulonimbus. The water content of the latter cloud type is the highest, but even so calculation of dynamical and microphysical factors leads to corrections (in the water content and temperature) amounting to only a few percent.

It should be noted in this regard that, given the present accuracy of measurements and studies of the physical mechanisms of cloud formation (as well as of all other meteorological fields), we can hardly hope for an accuracy of more than 10% when modeling atmospheric phenomena and processes. The divided approach to solving the system of equations is wholly necessary for cloudiness parametrization in problems of weather forecasting and constructing models of the general circulation of the atmosphere and the climate of the Earth. If such an approach is not used, the solution of the system of equations becomes so laborious that present-day computers could not possibly cope with it.

However, another serious difficulty crops up in cloud dynamics, namely that clouds are very sensitive to variations in the air temperature and humidity. Thus, such variations have to be measured and calculated with great precision.

An interesting conclusion was drawn by Buikov and Pirnach [46, 47], within the framework of a one-dimensional nonstationary model, concerning fluctuations of the cloud microstructure and intensity in time, and concerning their dependence on the mechanism of ice formation, the vertical velocity, and the turbulent exchange. It is shown that the modeled droplet spectra for mixed clouds are narrower than the spectra for water clouds. Taking microstructure into account has a very marked effect on precipitation intensity. But the amount of precipitation over a 10 or 20 hour period, estimated using microphysical and other methods (that of A. F. Dyubyuk, for instance) turns out to be practically the same.

Recently, Buikov and Khvorost'yanov [48, 315] modeled low clouds and fog. To the equations of heat and moisture transfer and the equation for the distribution function, they added the equations of steady uniform motion (along the horizontal), continuity, turbulent-energy balance, and long-wave radiation transfer. The numerical model derived from this system of equations reproduces a "statistical-mean" layer cloud, the properties of which have been studied experimentally in considerable detail [79]. It was concluded, interestingly enough, that the lower part of the cloud in the boundary layer can exist in the absence of saturation of the water vapor. It can also be concluded from the studies of Buikov that turbulent exchange and vertical movements play a major role in varying the temperature and water content of the cloud.

Tkachenko and his coworkers [139, 298-301] made a detailed study of the behavior of invariants Π and s, introduced in [179, 181], as well as of the lapse rate and the vertical specific-humidity gradient. They obtained relations for the rate of change of height with time of the cloud's lower boundary, which formed a basis for the short-range forecasting of clouds.

A few remarks about the studies of Buikov and his coworkers will be in order here. Their work has brought together, into a single system, equations describing processes with considerably different scales. On the one hand, these are the equations of heat and moisture transfer, describing the fields of temperature and humidity and the cloud-formation process (the water-content field), the scales of which measure hundreds or thousands of kilometers along the horizontal, hundreds of meters or kilometers along the vertical, and up to tens of hours in time. During numerical integration of these equations, the corresponding intervals are taken to be equal to about one tenth of the indicated scales. On the other hand, we have Maxwell's equation and the equation for the distribution function, describing the growth of an individual drop and the formation of the droplet-size spectrum. The scale of this process (the droplet size) ranges from fractions of a micron to several millimeters (if the problem is continued up to precipitation), that is, many orders of magnitude smaller than the scale of the cloud-formation process. It is clear that the numerical techniques for solving the equations of the first and second group will be quite different, and cannot in general be applied together.

Moreover, the structure of the equations of heat and moisture transfer is such that the rate and heat of condensation cannot be written in any form; they can always be eliminated from these equations, if invariants Π and s are introduced. Quantities characterizing droplet growth no longer enter into the equations for the invariants. Therefore, the water content, temperature, and humidity can be calculated to a higher degree of accuracy using these equations, and incomparably less computing time is needed.

The microphysical technique has, however, one illogicality:

Maxwell's equation describes the steady-state process of water-vapor condensation on a drop (the air humidity does not change in time), whereas in the equations for the distribution function and the transfer of heat and moisture, non steady-state terms are retained.

Although these shortcomings may affect some of the results (in particular, we might cite the excessively high variability in time of the water content, the precipitation intensity, etc.: in some instances, the temperature increments at levels which are close to one another may differ considerably), the modeling of cloud-formation processes with the aid of the microphysical equations has still, on the whole, helped to identify and elucidate a number of new regular features of cloud structure.

In this monograph, a great deal of attention will be given to the experimental material which has so far been gathered via radiosonde and aircraft probing of the atmosphere (with station networks and under expedition conditions). With the aid of airborne laboratories equipped with instruments for measuring cloud water content, air temperature and humidity, sizes and numbers of droplets, visibility, etc., numerous data have been obtained on such things as cloud-boundary heights, cloud thicknesses, vertical profiles of cloud water content, air temperature, and air humidity, the spatial distributions and time variations of these quantities, and the characteristics of turbulent exchange in clouds.

On the one hand, these data enable a more precise formulation of the problems of cloud formation: the order of magnitude of various terms in the initial system of equations can be evaluated, coefficients in the equations can be selected, boundary and initial conditions can be specified, etc. The quality of any given model is determined by comparing the theoretical results with experimental data. Clearly, a model will be acceptable only if it is able to explain the most important regularities of the phenomenon.

On the other hand, the model can be used to predict new regularities which have not yet been detected experimentally. In such cases, experiments should be designed to test the theoretical predictions.

Great interest is now being shown in statistical characteristics of clouds and fog, as well as of fields of temperature, humidity, and precipitation, established by analyzing large volumes of data from sounding of the atmosphere with aircraft, radiosondes, and radar, and also from satellite observations. In this monograph, the results of many such studies will be cited [1-4, 6, 11, 12, 14, 23, 24, 27, 35-38, 41, 42, 50, 51, 54-56, 59, 60, 66-68, 73-76, 78, 84-87, 90, 91, 94, 105, 107, 123, 124, 127, 131, 140-146, 156, 159, 166, 167, 195, 196, 203-205, 214-216, 218, 224-226, 235, 237, 243, 251-264, 266, 274, 277, 281, 282, 290, 291, 293-297, 302-304, 308, 320-322].

The vertical profiles of the fluxes and influxes of long-wave radiation in the presence of one or several cloud layers were considered by Feigel'son and his coworkers [248, 311, 312], and also by Herman and Goody [364]. In these papers, the radiation properties of clouds have been investigated in conjunction with the dynamics of their development: variation of boundary height, and vertical profiles of water content, temperature, and humidity. This became possible once the equations of radiation transfer had been supplemented by the equations for the flux of heat and moisture in a turbulent atmosphere, and once the author's method described in [179, 181, 394], had been invoked to solve them.

Dmitrieva-Arrago et al. [78, 93] evaluated the effect of cloudiness (for arbitrary amounts of clouds) on the fluxes and influxes of short-wave (solar) and long-wave (terrestrial) radiation. They made great use of the ideas behind the models of the general circulation of the atmosphere and climate of the Earth proposed by Smagorinsky, Manabe, and others, [388, 389, 421]. In these models, the radiation flux arriving at a given level z from above is the sum of the flux from the part of the atmosphere not covered by clouds and the fluxes from clouds located in upper, middle, and lower decks. The fluxes from clouds are assumed to be proportional to the amount of cloudiness visible from below, from level z. The transformation of fluxes in the layer between the lower cloud boundary and the given level is taken into account here.

The radiation flux arriving at level z from below is the sum of the flux from the Earth's surface (taking into account the portion of cloudless sky) and the fluxes from clouds, which are proportional to the amount of cloudiness visible from above, from level z. The radiation from clouds of the lower and middle decks is considered to be blackbody radiation, while that from clouds of the upper deck is assumed to be graybody radiation (emission coefficient of about 0.9). The distribution of the cloudiness at each deck is taken to be random (independent of the clouds of other decks). Given these assumptions, the radiative influx of heat can be estimated, provided that we know (for instance, with the aid of satellite observations) the total amount of cloudiness and the flux of departing (from the upper limit of atmosphere) long-wave radiation.

However, synoptic practice and the statistical data presented in Chapter 7 indicate that the assumption of a random (independent) distribution of the clouds of each deck is in conflict with the distribution actually observed. The main cloud types (nimbostratus, altostratus, cumulonimbus) are, as a rule, distributed over the vertical in two, and often three decks simultaneously. This means that the amounts of cloudiness in different decks are closely related to each other.

On the other hand, in all the models of the general circulation of the atmosphere worked out so far, the heights of the cloud boundaries and the cloud thicknesses have been assumed to be specified and invariable, while the cloud distributions over the height decks were calculated from the very old (1957) data of London [385]. Thus, it is clear that the problem of taking the cloudiness into account in models of the general circulation of the atmosphere and climate of the Earth is in need of considerable development and should be considered more deeply.

Smagorinsky [421] was the first to suggest using, in models of the general circulation, an additional relation, according to which the amount of cloudiness *n* of the lower and middle decks is proportional to the relative humidity of the air at the corresponding level. This relation is valid only statistically. For instance, one sampling gave the following relationship between the frequency of occurrence *f* (in %) of the clouds and the relative humidity:

Cloudiness *n*	*f*%					
	84–70	69–58	57–48	47–40	39–33	32–21
⩾8	86	72	52	46	32	14
5–8	8,5	10	11	12	16	8,1
2–5	3,3	7,2	14	13	13	22
<2	2,2	11	23	28	38	55
No. of cases	224	349	427	386	407	347

For high values of *f* (84-70%) it is seen that a cloudiness of 8 or greater is predominant (86% of cases). However, even for lower *f* (47-40%, for instance), the frequency of a cloudiness of 8 is quite high (46%). For f = 32-21%, low cloudiness (< 2) is observed only in 55% of the cases. The overall accuracy of Smagorinsky's relation is, according to the data of [92], 55-60%.

Smagorinsky, Manabe, and others [388, 389] were the first to include the heat of condensation and the amount of cloudiness in the model of the general circulation of the atmosphere. One of the most interesting conclusions is that such important synoptic objects as fronts and cyclones can be modeled only if the effects of water-vapor condensation are included in the calculations. If these effects are ignored, then only those (according to scale) features of the pressure, temperature, and velocity fields which are typical of the mean troposphere (for instance, for the 500 gPa*) level) can be modeled, that is, long-wave motion. Taking the amount of cloudiness into account, however, makes it possible to model features of the meteorological fields near sea level which are close to the observable ones (evolution of families of cyclones, fronts, and even their occlusion, etc.).

Soviet scientists [169, 170, 210] have focused considerable attention on modeling the general circulation of the atmosphere with the aid of numerical methods that include the fields of humidity and cloudiness. The most complete model of the general circulation and climate of the Earth was devised by Marchuk and his students [169, 170], who drew upon many years of experience in studying large-scale processes in the atmosphere and ocean. According to Marchuk [169], his ideas with regard to modeling the

* 1 gPa = 10^9 pascals. The pascal is a metric unit of pressure equal to one N/m^2. Note that 100 Pa = 100 N/m^2 = 1 mb. (Translator).

general circulation of the atmosphere and climate of the Earth were influenced considerably by "the work of N. E. Kochin, I. A. Kibel', J. G. Charney, Phillips, R. O. R. Thompson, E. N. Lorenz, A. M. Obukhov, E. N. Blinova, M. I. Budyko, K. Ya. Kondrat'ev, A. S. Monin, M. I. Yudin, Miyakoda, E. M. Feigel'son, M. E. Shvets, L. T. Matveev, and G. P. Kurbatkin". An understanding was gained of just what physical processes have to be described in the first place, in order to reproduce more or less accurately the main features of the structure of the atmosphere. Important contributions in this regard were made by Smagorinsky and Manabe, Mintz and Arakawa, Lewis, Casahar and Washington, Mason, and others.

A careful analysis has shown that, physically speaking, almost all the models of the general circulation of the atmosphere and ocean are nearly the same. Any differences between them usually involve the choice of coordinate systems, the transition from differential equations to difference equations, or the numerical procedures used to solve the difference equations and the application of these on the computer. In the model of Marchuk and his students, several important improvements were made, to complement the experience which had been gained beforehand (both with regard to choosing a more or less universal physical model and with regard to its mathematical implementation). When formulating a system of difference equations which did not deviate from the properties of the initial system, they made use of a number of properties, in particular the property of skew symmetry, since a skew-symmetrical nature of the system of equations makes it possible to base the stability of the entire calculation process on the property of the invariance of quadratic norms.

The method of splitting, which Marchuk provided with a theoretical basis for all stages of its application, enables replacing the very strict criterion of stability by a less stringent and more natural approximation criterion, and also makes it possible to introduce analogs of the stream functions which stabilize the calculation process substantially, when solving the equations describing the adaptation of the fields of the meteorological quantities.

From the physical point of view, the model of Marchuk and his students takes into account all the most significant features of the genesis of large-scale fields and processes in the atmosphere:

(a) The motion is (under the influence of the Coriolis force) close to geostrophic; although the deviations from geostrophic motion are slight, it is precisely these which are responsible for the evolution of meteorological fields. This brings us to the first difficulty when modeling the dynamics of the atmosphere: small quantities have to be found by eliminating large quantities.

(b) A westerly transport prevails in the atmosphere, intensifying with the height in the troposphere (the maximum velocity is near the tropopause). The westerly flow carries with it medium-scale (up to 1000 km in size) disturbances (long waves) and has a considerable effect on movements of large-scale planetary perturbations (superlong-wave perturbations).

(c) Processes observed in the atmosphere have a quite pronounced turbulent nature. Thus, we have to take into consideration the creation of atmospheric eddies when the primary motion becomes unstable and the subsequent cascade conversion of larger eddies into smaller perturbations, as well as the opposite mechanism, namely a development of turbulence whereby quasi-two-dimensional eddies grow in size (so-called negative eddy viscosity). Turbulence plays an important part in the interaction of the atmosphere with the ocean and the land, which are the main vehicles for transferring solar energy to atmospheric processes.

(d) The formation and evolution of cloud fields, which is the most important feature of the atmosphere.

The cloud cover, as we have shown, significantly influences the flux of solar radiation, and thus the amount of radiation absorbed by the ocean and land as well. The less cloudiness there is, the more solar radiation will be absorbed by the ocean. If slight cloudiness persists in some region for a long time (months or a season), then the ocean will heat up considerably there. The heated water (for instance, the Atlantic or Pacific ocean water) is then transported by currents to other (more northerly) regions. There, because of the low air temperatures, an unstable stratification is created (in the atmosphere and the ocean), which gives rise to powerful convective motions. The latter serve to transport stored heat from the ocean to the atmosphere. The consequent horizontal temperature differences (between the heated air mass from the ocean and its surroundings) promote the creation of cyclonic disturbances. The transfer of these via the planetary flow to the land is accompanied by a rise in temperature on the continents (with a substantial time lag). This is one way in which the cloudiness influences (with the ocean as its vehicle) the thermal regime of the continents, providing the possibility of long-range weather forecasting.

The cloudiness also influences the effective (long-wave) radiation of the Earth's surface. Whereas stratiform cloudiness can at present be modeled quite accurately, the same is not true for convective clouds, the importance of which for the transport of heat and moisture is also considerable.

In [100], some comments were made in favor of Smagorinsky's suggestion of a linear dependence of n on f. First, Lewis [382], and then Dushkin, Lomonosov, and Lunin [99], plotted empirical graphs relating the dew-point deficit Δ and the vertical velocity $\tau = dp/dt \approx -g\rho w$ to the amount of cloudiness. The accuracy of the estimate of the cloudiness using this method is about 75%; the success in determining the precipitation (for $\tau \leqslant -10$ gPa/12 hr and $\Delta < 2^0$C) ranges from 63 to 65%.

Let us make one remark here in connection with precipitation forecasting. The author shares the opinion of some that, by the time of the heaviest precipitation fallout in a cloud (or part of a cloud), the ascending motions are replaced by descending motions [136, 138, 177]. Most frequently these are mesoscale vertical motions which cannot be determined using synoptic data. Since heavy precipitation forms a sizeable part of the total precipitation, and since mesoscale vertical motions are a determining factor for such precipitation, it is clear that no close relationship between the amount of precipitation and vertical motions of large (synoptic) scale need be observed. The comparative results given above (and others as well) verify this conclusion.

The influence of cloudiness on the zonal temperature field is also taken into account in the models of the climate produced by Budyko, Sellers, and others. These models are based on the heat-balance equation and a number of empirical relations. Golitsyn and Mokhov [81] have discussed climatic stability vis-à-vis the amount of cloudiness, the albedo of the Earth's surface, and other factors.

In recent decades, attempts have been made to devise hydrodynamic (numerical) procedures for forecasting clouds, fog, and precipitation. The procedure for cloudiness forecasting developed by Lushev and the author [160, 161] is based on the equations of heat and moisture transfer, reduced to an invariant form. In this procedure, the velocity components are taken from a five-level (1000, 850, 700, 500, and 300 gPa) numerical model of pressure forecasting, worked out by Marchuk, Kurbatkin, and others [168, 170, 171]. Using this method, cloudiness has a total predictability of 80-85%.

Dymnikov and his coworkers [101, 102] devised a means of predicting cloudiness and precipitation based on a pressure-forecasting model using general equations, for higher accuracy in solving the finite-difference system of equations. The system of equations describing heat and moisture transfer is also reduced to invariant form. The invariants introduced by Dymnikov are somewhat different from ours. In the opinion of the authors of [101, 102] their invariants should remove the difficulty arising when determining the water content of a cloud from the difference between two large quantities. However, this difficulty persists in Dymnikov's method, too; it was merely shifted to the initial and boundary conditions.

Recently, a large series of studies dealing with the quantitative prediction of precipitation (steady as well as shower-type) have been carried out at the Gidromettsentr of the USSR. These studies were initiated by Dyubyuk [103], and later the following scientists have contributed significantly to the determination of numerical methods for precipitation prediction, often in conjunction with synoptic methods: Uspenskii, Lebedeva, Orlova, Mertsalov, Petrichenko, Galakhova, and Bagrov, to name a few [13, 72, 116, 153, 236, 246, 247, 259, 305-307]. The main results of these investigations are included in handbooks and methodological texts on the short-range forecasting of weather.

A procedure for forecasting shower-type precipitation, based on the slice method, has become widely known. Shishkin and his coworkers [156, 328, 329] have used this method to analyze the thermodynamic stability of the atmosphere. The results of radar studies of convective clouds, together with satellite and aerosynoptic information, are used in the prediction of showers [75, 82, 149, 212, 283, 288, 302, 303].

Studies have also been made in which precipitation is forecast on the basis of statistical relationships between the precipitation intensity and those dynamical-thermal characteristics of the atmosphere which have the greatest effect on cloud formation and evolution and which lead to precipitation fallout from the clouds so formed [42, 60, 242, 243, 263, 264, 288].

During recent decades, many studies have been made of wave and eddy motions in the atmosphere, and circulation in the form of cells, accompanied by the formation of the corresponding types of wave and convective clouds [26, 27, 36, 40, 56-58, 83, 95, 104, 111, 112, 132-134, 282, 302, 316, 324]. However, the majority of such studies have had as their goal the construction of models of the motion of the atmosphere, and they have not been applied to ascertain the conditions of cloud formation not to mention cloud characteristics such as water contents or boundaries).

Numerous models have been devised of individual convective clouds produced when an individual volume of air moves aloft [8, 9, 25, 80, 113, 115, 120, 130, 240, 310] or when a jet spreads out [16, 17, 65, 118]. In this monograph, we will discuss only topics in the dynamics of convective clouds pertaining to the formation of an ensemble (field) of such clouds [302, 304, 309, 327].*

We should also mention a series of studies of the moisture content of the atmosphere and the water reserves of clouds, carried out with the aid of radiometric equipment, both ground-based and mounted on aircraft and satellites [5, 7, 31-33, 59, 82, 94, 142, 217, 302, 321]. Models of the atmospheric moisture content and cloud water reserves have been constructed in [136, 138, 207].

Fedorov and Mamina [310] were the first to point out that there is a continual renewal of the liquid water in a precipitating cloud. The ratio of the amount of precipitation falling from a cloud (during a 2-3 day period) to the water mass contained in the cloud at a given moment fluctuates from 7 to 40. In [194] it was shown that these experimental data are in good agreement with the results of cloud and precipitation modeling discussed in Chapter 2 of this monograph.

Both Soviet and non-Soviet scientists have made many studies of fog as well as clouds, and they have focused in particular on modeling fog structure and the conditions of fog formation and on gathering the necessary experimental (laboratory and synoptic-

* Most of the non-Soviet works referenced also relate to studies of convective clouds.

aerological data [21, 22, 48, 66, 100, 109, 145, 237, 260, 261, 438, 439].

In conclusion, let us cite the paper of Obukhov and Golitsyn [232], in which the idea of the entrainment of drops by turbulent moles and the concept of the specific moisture content are invoked to explain cloud-formation conditions on Venus.

CHAPTER 1

EQUATIONS OF HEAT AND MOISTURE TRANSFER IN A TURBULENT ATMOSPHERE

Atmospheric motions are known to be turbulent. Motion in a viscous sublayer (a layer only a few millimeters thick close to the Earth's surface) constitutes an exception, as does motion in the near-Earth layer when there is a very slight breeze (up to 0.5–1 m/s) and a very stable (inversion) stratification.

1.1. Water-vapor influx (balance)

Let us consider an air particle (turbulent air parcel) in a flow. The specific humidity (mass of water vapor per unit mass of air) of such a particle moving with the flow can vary in time only as a result of molecular diffusion or condensation. Since the role of molecular diffusion in the atmosphere is negligible, it follows that for unsaturated moist air we can write

$$\frac{dq}{dt}=0 \quad \text{or} \quad \frac{\partial q}{\partial t}+u\frac{\partial q}{\partial x}+v\frac{\partial q}{\partial y}+w\frac{\partial q}{\partial z}=0, \qquad (1.1.1)$$

where d/dt indicates the individual (total) derivative with respect to time, characterizing the time rate of change of the quantity in an air particle moving with the flow; u, v, w are the components of the instantaneous velocity of the particle in the x, y, z directions of the Cartesian coordinate system; $\partial/\partial t$ indicates the local time derivative, characterizing the time rate of change of the quantity at a given point in space; and q is the specific humidity of the air.

The physical meaning of eqn. (1.1.1) is that the specific humidity in a moving air particle does not vary with time, that is, q is a conservative flow characteristic, prior to the beginning of condensation and in the absence of molecular diffusion (or neglecting it).

In view of the continuity equation,

$$\frac{\partial \rho}{\partial t}+\frac{\partial \rho u}{\partial x}+\frac{\partial \rho v}{\partial y}+\frac{\partial \rho w}{\partial z}=0 \qquad (1.1.2)$$

eqn. (1.1.1) can be written as

$$\frac{\partial \rho q}{\partial t}+\frac{\partial u(\rho q)}{\partial x}+\frac{\partial v(\rho q)}{\partial y}+\frac{\partial w(\rho q)}{\partial z}=0$$

or

$$\frac{\partial a}{\partial t}+\frac{\partial ua}{\partial x}+\frac{\partial va}{\partial y}+\frac{\partial wa}{\partial z}=0, \tag{1.1.3}$$

where $a = \rho q$ is the absolute air humidity (water-vapor density), and ρ is the air density. All the quantities appearing in eqns. (1.1.1)-(1.1.3) are instantaneous (unaveraged), and they characterize the state of the air particle at a certain (fixed) moment of time.

Since, as already mentioned, the motion of the atmosphere is predominantly turbulent, all the instantaneous quantities should be represented as sums: (1.1.4)

$$q=\bar{q}+q', \quad a=\bar{a}+a', \quad u=\bar{u}+u', \quad v=\bar{v}+v', \quad w=\bar{w}+w'.$$

Here the first terms ($\bar{q}$, $\bar{a}$,...) of the sums are the mean values of the quantities, while the second terms (q', a',...) are the deviations (fluctuations) from these mean values. The averaging is either over time, space, or over space-time.

Now, if we replace the instantaneous quantities in (1.1.3) by their sums according to relations (1.1.4), and take the average of the result, then in view of the postulates of the averaging operation (namely

$$\left(\overline{\varphi\pm f}=\bar{\varphi}\pm\bar{f},\ \overline{\frac{\partial f}{\partial t}}=\frac{\partial \bar{f}}{\partial t};\ \bar{\bar{\varphi}}=\bar{\varphi},\ \overline{\bar{\varphi} f}=\bar{\varphi}\bar{f},\ \bar{f'}=\bar{\varphi'}=0,\ \overline{\varphi f}=\bar{\varphi}\bar{f}+\overline{\varphi' f'}\right)$$

we get (1.1.5)

$$\frac{\partial \bar{a}}{\partial t}+\frac{\partial \bar{u}\bar{a}}{\partial x}+\frac{\partial \bar{v}\bar{a}}{\partial y}+\frac{\partial \bar{w}\bar{a}}{\partial z}=-\left(\frac{\partial \overline{u'a'}}{\partial x}+\frac{\partial \overline{v'a'}}{\partial y}+\frac{\partial \overline{w'a'}}{\partial z}\right).$$

The right-hand side of this equation is the turbulent influx of water vapor ε_V to a unit volume of air per unit time, the units of ε_V being kg(m^{3}s). On the basis of the relation

$$\varepsilon_v=-\left(\frac{\partial Q_x}{\partial x}+\frac{\partial Q_y}{\partial y}+\frac{\partial Q_z}{\partial z}\right),$$

between the water-vapor influx ε_V and the water-vapor flow $\mathbf{Q}$, we are justified in writing the x, y, z components of the turbulent flow of water vapor as

$$Q_x=\overline{u'a'}, \quad Q_y=\overline{v'a'}, \quad Q_z=\overline{w'a'}. \tag{1.1.6}$$

It can be assumed that eqn. (1.1.5) is the most general (prior to the beginning of condensation) equation describing the transfer of water vapor (or of any other substance) in a turbulent atmosphere (or, in the general case, in a turbulent medium). However, some problems arise when we try to use this equation, since its right-hand side contains the components of the turbulent flow expressed in terms of fluctuations of the velocity and the absolute humidity. These fluctuations can be determined (measured) only with the aid of special instruments at a limited number of points (at major observatories and on expeditions). Moreover, no complete data are available on the laws of variation (in time and space) of the fluctuations in the meteorological quantities. Consequently, for some decades now, beginning with the fundamental works of Taylor, Schmidt, Prandtl, and von Karman, the ideas of the semi-empirical theory of turbulence, concerning the relation between the fluctuations and mean values of the meteorological quantities, have been widely used to complete the system of equations.

Within the framework of the semiempirical theory, the components of the turbulent flow of water vapor (1.1.6) can be written as

$$Q_x = -\bar{\rho} k_x \frac{\partial \bar{q}}{\partial x}, \quad Q_y = -\bar{\rho} k_y \frac{\partial \bar{q}}{\partial y}, \quad Q_z = -\bar{\rho} k_z \frac{\partial \bar{q}}{\partial z}, \qquad (1.1.7)$$

where $\bar{q}$ is the mean value of the relative humidity, and k_x, k_y, k are the x, y, z components of the turbulence coefficient.

It should be noted that the turbulent flow of water vapor is proportional to the gradient of the specific (not absolute) humidity. Physically speaking, this is because the turbulent flows are proportional to the gradient of that characteristic of the substance (or property) which remains constant as the air particle moves. For water vapor, the specific humidity serves as the conservative characteristic.

Now, if eqn. (1.1.7) is substituted into eqn. (1.1.5), then the mean specific humidity $\bar{q}$ will appear in the latter as well as the mean absolute humidity $\bar{a}$. Clearly, we have to obtain, instead, an equation containing only one of these quantities. This, however, is not as simple an operation as might appear at first glance.

Actually, the mean values $\bar{a}$ and $\bar{q}$ are related to the fluctuations a' and q' as follows:

$$\bar{a} = \bar{\rho}\bar{q} + \overline{\rho' q'}, \quad a' = \bar{\rho} q' + \bar{q}\rho' + \rho' q' - \overline{\rho' q'}, \qquad (1.1.8)$$

since $a = \rho q$, $\bar{a} + a' = \bar{\rho}\bar{q} + \bar{\rho} q' + \bar{q}\rho' + \rho' q'$, and $\overline{\bar{a} + a'} = \bar{a} = \bar{\rho}\bar{q} + \overline{\rho' q'}$.

After averaging, we can write the continuity equation (1.1.2) as

$$\frac{\partial \bar{\rho}}{\partial t} + \sum_{1}^{3} \frac{\partial \bar{\rho} \bar{u}_i}{\partial x_i} = - \sum_{1}^{3} \frac{\partial \overline{\rho' u_i'}}{\partial x_i}. \qquad (1.1.9)$$

Here, and in the following, to keep the notation brief, we denote $x_1 = x$, $x_2 = y$, $x_3 = z$, and, correspondingly, $u_1 = u$, $u_2 = v$, $u_3 = w$. Substituting expressions (1.1.8) for $\bar{a}$ and a' into (1.1.5), and taking continuity equation (1.1.9) into account yields

$$\bar{\rho}\left(\frac{\partial \bar{q}}{\partial t}+\sum_{1}^{3}\bar{u}_i\frac{\partial \bar{q}}{\partial x_i}\right)=-\sum_{1}^{3}\left(\frac{\partial\bar{\rho}\overline{u_i'q'}}{\partial x_i}+\frac{\partial\bar{u}_i\overline{\rho'q'}}{\partial x_i}+\right.$$

$$\left.+\overline{\rho' u_i'}\frac{\partial \bar{q}}{\partial x_i}+\frac{\partial\overline{u_i'\rho' q'}}{\partial x_i}\right). \tag{1.1.10}$$

Thus, the transition from $\bar{a}$ to $\bar{q}$ is accompanied by the appearance, on the right-hand side of the equation for water-vapor transfer, of a number of additional terms. Here, what is most important is that some of these terms are completely new (in structure) - for instance, the last term in the sum on the right-hand side - and they contain triple products of the fluctuations, the so-called third moments $(\overline{u'\rho'q'})$ of the probability distribution. The first moments, let us recall, are the mean values $\bar{q}$, $\bar{\rho}$, $\bar{u}_i$, and the second moments are the averaged double products $u_i' q_i'$, $\rho_i' u_i'$, $\rho_i' q_i'$ This, in turn, means that to complete the system of equations it is no longer sufficient to invoke relations (1.1.7) between the second moments and the mean values. Instead, additonal relations or equations linking the third moments to the first and second moments are needed.

We will not consider further this difficult problem in statistical hydromechanics. Rather, let us adopt the hypothesis proposed as long ago as 1877 by Boussinesq. According to his hypothesis (approximation), the density fluctuations ρ are small in comparison with the mean value $\bar{\rho}$, that is $\rho/\bar{\rho} \ll 1$. As applied to the atmosphere, this condition is satisfied to a high degree of accuracy. Actually, from the equation of state it follows that

$$p'/\bar{p}=\rho'/\bar{\rho}+T'/\bar{T}.$$

Since the fluctuations of the pressure p in the Earth's atmosphere are negligible ($p'/\bar{p}$ is considerably less than both $\rho'/\bar{\rho}$ and $T'/\bar{T}$), therefore $-\rho'/\bar{\rho} \approx T'/\bar{T}$.

Experimental data indicate that fluctuations T' in the case of microscale and mesoscale turbulence vary from hundredths or tenths of a degree to several degrees Celsius. Therefore,

$$\left|\frac{\rho'}{\bar{\rho}}\right| \sim \left|\frac{T'}{\bar{T}}\right| \sim 10^{-4} \ldots 10^{-2},$$

whereas $|p'/\bar{p}| \sim 10^{-5}$.

It can be shown [53] that, if the hypothesis (approximation) of Boussinesq is valid, the density fluctuations can be set equal to zero in all terms of the equations of motion, continuity, and heat and moisture transfer, except for the term describing the effect of the buoyancy force (in thr third equation of motion).

Consequently, in Boussinesq's approximation, eqn. (1.1.10) takes the form

$$\bar{\rho}\frac{\partial \bar{q}}{\partial t} = -\left(\frac{\partial \bar{\rho}\overline{u'q'}}{\partial x} + \frac{\partial \bar{\rho}\overline{v'q'}}{\partial y} + \frac{\partial \bar{\rho}\overline{w'q'}}{\partial z}\right). \qquad (1.1.11)$$

Here $\bar{\rho}\overline{u'q'} = \overline{u'\bar{\rho}q'} = \overline{u'a'} = Q_x$ $\bar{\rho}\overline{v'q'} = Q_y$, $\bar{\rho}\overline{w'q'} = Q_z$— co-are the components of the turbulent flow of water vapor. Taking expressions (1.1.7) into account, we can rewrite eqn. (1.1.11) as

$$\bar{\rho}\frac{d\bar{q}}{dt} = \frac{\partial}{\partial x}\bar{\rho}k_x\frac{\partial \bar{q}}{\partial x} + \frac{\partial}{\partial y}\bar{\rho}k_y\frac{\partial \bar{q}}{\partial y} + \frac{\partial}{\partial z}\bar{\rho}k_z\frac{\partial \bar{q}}{\partial z}. \qquad (1.1.12)$$

The right-hand side of this equation is the turbulent influx of water vapor to a unit volume of air per unit time (rate of absolute influx). Now, designating the turbulent influx as ε_V, we obtain eqn. (1.1.12) in the form:

$$\frac{\partial \bar{q}}{\partial t} = -\left(\bar{u}\frac{\partial \bar{q}}{\partial x} + \bar{v}\frac{\partial \bar{q}}{\partial y}\right) - \bar{w}\frac{\partial \bar{q}}{\partial z} + \frac{\varepsilon_V}{\bar{\rho}}. \qquad (1.1.13)$$

According to this equation, at a fixed point in space, the specific humidity varies in time as a result of three factors:

(1) The advective influx of water vapor,

$$-\left(\bar{u}\frac{\partial \bar{q}}{\partial x} + \bar{v}\frac{\partial \bar{q}}{\partial y}\right);$$

This quantity will be positive if the vapor transfer is from a region of higher specific humidity to a region of lower specific humidity.

(2) The convective influx of water vapor,

$$-\bar{w}\frac{\partial \bar{q}}{\partial z};$$

since the specific humidity generally decreases with height ($\partial\bar{q}/\partial z<0$), therefore for an ascending motion ($\bar{w}>0$, as is usually observed in cyclones and troughs, at any rate in the lower half of the troposphere) the convective influx is positive, while for a descending motion ($\bar{w}<0$, as is observed in anticyclones and ridges) it is negative.

(3) The turbulent influx of water vapor (ε_V/ρ).

1.2. Heat influx (balance)

In the absence of heat influxes (radiant, molecular, condensational) the first law of thermodynamics for instantaneous (unaveraged) motion has the form

$$c_p \frac{dT}{dt} - \frac{RT}{p}\frac{dp}{dt} = 0.$$

If we introduce the potential temperature $\Theta = T(1000/p)^{R/c_p}$, then the last equation becomes

$$c_p \frac{d\theta}{dt} = 0 \quad \text{or} \quad \frac{\partial\theta}{\partial t} + u\frac{\partial\theta}{\partial x} + v\frac{\partial\theta}{\partial y} + w\frac{\partial\theta}{\partial z} = 0. \tag{1.2.1}$$

By combining this equation with continuity equation (1.1.2), we obtain

$$\frac{\partial\Phi}{\partial t} + \frac{\partial u\Phi}{\partial x} + \frac{\partial v\Phi}{\partial y} + \frac{\partial w\Phi}{\partial z} = 0, \tag{1.2.2}$$

where $\Phi = c_p\rho\Theta$ is the potential heat content of a unit volume of air.

After representing all the quantities as sums, as above:

$$\Phi = \overline{\Phi} + \Phi', \quad u = \bar{u} + u', \quad v = \bar{v} + v', \quad w = \bar{w} + w',$$

and then averaging equation (1.2.2), we obtain

$$\frac{\partial\overline{\Phi}}{\partial t} + \frac{\partial\bar{v}\overline{\Phi}}{\partial x} + \frac{\partial\bar{v}\overline{\Phi}}{\partial y} + \frac{\partial\bar{w}\overline{\Phi}}{\partial z} = -\left(\frac{\partial\overline{u'\Phi'}}{\partial x} + \frac{\partial\overline{v'\Phi'}}{\partial y} + \frac{\partial\overline{w'\Phi'}}{\partial z}\right). \tag{1.2.3}$$

The right-hand side of this equation is the turbulent influx of heat to a unit volume of air per unit time (rate of absolute heat influx)

$$\varepsilon_T = -\left(\frac{\partial\overline{u'\Phi'}}{\partial x} + \frac{\partial\overline{v'\Phi'}}{\partial y} + \frac{\partial\overline{w'\Phi'}}{\partial z}\right). \tag{1.2.4}$$

The components of the turbulent heat flow are:

$$Q_x = \overline{u'\Phi'}, \quad Q_y = \overline{v'\Phi'}, \quad Q_z = \overline{w'\Theta'}. \tag{1.2.5}$$

If we introduce into eqn. (1.2.5) the mean values of the

potential temperature $(\Theta=\bar{\Theta}+\Theta')$ and the air density $(\rho=\bar{\rho}+\rho')$, then we have

$$\Phi=c_p\rho\Theta=c_p(\bar{\rho}\bar{\Theta}+\rho'\bar{\Theta}+\bar{\rho}\Theta'+\rho'\Theta'),$$

from which

$$\bar{\Phi}=c_p(\bar{\rho}\bar{\Theta}+\overline{\rho'\Theta'}),\quad \Phi'=c_p(\rho'\bar{\Theta}+\bar{\rho}\Theta'+\rho'\Theta'-\overline{\rho'\Theta'}).$$

After substituting these expressions into eqn. (1.2.3), taking the averaged continuity equation (1.1.9) into account, we get

$$c_p\bar{\rho}\left(\frac{\partial\bar{\Theta}}{\partial t}+\sum_1^3\bar{u}_i\frac{\partial\bar{\Theta}}{\partial x_i}\right)=-\sum_1^3 c_p\left(\frac{\partial\bar{\rho}\overline{u_i'\Theta'}}{\partial x_i}+\frac{\partial\bar{u}_i\overline{\rho'\Theta'}}{\partial x_i}+\right.$$

$$\left.+\overline{\rho' u_i'}\frac{\partial\bar{\Theta}}{\partial x_i}+\frac{\partial\overline{u_i'\rho'\Theta'}}{\partial x_i}\right). \tag{1.2.6}$$

This equation describing the heat influx for averaged motion is quite complicated. Only in the Boussinesq approximation, where the density fluctuations are neglected, does the heat-influx equation assume the simpler form in which it is usually expressed:

$$c_p\bar{\rho}\left(\frac{\partial\bar{\Theta}}{\partial t}+\bar{u}\frac{\partial\bar{\Theta}}{\partial x}+\bar{v}\frac{\partial\bar{\Theta}}{\partial y}+\bar{w}\frac{\partial\bar{\Theta}}{\partial z}\right)=-c_p\left(\frac{\partial\bar{\rho}\,\overline{u'\Theta'}}{\partial x}+\right.$$

$$\left.+\frac{\partial\bar{\rho}\overline{v'\Theta'}}{\partial y}+\frac{\partial\bar{\rho}\overline{w'\Theta'}}{\partial z}\right). \tag{1.2.7}$$

Within the framework of the semiempirical theory, the components of the turbulent heat flow can be represented as

$$c_p\bar{\rho}\overline{u'\Theta'}=-c_p\bar{\rho}k_x\frac{\partial\bar{\Theta}}{\partial x},\quad c_p\bar{\rho}\overline{v'\Theta'}=-c_p\bar{\rho}k_y\frac{\partial\bar{\Theta}}{\partial y},$$

$$c_p\bar{\rho}\overline{w'\Theta'}=-c_p\bar{\rho}k_z\frac{\partial\bar{\Theta}}{\partial z}, \tag{1.2.8}$$

while the heat-influx equation becomes,

$$\bar{\rho}\left(\frac{\partial\bar{\Theta}}{\partial t}+\bar{u}\frac{\partial\bar{\Theta}}{\partial x}+\bar{v}\frac{\partial\bar{\Theta}}{\partial y}+\bar{w}\frac{\partial\bar{\Theta}}{\partial z}\right)=$$

$$=\frac{\partial}{\partial x}\bar{\rho}k_x\frac{\partial\bar{\Theta}}{\partial x}+\frac{\partial}{\partial y}\bar{\rho}k_y\frac{\partial\bar{\Theta}}{\partial y}+\frac{\partial}{\partial z}\bar{\rho}k_z\frac{\partial\bar{\Theta}}{\partial z}.$$

(1.2.9)

This equation is usually also simplified somewhat. Since

$$\frac{1}{\bar{\Theta}}\frac{\partial\bar{\Theta}}{\partial x}=\frac{1}{\bar{T}}\frac{\partial\bar{T}}{\partial x}-\frac{R}{c_p}\frac{1}{\bar{p}}\frac{\partial\bar{p}}{\partial x},$$

where the second term on the right-hand side is about 1/10 as great as the first, and

$$\frac{1}{\bar{\Theta}}\frac{\partial\bar{\Theta}}{\partial z}=\frac{1}{\bar{T}}(\gamma_a-\gamma),$$

where $\gamma = \partial T/\partial z$ is the vertical temperature gradient [lapse rate] $\gamma_a = g/c_p$ is the dry-adiabatic gradient, and $\bar{T}/\bar{\Theta}\approx 1$, it follows that

$$\frac{\partial\bar{T}}{\partial t}=-\left(\bar{u}\frac{\partial\bar{T}}{\partial x}+\bar{v}\frac{\partial\bar{T}}{\partial y}\right)-\bar{w}(\gamma_a-\gamma)+k_x\frac{\partial^2\bar{T}}{\partial x^2}+$$

$$+k_y\frac{\partial^2\bar{T}}{\partial y^2}+\frac{1}{\bar{\rho}}\frac{\partial}{\partial z}\bar{\rho}k_z\left(\frac{\partial\bar{T}}{\partial z}+\gamma_a\right).$$

(1.2.10)

Here, we have neglected the dependences of $\bar{\rho}k_x$ and $\bar{\rho}k_y$ on x and y, and we have taken into account that the foregoing relation for for $\partial\bar{\Theta}/\partial x$ is also valid for the derivatives with respect to y and t.

1.3. Influxes of heat and water vapor in moist saturated air (in a cloud)

The transition of water from one state of aggregation to another (phase transitions of water) constitutes the most important feature of processes in moist saturated air, distinguishing them from processes in dry air or moist unsaturated air. In contrast to other gases (nitrogen, oxygen, argon), water vapor at the temperatures observed in the Earth's atmosphere can change into either the liquid (droplets) or solid (crystals) state. The condition $T<T_{cr}$ (where T_{cr} is the critical temperature of the gas) is necessary but not sufficient for the transition of water vapor (just

like any other gas) to the liquid state. The major purpose of this book will be to describe the conditions of water-vapor condensation and to evaluate quantitatively the roles of various factors in the formation of clouds and fogs. It should be noted that the theory describing processes accompanied by phase transitions is one of the most complicated branches of heat physics and hydrodynamics.

Let us designate as ν the absolute rate of water-vapor condensation; that is, the mass of water vapor condensing on drops in a unit volume of cloud air per unit time. The absolute air humidity will diminish by this amount. Thus, the equation describing the influx of water vapor in saturated moist air, taking (1.1.12) into account, can be written as

$$\bar{\rho}\frac{d\bar{q}}{dt}=\frac{\partial}{\partial x}\bar{\rho}k_x\frac{\partial\bar{q}}{\partial x}+\frac{\partial}{\partial y}\bar{\rho}k_y\frac{\partial\bar{q}}{\partial y}+\frac{\partial}{\partial z}\bar{\rho}k_z\frac{\partial\bar{q}}{\partial z}-\nu. \qquad (1.3.1)$$

If we neglect the x and y dependences of $\bar{\rho}k$ and $\bar{\rho}k$, then we can write (1.3.1) as

$$\frac{d\bar{q}}{dt}=k_x\frac{\partial^2\bar{q}}{\partial x^2}+k_y\frac{\partial^2\bar{q}}{\partial y^2}+\frac{1}{\bar{\rho}}\frac{\partial}{\partial z}\bar{\rho}k_z\frac{\partial\bar{q}}{\partial z}-\frac{\nu}{\bar{\rho}}. \qquad (1.3.2)$$

As water vapor condenses, latent heat is released. If L is the latent heat of condensation, then an amount of heat equal to L will be liberated per unit time in a unit volume of cloud air. The equation for the heat influx in saturated moist air, taking (1.2.9) into account, thus becomes,

$$c_p\bar{\rho}\frac{d\bar{\Theta}}{dt}=c_p\left(\frac{\partial}{\partial x}\bar{\rho}k_x\frac{\partial\bar{\Theta}}{\partial x}+\frac{\partial}{\partial y}\bar{\rho}k_y\frac{\partial\bar{\Theta}}{\partial y}+\frac{\partial}{\partial z}\bar{\rho}k_z\frac{\partial\bar{\Theta}}{\partial z}\right)+L\nu+\varepsilon_r, \qquad (1.3.3)$$

where the additive term ε_r on the right-hand side is the absolute rate of radiative (radiant) heat influx.

If the simplifications mentioned at the end of Section 1.2 are used, then eqn. (1.3.3) becomes

$$\frac{d\bar{T}}{dt}+\gamma_a\bar{w}=k_x\frac{\partial^2\bar{T}}{\partial x^2}+k_y\frac{\partial^2\bar{T}}{\partial y^2}+\frac{1}{\bar{\rho}}\frac{\partial}{\partial z}\bar{\rho}k_z\left(\frac{\partial\bar{T}}{\partial z}+\gamma_a\right)+$$

$$+\frac{L\nu}{c_p\bar{\rho}}+\frac{\varepsilon_p}{c_p\bar{\rho}}. \qquad (1.3.4)$$

Equations (1.3.1)-(1.3.4) serve as initial equations when posing the majority of the problems discussed in this monograph.

When studying stratiform clouds, as opposed to convective

clouds (that is, when studying frontal Ns-As-Cs or internal St-Sc--Ac-Cc), the horizontal dimensions of which are considerably larger than the vertical dimensions, evaluations show that the heat and moisture influxes due to horizontal turbulence can safely be neglected. Physically speaking, it is clear that the horizontal exchange [mixing] influences the cloudiness very significantly only close to the horizontal cloud boundaries.

Consequently, with a sufficient degree of accuracy, the equations describing the formation and evolution of stratiform clouds can be written as

$$\frac{\partial \bar{q}}{\partial t} = -\left(\bar{u}\frac{\partial \bar{q}}{\partial x} + \bar{v}\frac{\partial \bar{q}}{\partial y}\right) - \bar{w}\frac{\partial \bar{q}}{\partial z} + \frac{1}{\bar{\rho}}\frac{\partial}{\partial z}\bar{\rho}k_z\frac{\partial \bar{q}}{\partial z} - \frac{\nu}{\bar{\rho}}, \tag{1.3.5}$$

$$\frac{\partial \bar{\theta}}{\partial t} = -\left(\bar{u}\frac{\partial \bar{\theta}}{\partial x} + \bar{v}\frac{\partial \bar{\theta}}{\partial y}\right) - \bar{w}\frac{\partial \bar{\theta}}{\partial z} + \frac{1}{\bar{\rho}}\frac{\partial}{\partial z}\bar{\rho}k_z\left(\frac{\partial \bar{T}}{\partial z} + \gamma_a\right) + \frac{L\nu}{c_p\bar{\rho}} + \frac{\varepsilon_p}{c_p\bar{\rho}}. \tag{1.3.6}$$

The left-hand side of eqn. (1.3.5) and the first and second terms on the right-hand side of this equation are generally thought to be of the same order of magnitude under actual atmospheric conditions (this conclusion also applies to the heat-influx equation). However, there is less agreement about the role of the heat and moisture influxes due to vertical turbulent mixing: in most works it is stated, without any discussion, that outside of the boundary layer these influxes are negligible (compared to, for example, the influx due to ordered vertical currents). On the other hand, elementary evaluations made on the basis of conventional ideas of dimensional analysis indicate that the role of vertical turbulent exchange in heat and moisture transfer is comparable to the roles of all the other factors, not just in the boundary layer but throughout the entire troposphere. Actually if we introduce characteristic values of the height (H), the vertical velocity ($\tilde{w}$), and the turbulence coefficient ($\tilde{k}$), then the third and fourth terms on the right-hand side of (1.3.5) become

$$\frac{\tilde{w}}{H} w_1 \frac{\partial \bar{q}}{\partial z_1} \quad \text{and} \quad \frac{\tilde{k}}{H^2} k_1 \frac{\partial^2 \bar{q}}{\partial z_1^2},$$

where $w_1 = \bar{w}/\tilde{w}$, $k_1 = k_z/\tilde{k}$, $z_1 = z/H$ are dimensionless quantities of the order of unity.

Now, if the dimensionless factors in these terms are assumed to be of the same order of magnitude, then the characteristic value of the turbulence coefficient for which the turbulent influx of moisture plays the same role as the transfer via vertical currents must be

$$\tilde{k} = \tilde{w} H.$$

For $H = 10^4$ m (the thickness of the troposphere) and $\tilde{w} = 0.1, \ldots, 1$ cm/s, we get $\tilde{k} = 10, \ldots, 100$ m^2/s.

Many calculations have been carried out during recent decades on the basis of data on drag-to-weight ratios for aircraft and the results of temperature-wind soundings of the atmosphere. These have yielded values of the turbulence coefficient in the free atmosphere which were the same as those stated above. For instance, according to [68, 181, 183], the frequency maximum of the turbulence coefficient in clouds is reached at 50 m^2/s, and outside clouds at 20 to 30 m^2/s. Often, the turbulence coefficient in the free atmosphere exceeds 100 m^2/s (and in jet streams it can be 1000 to 1500 m^2/s).

However, the equation of $\tilde{k}$ based on the classical concepts of dimensional analysis are on the high side. Actually, if the specific humidity is represented as follows, as a first approximation,

$$\bar{q} = q_0 \exp(-\beta z) \quad \text{or} \quad \bar{q} = q_0 \exp(-\beta H z_1),$$

where βH is equal to about 4 according to experimental data, then the first and second derivatives of the specific humidity will be

$$\frac{\partial \bar{q}}{\partial z_1} = -q_0 \beta H \exp(-\beta H z_1),$$

$$\frac{\partial^2 \bar{q}}{\partial z_1^2} = q_0 (\beta H)^2 \exp(-\beta H z_1).$$

Consequently, in order for the moisture influxes due to vertical currents and turbulent mixing to be of the same order of magnitude, it is sufficient that

$$\frac{\tilde{w}}{H} = \frac{\tilde{k}}{H^2} \beta H,$$

from which

$$k = \frac{1}{\beta H} \tilde{w} H.$$

For $\tilde{w} = 0.1, \ldots, 1$ cm/s, we obtain the following characteristic value of the turbulent coefficient: $\tilde{k}$ = 2,...25 m^2/s.

These considerations make it clear that the vertical turbulent influx of heat and moisture is of just as much importance as the other factors. Moreover, specific calculations carried out, first in [189] and then by other investigators, confirm the following conclusion: cloudiness of great vertical extent often forms under the influence of turbulent mixing alone. It should be stressed that here we are referring to ordinary small-scale turbulence, which is generally assumed to affect the redistribution of heat, moisture, and momentum in the boundary layer of the atmosphere.

1.4. Equations for temperature and condensation rate of water vapor

In order to demonstrate the complexity of the equations of heat and water-vapor influx, as well as the seriousness of the difficulties encountered when solving them with the aid of direct methods, let us derive the equations for the temperature and the condensation rate v. System (1.3.5), (1.3.6) for a cloud, where the water vapor is in a saturated state, is supplemented by a relation for the specific humidity

$$q = q_m = 0{,}622\,\frac{E(T)}{p}, \tag{1.4.1}$$

where $E(T)$ is the pressure of the saturated water vapor.

To these we add:

(1) the equation of statics

$$-\frac{1}{p}\frac{\partial p}{\partial z} = \frac{1}{H}, \tag{1.4.2}$$

where $H = RT/g$ is the height of the homogeneous atmosphere;

(2) the equation of state for air

$$p = R\rho T; \tag{1.4.3}$$

(3) the Clausius-Clapeyron equation

$$\frac{E'}{E} = \frac{L}{R_v T^2}, \tag{1.4.4}$$

where $E' = dE/dT$; and R_v is the specific gas constant of the water vapor.

The successive differentiation of formula (1.4.1) now gives

$$\frac{\partial q_m}{\partial z}=\frac{0,622}{p}\left(E'\frac{\partial T}{\partial z}+\frac{E}{H}\right),$$

$$\frac{1}{\rho}\frac{\partial}{\partial z}\rho k\frac{\partial q_m}{\partial z}=\frac{0,622}{p}\left[E'\frac{\partial}{\partial z}k\frac{\partial T}{\partial z}+k\left(E''-\frac{E'}{T}\right)\left(\frac{\partial T}{\partial z}\right)^2+\right.$$

$$\left.+\frac{k}{H}\left(E'-\frac{2E}{T}\right)\frac{\partial T}{\partial z}+\frac{E}{H}\frac{\partial k}{\partial z}\right], \tag{1.4.5}$$

where

$$E''=\frac{d^2E}{dT^2}=\frac{LE}{R_v T^3}\left(\frac{L}{R_v T}-2\right).$$

Similarly, by using the relation $\Theta=T(1000/p)^{(\varkappa-1)/\varkappa}$, we obtain

$$\frac{1}{\rho}\frac{\partial}{\partial z}\rho k\frac{\partial\theta}{\partial z}=\frac{\theta}{T}\left[\frac{\partial}{\partial z}k\frac{\partial T}{\partial z}+\gamma_a\frac{\partial k}{\partial z}+\right.$$

$$\left.+\frac{k}{T}\left(\gamma_a+\frac{\partial T}{\partial z}\right)\left(\frac{\partial T}{\partial z}-\frac{g}{\varkappa R}\right)\right]. \tag{1.4.6}$$

If relation (1.4.1) and the formula for Θ are differentiated with respect to time, then we have

$$\frac{d\theta}{dt}=\frac{\theta}{T}\frac{dT}{dt}-\frac{\varkappa-1}{\varkappa}\frac{\theta}{p}\frac{dp}{dt},\quad \frac{dq_m}{dt}=\frac{0,622}{p}\left(E'\frac{dT}{dt}-\frac{E}{p}\frac{dp}{dt}\right). \tag{1.4.7}$$

Next, eqn. (1.3.5) is multiplied by L/c_p and added to (1.3.6). Taking relations (1.4.5)-(1.4.7) into account, we get

$$(1+c)\frac{dT}{dt}-b\frac{dp}{dt}=\frac{\partial}{\partial z}k\frac{\partial T}{\partial z}+\gamma_a\frac{\partial k}{\partial z}+\frac{k}{T}\left(\gamma_a+\frac{\partial T}{\partial z}\right)\times$$

$$\times\left(\frac{\partial T}{\partial z}-\frac{g}{\varkappa R}\right)+\frac{c}{E'}\left[E'\frac{\partial}{\partial z}k\frac{\partial T}{\partial z}+k\left(E''-\frac{E'}{T}\right)\left(\frac{\partial T}{\partial z}\right)^2+\right.$$

$$\left.+\frac{k}{H}\left(E'-\frac{2E}{T}\right)\frac{\partial T}{\partial z}+\frac{E}{H}\frac{\partial k}{\partial z}\right]. \tag{1.4.8}$$

Here

$$c = 0{,}622\frac{LE'}{c_p p},\quad b = \frac{\varkappa - 1}{\varkappa}\frac{T}{p} + 0{,}622\frac{LE}{c_p p^2}\frac{T}{\theta}.$$

If derivative dT/dt is found from the latter equation and substituted into eqn. (1.3.6), then we obtain an equation for $\mathbf{v}$ (not given here). Equation (1.4.8) gives us an idea of how complicated the equation describing the heat influx of a cloud is: it is nonlinear with respect to the temperature, and it has variable coefficients and a complex right-hand side. The equation for $\mathbf{v}$ has an equally complicated form.

The boundary conditions to which the unknown functions must conform correspond to moving cloud boundaries, and this complicates the solution of eqn. (1.4.8) even more.

There is another (major) difficulty which crops up when studying the cloud-formation process, aside from those mentioned above, which are associated with the complex form of the differential equations and the confitions at the moving boundaries. This new difficulty can be stated as follows: the moment of onset of cloud formation, and thus the transition to system (1.3.5), (1.3.6) from the system describing the transfer of heat and water vapor in a cloudless atmosphere, are unknown. Ascertaining the moment of onset is an important theoretical problem.

1.5. Cloud-droplet distribution functions

Like the water content, the cloud-droplet spectrum is an important cloud characteristic. The function $f(r_d, \mathbf{x}, t)$ the density of the droplet-radius r_d distribution at a time t in a physically infinitesimal volume dV containing a variable point $\mathbf{x}$, is a quantitative measure of the spectrum. Function f is usually normalized to unity:

$$\int_0^\infty f(r_d, \mathbf{x}, t)\, dr_d = 1.$$

By definition, $f(r_d, \mathbf{x}, t)\, dr_d dV$ is the probability of droplet fall with a radius lying in the interval $(r_d, r_d + dr_d)$ at a time t in a small volume dV containing the point with a radius vector $\mathbf{x}$.

Some other variable, l, is often used instead of the droplet radius r_d. If $\varphi(l, x, t)$ is the density of the droplet distribution, reckoned in terms of l, then f and φ are related as follows

$$\varphi(l, \mathbf{x}, t) = f(r_d, \mathbf{x}, t)\frac{dr_d}{dl}.$$

In particular, if $l = r_d^3$, that is, if it is proportional to the drop volume, then

$$\varphi_2(r_d^3, \mathbf{x}, t) = \frac{1}{3r_d^2} f(r_\text{к}, \mathbf{x}, t);$$

if $l = r_d^2$ that is, if it is proportional to the drop area, then

$$\varphi_1(r_d^2, \mathbf{x}, t) = \frac{1}{2r_d} f(r_d, \mathbf{x}, t).$$

In recent decades, many studies have been published in which the problem of droplet growth has been discussed, and, in particular, that of the formation of the cloud-droplet-size spectrum. During the initial stage of condensation, when the droplet spectrum is influenced directly by the spectrum of condensation nuclei, the determining factors will be such as the surface forces of a nucleus surrounded by water, the content of soluble hygroscopic substances in the nucleus, discontinuities in the water-vapor density and temperature at the drop surface, and electrical charges. This problem is well on the way towards being solved, and it is only with regard to the important question of the value of the condensation coefficient that there is still a lack of convincing data.

The further evolution of the cloud-droplet spectrum is influenced by condensation and coalescence, with a gradual predominance of the latter process over the former. The theory of regular droplet growth by condensation and coalescence proved unable to explain the broadening with time of the droplet spectrum during the stage when condensation processes are dominant, a broadening which has been established by numerous observations. According to this theorey, the droplet spectrum should become narrower as time goes by.

Belyaev [19] was apparently the first to point out that turbulent fluctuations of the air-flow velocity can lead to fluctuations of the specific density close to a droplet, and, in addition, to irregular (random) fluctuations of the rate of droplet growth. Subsequent theoretical advances by Sedunov [271], Mazin [163-165], Levin and Sedunov [154, 155], Stepanov [284, 285] and Voloshchuk and Sedunov [62-64] showed that it is precisely turbulence which is responsible for the broadening of the cloud--droplet spectrum.

1.5.1. Droplet growth by condensation in a turbuent medium

Voloshchuk and Sedunov [64] obtained an equation for the droplet distrubution function in a turbulent cloud during the initial stage of its development, when the factor determining droplet growth is water-vapor condensation. The initial equation describing the variation in the number of droplets in an undisturbed medium has the form

$$\frac{\partial f}{\partial t} + \operatorname{div}(\mathbf{v}f) + \frac{\partial}{\partial r_d}(\dot{r}_d f) = 0, \qquad (1.5.1).$$

where $\mathbf{v}$ is the instantaneous droplet velocity, which differs from the velocity $\mathbf{u}$ of the air by a term taking into account the fall (sedimentation) of droplets under the influence of gravity, and $\operatorname{div}\mathbf{v} = \operatorname{div}\mathbf{u}$. The rate of growth by condensation of an individual droplet, $\dot{r}_d = dr_d/dt$ which appears in eqn. (1.5.1), can be represented by Maxwell's formula

$$\dot{r}_d = \frac{D_e}{r_d \rho_d}\left[\rho_v - \rho_m(1+\Delta)\right], \tag{1.5.2}$$

where D_e is some effective diffusion coefficient, taking into account the effects of turbulence and of discontinuities in the water-vapor density and temperature at the droplet surface, as well as of the psychrometric correction, etc. Quantities ρ_d, ρ_v, and ρ_m in eqn. (1.5.2) are the densities of water (droplets), water vapor, and saturated water vapor over a plane surface of pure water, respectively, and Δ is a correction for the density of saturated water vapor, allowing for the effects which surface forces, salt content, electrical charges, and so forth, have on the vapor density at the droplet surface.

The rate of water-vapor condensation, which is equal to the rate of increase of the absolute water content of the cloud in time $(d\delta^*/dt)$, can be written as

$$\frac{d\delta^*}{dt} = \frac{4}{3}\pi\rho_d n \int_0^\infty \frac{dr_d^3}{dt} f\,dr_d = 4\pi\rho_d n \int_0^\infty r_d^2 f \dot{r}_d\,dr_d$$

or, taking eqn. (1.5.2) into account, as

$$\frac{d\delta^*}{dt} = (\rho_v - \rho_m)F_1 - \rho_m F_2. \tag{1.5.3}$$

Here, n is the number of droplets per unit volume,

$$F_1 = 4\pi n \int_0^\infty D_e r_d f\,dr_d, \quad F_2 = 4\pi n \int_0^\infty D_e r_d f \Delta dr_d. \tag{1.5.4}$$

The given system of equations is supplemented by the equation of conservation of the water-vapor mass

$$\dot{\rho}_v + \rho_v \operatorname{div}\mathbf{u} + \dot{\delta}^* = 0, \tag{1.5.5}$$

the heat-influx equation (first law of thermodynamics)

$$\rho c_p \dot{T} = \dot{p} + L\dot{\delta}^* + \rho\varepsilon; \tag{1.5.6}$$

and also the equation of state for air ($p = R\rho T$) and the equation of statics $\left(\dot{p} \approx w\frac{\partial p}{\partial z} = -g\rho w\right)$. In these equations, ε is the radiative and molecular heat influx, and w is the vertical velocity component.

From system (1.5.1)-(1.5.6) it follows that

$$\dot{\rho}_v = -\frac{\rho_v}{T}\left(\dot{T} + \frac{g}{R} w\right) - (\rho_v - \rho_m) F_1 + \rho_m F_2, \tag{1.5.7}$$

$$\dot{T} = -\frac{g}{c_p} w + \frac{L}{\rho c_p} [(\rho_v - \rho_m) F_1 - \rho_m F_2] + \frac{\varepsilon}{c_p},$$

$$\operatorname{div} \mathbf{u} = \frac{1}{T}\left(\dot{T} + \frac{g}{R} w\right).$$

This system can be simplified considerably. The supersaturation in a cloud at every stage of its development is very slight, because of the quite rapid molecular transfer of water vapor to condensation nuclei and also because of slow processes causing the water-vapor density to deviate from the saturated value. These processes are slow compared with phase-relaxation processes, which have a characteristic time $\tau_0 = 1/F_1$. Therefore assuming that $|\Delta| \ll 1$, and limiting ourselves to the first (linear) approximation (when expanding the quantities in terms of Δ), we obtain

$$\rho_v \approx \rho_m \left\{1 + \frac{1}{F_1}\left[F_2 - \frac{1}{T}\left(\dot{T} + \frac{g}{R} w\right) - \frac{\dot{\rho}_m}{\rho_m}\right]\right\}. \tag{1.5.8}$$

Next, by invoking the Clausius-Clapeyron equation

$$\dot{\rho}_m = \frac{\rho_m}{T}\left(\frac{L}{R_v T} - 1\right)\dot{T}, \tag{1.5.9}$$

on the basis of (1.5.2), (1.5.7), and (1.5.8) we obtain:

$$\dot{r}_d = \frac{D_э}{r_d F_1} \frac{\rho_m}{\rho_d} (a_1 w - a_2 \varepsilon + F_2 - F_1 \Delta), \tag{1.5.10}$$

$$\dot{T} = -\gamma_a' w + a_3 \varepsilon, \tag{1.5.11}$$

$$\operatorname{div} \mathbf{u} = b_2 w + \frac{a_3}{T} \varepsilon. \tag{1.5.12}$$

Here

$$a_1 = \frac{c_p g}{RT}\left(\frac{L}{c_p T} \frac{R}{R_v} - 1\right) a_3,$$

$$a_2 = \frac{L}{R_v T^2} a_3, \quad a_3^{-1} = c_p\left(1 + \frac{L}{c_p T} \frac{L}{R_v T} \frac{\rho_m}{\rho_к}\right),$$

$$\gamma_a' = g\left(1 + \frac{L}{RT} \frac{\rho_m}{\rho}\right) a_3, \quad b_2 = \frac{1}{T}\left(\frac{g}{R} - \gamma_a'\right),$$

γ_a' being the moist-adiabatic gradient.

The systems of equations (1.5.1), (1.5.10)-(1.5.12) also forms a basis for the description of condensation processes in a turbulent medium. Since these equations were derived for a time considerably longer than the time of phase relaxation ($t \gg \tau_0$), they cannot be used to describe a process at the initial period of time ($t \leqslant \tau_0$) or in thin layers close to cloud boundaries. On the other hand, with the aid of the above system it is possible to take into account, firstly, the effect of turbulence with a very small scale (for instance, through coefficient D_e) and, secondly, turbulent fluctuations with lifetimes τ_* which are considerably longer than both τ_0 ($\tau_* \gg \tau_0$) and the droplet relaxation time τ_v ($\tau_* \gg \tau_v$). As mentioned above, the presence of vertical-velocity fluctuations gives rise to fluctuations in the water-vapor density and the condensation rate, as well as to a broadening of the droplet spectrum.

In order to allow for the effect of turbulent velocity fluctuations on the droplet distribution, let us express all the quantities as sums of the mean (regular) value and the deviation (fluctuation) from it:

$$f = \bar{f} + f', \quad w = \bar{w} + w', \quad r_d = \bar{r}_d + r_d' \quad \text{etc.}$$

where $\mathbf{v}' = \mathbf{u}'$, since, for $\tau_* \gg \tau_v$, the fluctuations in velocity and the droplet fluctuations are identical (complete entrainment of droplets by turbulent air parcels).

If the quantities, written in this form, are now substituted into eqns. (1.5.1) and (1.5.10)-(1.5.12) and these are averaged, after some tedious (albeit not complicated) operations we obtain the following equation for the droplet-size distribution function:

$$\frac{df}{dt} - v_s \frac{\partial f}{\partial z} + \frac{\rho_m}{F_1 \rho_d} \frac{\partial}{\partial r_d} \frac{D_e}{r_d} \Big\{ a_1 w + a_2 \varepsilon + F_2 - F_1 \Delta +$$
$$+ a_1 k_z \Big[\frac{d \ln \rho_m}{dT} \Big(\frac{\partial T}{\partial z} + \gamma_a' \Big) + \frac{1}{F_1} \Big(\frac{\partial F_1}{\partial z} + \frac{a_1}{F_1} \frac{\rho_m}{\rho_\kappa} F_* \Big) \Big] \Big\} f =$$
$$= \frac{\partial}{\partial z} k_z \frac{\partial f}{\partial z} + \frac{a_1}{F_1} \frac{\rho_m}{\rho_d} \frac{\partial}{\partial r_d} \frac{D_e}{r_d} \times$$
$$\times k_z \Big(\frac{\partial}{\partial z} + b_2 + \frac{a_1}{F_1} \frac{\rho_m}{\rho_d} \frac{\partial}{\partial r_d} \frac{D_e}{r_d} \Big) f. \qquad (1.5.13)$$

Here, all the quantities are mean values (the averaging bar is dropped). In (1.5.13)

$$df/dt = \partial f/\partial t + u\,\partial f/\partial x + v\,\partial f/\partial y + w\,\partial f/\partial z,$$

$$F_* = 4\pi \int_0^\infty D_e r_d \frac{\partial}{\partial r_d} \Big(\frac{D_e}{r_d} f \Big) dr_\kappa,$$

where v_s is the fall velocity, and k_z is the turbulence coefficient in the vertical direction.

While deriving eqn. (1.5.13), we made wide use of the concepts and relations of the semiempirical theory of turbulence. With the help of these, the second moments of the fluctuations could be related to the derivatives, with respect to the coordinates, of the mean values of the quantities. Although in this form eqn. (1.5.13) is very complicated, it contains only those terms which are attributable to the fluctuations of the vertical component of the air velocity (which is in turn responsible for the vertical turbulent mixing). If the fluctuations of the horizontal velocity components are taken into account, then eqn. (1.5.13) will include additional terms containing k_x and k_y.

Given the above assumptions, the equations for T and the relative supersaturation Δ become

$$\frac{dT}{dt} + \gamma_a' w = \frac{\partial}{\partial z} k_z \left(\frac{\partial T}{\partial z} + \gamma_a'\right) + a_3 \varepsilon, \quad (1.5.14)$$

$$\frac{d\Delta}{dt} = \frac{\partial}{\partial z} k_z \left(\frac{\partial \Delta}{\partial z} + b_2\right). \quad (1.5.15)$$

We recall that, when deriving all these equations, the entrainment of cloud droplets by turbulent moles is assumed to be complete, and Δ was assumed to be small in comparison with unity.

The authors of [64] emphasize that vertical turbulent mixing still plays a role even when there is no supersaturation of water vapor ($\Delta = 0$) and when the fall velocity of the droplets is negligible ($v_s \approx 0$). Equations (1.5.13) and (1.5.14) describe, in this case, the evolution of the temperature field and droplet spectrum from an arbitrary initial state to the steady state, for which in a closed cloud the following conditions are valid,

$$\frac{\partial T}{\partial z} + \gamma_a' = 0; \quad \frac{\partial f}{\partial z} + b_2 f + \frac{a_1}{F_1} \frac{\rho_m}{\rho_d} \frac{\partial}{\partial r_d}\left(\frac{D_e}{r_d} f\right) = 0, \quad (1.5.16)$$

signifying that neither the temperature nor the droplet spectrum varies with time. If the cloud is not closed, then at its boundaries neither of conditions (1.5.16) is satisfied, and in this case eqns. (1.5.13) and (1.5.14) describe the evolution of the temperature field and the droplet spectrum prior to the steady state, determined by the transfer of either heat or droplets through the cloud boundaries.

Even in the form given by (1.5.13), the equation for the function f has not so far been utilized in the modeling of cloud-formation processes. How justified are the premises adopted when deriving the equation for f (relations between scales τ_0, τ_*, τ_v, etc.), and whether this equation was simplified legitimately, are questions which, like some others, must await comprehensive experimental investigations.

1.5.2. Droplet growth by coalescence in a turbulent medium

Like condensation, coalescence is also an important factor in the formation of the droplet spectrum. Droplets in clouds collide and merge together for a variety of reasons: the difference in the falling speeds of droplets of different sizes in the gravitational field (gravitational coalescence); the result of electrical and induction forces (electrostatic coalescence); because droplets are drawn into the hydrodynamic wakes of other droplets; the random molecular fluctuations of the density and mean velocity of air molecules (Brownian coalescence); differences in temperature and pressure; etc.

Turbulence has a significant and unique effect on the coalescence of droplets. If the internal scale of turbulence (according to Kolmogorov) is less than or comparable to the particle sizes, then the turbulence will cause particles to migrate, facilitating collisions between them (so-called direct turbulent coalescence). However, in the atmosphere, the internal scale of turbulence is usually not less than 0.1 cm. Consequently, the direct effect of turbulence is exerted on very large particles (snowflakes or raindrops).

If, on the other hand, the internal scale of turbulence is much greater than the particle size, then the turbulence will intensify the coalescence caused by other factors. For instance, the presence of turbulence enhances the entrainment of droplets into the hydrodynamic wakes of other droplets or into the spheres of action of electrical forces, etc. Turbulence creates inhomogeneities of the air-velocity field, which in turn make it easier for droplets to approach one another.

Since the distances between particles are considerably greater than their sizes, we can limit ourselves to just paired collisions (the probability of a collision between three or more particles is negligible compared to the probability of a two-particle collision).

The equation for the droplet-volume v_d distribution function $\varphi_2(v_d, t)$, allowing for the variation in the droplet number due just to coalescence (paired collisions), has the forllowing form:

$$\frac{\partial\varphi_2}{\partial t} + \operatorname{div} \mathbf{v}\varphi_2 = \frac{1}{2}\int_0^{v_d} P(v_d', v_d - v_d')\varphi_2(v_d - v_d')\varphi_2(v_d')\,dv_d' - \int_0^{\infty} P(v_d, v_d')\varphi_2(v_d)\varphi_2(v_d')\,dv_d', \tag{1.5.17}$$

where **v** is the droplet velocity, and $P(v_d, v'_d)$ is the probability of a collision between droplets of volume v_d and v'_d.

Equation (1.5.17) has been widely used to describe the coalescence of aerosol particles for quite a long time, beginning with the works of Smolukhovskii (1936), who gave this equation in discrete form, and Müller (1928), who was the first to establish the form of this equation for a continuous variation of the independent variables (radius, volume, etc.). The monograph of Voloshchuk and Sedunov [63] includes a detailed discussion of the physical premises, the derivation of the coalescence equation, and the statistical basis, as well as the various forms of the coalescence equation. Following Voloshchuk [62], let us consider the effect of turbulence on the coalescence of cloud droplets.

If, as previously, we represent φ_2 as the sum of the mean value $(\bar{\varphi}_2)$ and the deviation from it (φ'_2), and substitute this sum into eqn. (1.5.17) and average it, we obtain an equation for $\bar{\varphi}_2$. By subtracting the obtained equation (for $\bar{\varphi}_2$) from eqn. (1.5.17), we obtain an equation for φ'_2. In general, however, these equations are extremely complicated, and so we will not give them here. Only in certain extreme cases will the equation for $\bar{\varphi}_2$ have a usable form. Let us assume that:

(a) turbulent fluctuations of $P(v_d, v'_d)$ can be neglected; this assumption is based on the very likely conjecture that the small-scale turbulence, which can affect $P(v_d, v'_d)$ correlates slightly with the large-scale turbulence producing the fluctuations of the cloud-droplet spectrum;

(b) the influence of the compressibility of the medium on the cloud-droplet spectrum is only slight; for quite wide assumptions it can be taken that

$$\operatorname{div} \mathbf{v} \approx \frac{1}{\lambda} \bar{w}, \quad \operatorname{div} \mathbf{v}' \approx \frac{1}{\lambda} w',$$

where $\bar{w}$ is the mean value of the vertical velocity, w' is its fluctuation, and λ is a parameter which is slightly temperature-dependent and is of the order of 10^4 m in value;

(c) the typical lifetime of a turbulent air parcel is considerably shorter than the characteristic time of variation of the droplet spectrum due to coalescence, which is of the order of 10^3–10^4 s.

If, in addition, we use the basic relations of the semi-empirical theory of turbulence (the scheme of nonlocal parametrization worked out by Voloshchuk), then we can put the equation for the mean value of the distribution function $\bar{\varphi}_2(v_d, t)$ of passive-droplet volumes during the coalescence stage into the following form:

$$\frac{\partial\bar{\varphi}_2}{\partial t}+\bar{v}_i\frac{\partial\bar{\varphi}_2}{\partial x_i}=\frac{\partial}{\partial x_i}k_{ij}\frac{\partial\bar{\varphi}_2}{\partial x_j}+\frac{1}{2}\int_0^{v_d}P(v_d', v_d-v_d')\bar{\varphi}_2(v_d-v_d')\times$$

$$\times\bar{\varphi}_2(v_d')dv_d'+\frac{1}{2}\int_0^{v_d}\lambda_{ij}P(v_d', v_d-v_d')\frac{\partial\bar{\varphi}_2(v_d-v_d')}{\partial x_i}\frac{\partial\bar{\varphi}_2(v_d')}{\partial x_j}dv_d'-$$

$$-\int_0^{\infty}P(v_d, v_d')\bar{\varphi}_2(v_d)\bar{\varphi}_2(v_d')dv_d'-$$

$$-\int_0^{\infty}\lambda_{ij}P(v_d, v_d')\frac{\partial\bar{\varphi}_2(v_d)}{\partial x_i}\frac{\partial\bar{\varphi}_2(v_d')}{\partial x_j}dv_d'. \quad (1.5.18)$$

Here, $\bar{v}_i$ is the component of the mean velocity of the air in the x_i-direction, k_{ij} are the tensor components of the turbulence coefficient, $\lambda_{ij} = L_iL_j$ (where L_j is the effective turbulence scale of the jth direction), and the recurring subscripts in the products indicate summation over the index (from 1 to 3).

Although a number of important limiting conditions were introduced during the derivation of eqn. (1.5.18), this is still so complicated that neither a qualitative analysis of it nor a quantitative evaluation of $\bar{\varphi}_2$ on the basis of this equation can be obtained.

1.6. Equations of stochastic condensation

In a series of works [77, 213, 284, 285], Stepanov and his coworkers attempted to construct a fairly complete system of equations describing the macrophysical conditions of cloud formation and the microphysical processes of droplet growth. Using, as a basis, equations for the temperature, specific moisture content, and water content of a cloud, together with the droplet-size (surface area) distribution function, and then averaging these, Stepanov derived equations for the mean values of the temperature T, specific water content δ, droplet number n, distribution function φ_1 and water-vapor supersaturation Δ. The initial equations are valid if the product $nr_d^3 \ll 1$, and if the characteristic distance l, and characteristic time t, of the variation in T, φ_1, ρ_v, and w are much greater than $n^{-1/3}$ and $n^{-2/3}D^{-1}$, respectively. In this case T, φ_1, ρ_v, and w at a point will refer to characteristics of a single real system, averaged over volumes much smaller than l^3. On the other hand, the physically small volume has to be so selected that all the properties of the system, described by the distribution function l^3, are realized. As applied to real clouds, this means that the characteristic time and distance have lower limits of the order of 10^{-2} s and 10^{-1} cm, respectively. The field of turbulent motions is characterized by the velocity $\boldsymbol{w}$ which is unaffected by the fields of T, ρ_V, and δ. This means that the eddies have an upper size limit dictated by a scale associated with the buoyancy force, the magnitude of which, in clouds, is several hundred meters.

Under actual atmospheric conditions the time of phase relaxation τ_0 is much less than the Lagrangian integral time scale of turbulence t_L (that is, $\tau_0 \ll t$). For this condition the approximation $\Delta \ll \delta$ is valid with a high degree of accuracy (the supersaturation of the vapor is considerably less than the cloud water content). The following assumptions are also made:

(1) terms of the order of τ_0/t_L, τ_0/t, and t_L/t can be dropped;

(2) the motion of turbulent air parcels is adiabatic;

(3) the diffuse approximation can be used (the space-time scales of the variation of the mean values are much greater than the typical scales of the fluctuating quantities);

(4) the relation $\overline{v_i'(t)v_i'(t)}=t_L k_{ij} \exp(-|t-t_0|/t)$ can be used to correlate the velocity fluctuations $v\ (t)$, where k_{ij}, a tensor, is the turbulence coefficient;

(5) when calculating the turbulent fluctuations, the quantities τ_0 and $\tilde{\rho m}$ can be put into the form $\tau_0 = \tau_0(\delta, n, c, t)$ and $\tilde{\rho m} = \tilde{\rho m}(\delta^*, n, T, c, t)$, where c is the specific salinity of the condensation nuclei. The explicit dependence of τ_0 and $\tilde{\rho m}$ on time is introduced in order to allow for the effect on τ_0 and $\tilde{\rho m}$ of the modification of the droplet spectrum under steady-state conditions;

(6) by the initial moment, the droplet spectrum has "separated" from the spectrum of dry condensation nuclei, that is, by the initial moment, drops have formed on practically all the active condensation nuclei and have grown so much that there are essentially no drops left with radii close to those of the condensation nuclei (r_c); thus, the activation of new condensation nuclei can be neglected.

If we make the foregoing assumptions, the system of equations describing the dynamics of cloud formation and droplet growth under the influence of stochastic (turbulent) condensation becomes:

$$\left(\frac{\partial}{\partial t}+u\frac{\partial}{\partial x}+v\frac{\partial}{\partial y}+w\frac{\partial}{\partial z}-\frac{\partial}{\partial z}k\frac{\partial}{\partial z}\right)\mathbf{A}=0, \tag{1.6.1}$$

$$\mathbf{A}=\left(q_m+\delta,\ T-\frac{L}{c_p}\delta,\ n,\ c\right), \tag{1.6.2}$$

$$\frac{\partial\delta}{\partial t}+u\frac{\partial\delta}{\partial x}+v\frac{\partial\delta}{\partial y}+w\frac{\partial\delta}{\partial z}=\frac{\partial}{\partial z}k\frac{\partial\delta}{\partial z}+\frac{\Delta_1}{\rho\tau}-$$

$$-\frac{c_p}{L}\frac{\partial}{\partial z}k\frac{\partial b}{\partial z}+d,\quad \frac{\partial\varphi_1}{\partial t}+u\frac{\partial\varphi_1}{\partial x}+v\frac{\partial\varphi_1}{\partial y}+w\frac{\partial\delta}{\partial z}+ \tag{1.6.3}$$

$$+\frac{\partial}{\partial F}\left\{\frac{8\pi D}{\rho_d}\left[\rho_v-\rho_m(1+\Delta)\right]+\lambda\right\}\varphi_1=$$

$$=\left(\frac{\partial}{\partial z}+\frac{\partial\rho^*}{\partial z}\frac{\partial}{\partial F}\right)\left(k\frac{\partial\varphi_1}{\partial z}+k\frac{\partial\rho^*}{\partial z}\frac{\partial\varphi_1}{\partial F}\right)+I. \tag{1.6.4}$$

Here, $F = 4\pi r_d^2$ is the surface area of a droplet of radius ,

$$\Delta_1=\rho_v-\rho_m\left(1+\tilde{\Delta}\right),$$

$$\tilde{\Delta}=\overline{r_d\varphi_1\Delta}/\overline{r_d\varphi_1},$$

$$\frac{\partial b}{\partial z}=\frac{\rho_d}{8\pi D\tau_0\rho}\frac{L}{c_p}\frac{\partial\rho^*}{\partial z},$$

$$\frac{\partial\rho^*}{\partial z}=-\frac{8\pi D\tau_0}{\rho_d}\frac{g}{c_p\gamma_1}\tilde{\rho}_m\left(\frac{L}{c_pT}\frac{L}{R_vT}-\frac{L}{RT}\right),$$

$$\tilde{\rho}_m=\rho_m\left(1+\tilde{\Delta}\right),$$

$$\gamma_1=1+\frac{\partial\tilde{\rho}_m/\rho}{\partial\delta}+\frac{L}{c_pT}\frac{L}{R_vT}\frac{\tilde{\rho}_m}{\rho},$$

$$I\varphi_1=2k\frac{\partial}{\partial F}\int_0^t dt_1\left\{N(t,t_1)\left[\frac{\partial\varphi_1(t_1)}{\partial z}+\frac{\partial\rho^*}{\partial z}\frac{\partial\varphi_1(t_1)}{\partial F}\right]+\right.$$

$$\left.+\int_0^t dt_2 N(t,t_2)\frac{\partial}{\partial F}N(t_1,t_2)\right\}\varphi_1(t_1),$$

$$N(t,t_1)=-\frac{1}{\gamma_1}\frac{8\pi D}{\rho_d}\frac{L}{c_p}\frac{\partial\rho_m}{\partial T}\left(\tilde{\Delta}-\Delta\right)\left\{\frac{\partial}{\partial z}\frac{\rho\delta+\rho_m}{\rho}\times\right.$$

$$\times(t_1)-\frac{\partial(\tilde{\rho}_m/\rho)}{\partial n}\frac{\partial}{\partial z}n(t_1)-\frac{\partial(\tilde{\rho}_m/\rho)}{\partial(c/\rho)}(t)\frac{\partial(c/\rho)}{\partial z}(t_1)+$$

$$\left.+\left[1+\frac{\partial(\tilde{\rho}_m/\rho)}{\partial\delta}(t)\right]-\left[\frac{c_p}{L}\frac{\partial\theta(t_1)}{\partial z}-\frac{\partial\delta(t_1)}{\partial z}\right]\right\},$$

$$\lambda=\overline{r_dFI\varphi_1}/\overline{r_d\varphi_1},\quad \tau_0^{-1}=4\pi D\overline{r_d\varphi_1},$$

$$d=-k\tau_0\rho\frac{c_p}{L}\frac{\partial b}{\partial z}\left[\frac{\partial}{\partial z}\left(\frac{1}{\tau_0\rho}\right)-\frac{c_p}{L}\frac{\partial b}{\partial z}\frac{\partial}{\partial\delta}\left(\frac{1}{\tau_0\rho}\right)\right];$$

a bar indicates averaging with respect to droplet size (area) and with respect to the content of salts.

The rate of increase of the droplet surface area F is found using Maxwell's formula

$$\frac{dF}{dt} \doteq \frac{8\pi D}{\rho_d}[\rho_v - \rho_m(1+\Delta)]. \qquad (1.6.5)$$

All the quantities in the system (1.6.1)-(1.6.5) are averaged over the turbulent fluctuations. It is clear that the first and second equations of system (1.6.1) are the equations for the invariants introduced by us, namely the specific moisture content s and function Π^*. The third equation (for the number of droplets n per unit air mass) is a corollary of the assumption that no new droplets are being formed (they have all formed on condensation nuclei by the initial moment): naturally, under these conditions n can vary only as a result of transfer and turbulence.

From the system (1.6.1)-(1.6.3), we can obtain an expression for the excess of the water-vapor density in cloud air over the vapor density at the droplet surface, that is, for the difference $\rho_v - \rho m(1 + \Delta)$. This equation turns out to be algebraic. Consequently, for $t \gg \tau_0$ the initial problem for vapor supersaturation cannot be formulated.

The system of equations (1.6.1)-(1.6.5) possesses more generality in comparison with the system presented above (that of Sedunov and Voloshchuk). The system (1.6.1)-(1.6.5) was derived without assuming that the relative supersaturation $\Delta = (\rho_v - \rho m)/$ $/\rho m$ is a small quantity. During the initial condensation stage, it may differ considerably from unity. For instance, taking into account the surface tension σ and the content of salts, the quantity Δ can be represented as

$$\Delta = \frac{2\sigma}{R_v \rho_d T r_d} - \frac{a_0 v_c}{v_d - b_0 v_c},$$

where v_c is the volume of a condensation nucleus, v_d is the volume of a droplet, and a_0, b_0 are constants for a given nucleus (for NaCl $a_0 = 1.215$ and $b_0 = 1.815$; for $CaCl_2$ $a_0 = 0.91$ and $b_0 = 3.16$).

In the studies discussed above, formulas were obtained for the mean surface area of a droplet $\bar{F}$ and the variance of this quantity. In these studies, the effects of various factors (nonuniformity of the fields of potential and pseudopotential temperature, specific moisture content, droplet number, local disturbances, and nonuniformity of the field of condensation nuclei) on the changes in cloud characteristics (water content and droplet spectrum) have been considered.

For constant gradients of the pseudopotential temperature, droplet number, and specific moisture content or concentration of condensation nuclei, the droplet spectrum will broaden with time because of turbulence, and it will shift into the region of smaller sizes.

CHAPTER 2

MODELS OF STRATIFORM-CLOUD FORMATION

2.1. Solution of the system of equations for heat and moisture exchange in a turbulent atmosphere

Using the results of Chapter 1 as a basis, let us write, in the following form, the initial equations describing the formation of clouds in the atmosphere:

$$\frac{d\theta}{dt} = \frac{\partial}{\partial z} k \frac{\partial \theta}{\partial z} + \frac{L\nu}{c_p \rho}, \tag{2.1.1}$$

$$\frac{dq}{dt} = \frac{\partial}{\partial z} k \frac{\partial q}{\partial z} - \frac{\nu}{\rho}. \tag{2.1.2}$$

This system differs from the system (1.3.5) and (1.3.6) only in that we have neglected the dependence of density on height. This is quite justified, since the second factor of the product ρk has not been studied sufficiently (especially above the ground layer). The errors arising due to the uncertainty of the dependence of k on z will naturally be compensated by the inaccuracies introduced by assuming the density to be constant with height. In the system (2.1.1) and (2.1.2), the previously introduced turbulence coefficient k_z is designated as k, while the radiative heat influx is dropped.

Since, as mentioned above, the direct solution of system (2.1.1), (2.1.2), involves serious difficulties (technical as well as theoretical), investigators have concentrated on devising methods which tend to eliminate the most important of these difficulties.

The main physical premise of the method proposed by the author [179, 181] is, as stated above, the assumption that cloud water droplets and ice crystals are more-or-less *completely entrained* by the air particles (moles) which participate in the turbulent mixing. Let us now present an argument in favor of this hypothesis of total entrainment. If at the initial moment a droplet of radius r_d has a velocity v_0 (relative to the air particle in which it is located), then the maximum distance (the so-called inertial traversed path) l_i traversed by the droplet in the air particle is given by the formula (see [192]):

$$l_i = v_0 \tau = \frac{2 v_0 r_d^2 \rho_d}{9 \eta},$$

where τ is the relaxation time, ρ_d is the droplet (water) density and $\eta = 1.4 \cdot 10^{-4}$ to $1.9 \cdot 10^{-4}$ g/(cm.s) is the molecular viscosity of air. Table 2.1 gives values of l_i for various values of v_0 and r_d (for $\eta = 1.72 \cdot 10^{-4}$ g/(cm.s) and $\rho_d = 1$ g/cm^3).

Table 2.1. Inertial traversed path of droplet (cm)

v_0, cm/s	r_d μm				
	1	5	10	50	100
1	$1,29 \cdot 10^{-5}$	$3.23 \cdot 10^{-4}$	$1,29 \cdot 10^{-3}$	$3.23 \cdot 10^{-2}$	$1.29 \cdot 10^{-1}$
5	$6,46 \cdot 10^{-5}$	$6.62 \cdot 10^{-3}$	$6.46 \cdot 10^{-3}$	$1,62 \cdot 10^{-1}$	$6.46 \cdot 10^{-1}$
10	$1,29 \cdot 10^{-3}$	$3,23 \cdot 10^{-3}$	$1.29 \cdot 10^{-2}$	$3,23 \cdot 10^{-1}$	1.29
50	$6,46 \cdot 10^{-4}$	$1,62 \cdot 10^{-2}$	$6.46 \cdot 10^{-2}$	1,62	6.46
100	$1.29 \cdot 10^{-3}$	$3,23 \cdot 10^{-2}$	$1,29 \cdot 10^{-1}$	3,23	12.92

Traversing the path l_i, the droplet begins to move together with the air particle (is completely entrained by it). Since the size of a turbulent particle L_T is considerably greater than l_i (according to estimates, L_T usually ranges from tens of centimeters to hundreds of meters, or even several kilometers), it is obvious that drops with radii less than 50-100 μm will almost all participate in the turblent motions of air particles (moles).

Assuming complete entrainment of cloud droplets and ice crystals by turbulent air parcels, we can write the equation for the specific water content δ of a cloud as

$$\frac{\partial \delta}{\partial t} + u\frac{\partial \delta}{\partial x} + v\frac{\partial \delta}{\partial y} + w\frac{\partial \delta}{\partial z} = \frac{\partial}{\partial z} k \frac{\partial \delta}{\partial z} + \frac{\nu}{\rho}. \tag{2.1.3}$$

System (2.1.1)-(2.1.3) for a cloud whose water vapor is in the saturated state is supplemented by the relation

$$q = q_m = 0,622\,\frac{E(T)}{p}, \tag{2.1.4}$$

where $E(T)$ is the saturation vapor pressure (WMO formulas are used to calculate this quantity).

Now, if we introduce the specific moisture content

$$s = q + \delta, \tag{2.1.5}$$

the addition of eqns. (2.1.2) and (2.1.3) then gives

$$\frac{\partial s}{\partial t} + u\frac{\partial s}{\partial x} + v\frac{\partial s}{\partial y} + w\frac{\partial s}{\partial z} = \frac{\partial}{\partial z} k \frac{\partial s}{\partial z}. \tag{2.1.6}$$

A second differential equation that does not contain the condensation rate ν can be obtained by adding eqns. (2.1.1) and (2.1.2), after first multiplying (2.1.2) by L/c_p:

$$\frac{\partial \Pi}{\partial t}+u\frac{\partial \Pi}{\partial x}+v\frac{\partial \Pi}{\partial y}+w\frac{\partial \Pi}{\partial z}=\frac{\partial}{\partial z}k\frac{\partial \Pi}{\partial z}, \tag{2.1.7}$$

where

$$\Pi=\Theta+\frac{L}{c_p}q. \tag{2.1.8}$$

In addition to invariant Π, let us introduce a second invariant quantity

$$\Pi^*=\Theta-\frac{L}{c_p}\delta,$$

the equation of which is obtained by subtracting eqn. (2.1.3) from (2.1.1). The equation for Π^* has the same form as the one for Π, and function Π^* is a linear combination of Π and s:

$$\Pi^*=\Pi-\frac{L}{c_p}s.$$

Quantities s and Π possess a very important property: they are invariant with respect to the cloud-formation process, that is, differential equations (2.1.6) and (2.1.7) retain the very same from prior to the onset of water-vapor condensation and after it.

If there is no turbulent mixing, then according to (2.1.6) and (2.1.7) the individual derivatives ds/dt and $d\Pi/dt$ will be equal to zero. Physically speaking, this means that in a moving mass of air, functions s and Π do not vary with time (s = const, Π = const), regardless of whether water vapor condenses or not. Consequently, here we have two invariants (in the absence of turbulent mixing) which remain constant (in time) both before water vapor starts to condense and during the formation and evolution of the cloud. As equations (2.1.6) and (2.1.7) indicate, s and Π will vary in a moving air mass under the influence of turbulent mixing (but not water-vapor condensation).

The system of differential equations (2.1.6), (2.1.7), is considerably simpler than the system (2.1.1), (2.1.2), in which fundamental complications arise in connection with the quantity ν.

Since it is tropospheric clouds which are the object of study (above the tropopause nacreous and noctilucent clouds are observed comparatively rarely, and they are of theoretical interest only), eqns. (2.1.6) and (2.1.7) are to be solved for a bounded (but not semi-infinite) layer, that is, for the troposphere. This simplifies the problem considerably from the mathematical point of view, especially the calculations. Moreover, the boundary conditions pertaining to a finite height (the tropopause) can be verified experimentally, which is impossible for a condition at infinity.

The boundary conditions for the unknown functions (we will give them only for s) are written in the form

$$-\frac{\partial s}{\partial z} = \beta' (s - s_1) + s_3 \text{ for } z = 0 \text{ (Earth's surface)}, \quad (2.1.9)$$

$$-\frac{\partial s}{\partial z} = \beta'' (s - s_2) + s_4 \text{ for } z = H \text{ (tropopause)}, \quad (2.1.10)$$

where $s_1(x, y, t)$, $s_3(x, y, t)$, $s_4(x, y, t)$ are known functions, and β', β'' are coefficients of moisture exchange.

From (2.1.9) and (2.1.10), as particular cases, other forms of the boundary conditions can be obtained. For instance, if $\beta' = \beta'' = 0$, these equations become

$$-\frac{\partial s}{\partial z} = s_3(x, y, t) \text{ for } z = 0, \quad -\frac{\partial s}{\partial z} = s_4(x, y, t) \text{ for } z = H, \quad (2.1.11)$$

that is, at the boundaries of the troposphere the water-vapor transport is known. If $1/\beta' = 1/\beta'' = 0$, then

$$s = s_1(x, y, t) \text{ for } z = 0, \quad s = s_2(x, y, t) \text{ for } z = H, \quad (2.1.12)$$

that is, the specific moisture contents at the Earth's surface and at the tropopause are known.

In cases where the rate of horizontal transfer varies little with height (u, v = const), new independent invariables $\zeta = x - ut$ and $\eta = y - vt$ (variables z and t remain as before) can be introduced, and equations (2.1.6) and (2.1.7) become

$$\frac{\partial s}{\partial t} + w \frac{\partial s}{\partial z} = \frac{\partial}{\partial z} k \frac{\partial s}{\partial z}, \quad (2.1.13)$$

$$\frac{\partial \Pi}{\partial t} + w \frac{\partial \Pi}{\partial z} = \frac{\partial}{\partial z} k \frac{\partial \Pi}{\partial z}. \quad (2.1.14)$$

Solutions of the system (2.1.4)-(2.1.8) will be presented and discussed in this and the following chapters. Here, let us express them in the form of certain known functions of the independent variables:

$$s(x, y, z, t) = f_1(x, y, z, t), \quad (2.1.15)$$

$$\Pi(x, y, z, t) = f_2(x, y, z, t). \quad (2.1.16)$$

Let us show how, knowing the form of functions f_1 and f_2, we can specify the quantities most important for the pheonomena in question: the water content and temperature of the cloud.

Inside the cloud, water vapor is in the saturated state, so that the specific humidity is related to the temperature by relation (2.1.4). The specific water content will be, according to (2.1.5), equal to the difference

$$\delta = s - q_m. \tag{2.1.17}$$

Inside the cloud relation (2.1.8) becomes, in view of (2.1.4) and (2.1.16),

$$\Theta + \frac{L}{c_p} q_m = f_2. \tag{2.1.18}$$

Now let us use the Clausius-Clapeyron equation (in integral form)

$$\ln \frac{E}{E_0} = \frac{L}{R_v}\left(\frac{1}{T_0} - \frac{1}{T}\right), \tag{2.1.19}$$

where E_0 is the saturation vapor pressure at a temperature T_0, and the formula $\Theta = T(1000/p)^{(\varkappa-1)/\varkappa}$, to transform equation (2.1.18) into the form

$$c_1 \frac{q_m}{q_m(0)} - \frac{c_2\left(\frac{1000}{p}\right)^{\frac{\varkappa-1}{\varkappa}}}{\ln \frac{q_m}{q_m(0)} + \ln \frac{p}{1000} - \frac{c_2}{T_0}} = f_2(x, y, z, t) \tag{2.1.20}$$

or

$$T\left(\frac{1000}{p}\right)^{\frac{\varkappa-1}{\varkappa}} + c_1 \frac{1000}{p} \exp\left[c_2\left(\frac{1}{T_0} - \frac{1}{T}\right)\right] = f_2(x, y, z, t), \tag{2.1.20'}$$

where $c_1 = Lq_m(0)/c_p$, $c_2 = L/R_v$, $\varkappa = c_p/c_v$ are constants, and $q_m(0)$ is the saturation specific humidity for a temperature T_0 and a pressure of 1000 gPa.*) The calculation of q_m and T according to formulas (2.1.20) and (2.1.20') is carried out with the help of a special nomogram, the coordinates of which are the function Π and the logarithm of the pressure p. The two families of curves on this nomogram portray isolines of the saturation specific humidity and isotherms [187].

There is another (numerical) means of determining the temperature according to the known function Π, which is especially convenient when the system of equations is integrated using numerical methods. Actually, if at some ith step (in time) the increment of Π is $\Delta\Pi_i = \Pi_i - \Pi_{i-1}$ then the temperature increment $\Delta T_i = T_i - T_{i-1}$

---

*) Recall that 1 Pa = 1 N/m^2 = .01 mb (Translator).

in a cloud at a level with a pressure p can be calculated from the relation

$$\Delta\Pi_i = \left[\left(\frac{1000}{p}\right)^{\frac{\varkappa-1}{\varkappa}} + 0{,}622\frac{L^2}{c_p p}\frac{E_{i-1}}{R_n T_{i-1}^2}\right]\Delta T_i,$$

where E_{i-1} is the saturation vapor pressure at a temperature T_{i-1} To obtain this relation, we used formula (2.1.8), differentiated with respect to time, expression (2.1.4), and the Clausius-Clapeyron equation. After calculating ΔT_i using the latter relation, we find the temperature in the cloud at the ith step of integration:

$$T_i = T_{i-1} + \Delta T_i.$$

If the above method is used to find the specific moisture content s and the saturation specific humidity q_m then first of all it should be possible to answer the question, so important in practice, of whether at a given moment there will be clouds in the specified layer, or not. According to the foregoing, clouds will be observed in those layers where the inequality $s \geqslant q_m$ holds true, and they will be absent in the layer if $s < q_m$

The method also makes it possible to calculate the quantitative characteristics of developing cloudiness, and, in particular, formula (2.1.17) can be applied to find its specific water content. The calculation is carried out for all levels where the water content is positive. As soon as negative values of the water content are obtained, corresponding to levels under the lower cloud boundary and above the upper boundary, the calculation of δ is terminated. The heights of the lower and upper cloud boundaries are found from the equation

$$f_1(x, y, z, t) = q_m,$$

since at these levels the cloud water content is zero.

For levels situated outside the cloud, the potential temperature (and the kinetic temperature as well) is calculated using the relation

$$\Theta(x, y, z, t) = f_2(x, y, z, t) - \frac{L}{c_p} f_1(x, y, z, t).$$

By carrying out calculations for various times (for example, 6, 12, 18, and 24 h after the initial time), we are able to monitor the evolution of the fields of water content, humidity, and temperature, as well as the variations of the cloud boundaries with time at different points.

2.1.1. *Taking into account the fall of cloud elements and the fallout of precipitation*

If the water droplets and ice crystals become large enough (tens or hundreds of microns in radius), so that precipitation starts to fall from the cloud, then it is necessary to take into account the transport of moisture caused by the fall of cloud elements under the influence of gravity. Let us designate as Q_d the transport of drops [droplets] and crystals due to gravity (its units are kg/m^{2}s)) and let us carry out the usual operations. Then the equation for the transfer (balance) of moisture, that is, (2.1.6), can be written as

$$\frac{ds}{dt}=\frac{\partial}{\partial z}k\frac{\partial s}{\partial z}-\frac{1}{\rho}\frac{\partial Q_d}{\partial z}. \tag{2.1.21}$$

Let us discuss the derivation of the relation for Q_d. If $v(r_d)$ is the rate of fall of a drop of radius r_d and if $f(r_d)$ is the drop-radius distribution function, then as a result of the falling of drops with radii between r_d and r_d+dr_d, the following mass of liquid water will pass through a horizontal area of 1 m^2 in 1 s:

$$nf(r_d)\frac{4}{3}\pi\rho_d r_d^3 v(r_d)\,dr_d,$$

where n is the number of droplets in 1 m^3 of cloudy air, and ρ_d is the droplet density.

By summing the latter expression over all radii, we obtain a formula for the transport of water droplets and ice crystals:

$$Q_d=-\frac{4}{3}\pi\rho_d n\int_0^\infty r_d^3 f(r_d)\,v(r_d)\,dr_d. \tag{2.1.22}$$

Since the flow Q_d is directed vertically downward, it is negative ($Q_d < 0$).

On the other hand, the relation for the specific water content of the cloud can be written as

$$\delta=s-q_m=\frac{n}{\rho}\int_0^\infty\frac{4}{3}\pi\rho_d r_d^3 f(r_d)\,dr_d, \tag{2.1.23}$$

where n/ρ is the number of droplets and crystals per unit mass of air. A comparison of (2.1.22) and (2.1.23) gives

$$Q_d=-\rho(s-q_m)\tilde{v}, \tag{2.1.24}$$

where $\tilde{v}$ signifies the weighted-mean (over the mass) falling velocity of the cloud elements:

$$\tilde{v} = \frac{\int_0^\infty r_d^3 f(r_d) v(r_d) dr_d}{\int_0^\infty r_d^3 f(r_d) dr_d}. \qquad (2.1.25)$$

Now, taking relation (2.1.24) into account, we can rewrite eqn. (2.1.21) for the moisture balance in the cloud as

$$\frac{ds}{dt} = \frac{\partial}{\partial z} k \frac{\partial s}{\partial z} + \frac{\partial \tilde{v} (s - q_m)}{\partial z}, \qquad (2.1.26)$$

where once again we have neglected the dependence of the air density on height, since the inaccuracies in determining $\tilde{v}$ tend to compensate this effect.

The foregoing considerations, together with the form of equation (2.1.26), indicate that for problems in the dynamics of clouds and precipitation the most important (and practically the only) quantity which is to be taken from microphysical studies is the weighted-mean value (over the mass) of the falling velocity of the cloud elements. In general, as we see from (2.1.25), to determine $\tilde{v}$ we must know the distribution function $f(r_d)$ and the falling velocity $v(r_d)$ over the entire range of r_d In view of the difficulties noted in Chapter 1 with regard to ascertaining the droplet-size distribution functions, Lebedev [150] and the author [191] suggested approximating the velocity $\tilde{v}$ with the aid of certain analytical expressions.

For instance, in [191] the author, basing his analysis on the experimental data of Gunn and Kintzer, showed that the falling velocity of drops can be expressed as a function of drop radius, with a quite acceptable accuracy, by the formula

$$v(r_d) = v_\infty [1 - \exp(-a r_d)], \qquad (2.1.27)$$

where $v_\infty = 995$ cm/s and $a = 1.2 \cdot 10^{-3}$ μm^{-1} are constants determined using the method of least squares. It should be noted that the falling velocity of water drops in air, unlike that of solid particles, does not increase indefinitely with the drop radius. Rather, it approaches a limit v_∞, due to a flattening of the drops. Moreover, large drops (with radii greater than 2.5-3 mm) begin to break up (shatter).

Now, if we take, for the function $f(r_d)$ a gamma distribution (D'yachenko)

$$f(r_d) = A r_d^b \exp(-c r_d) \qquad (2.1.28)$$

which agrees well with experimental data, or a log-normal distribution (Levin)

$$f(r_d) = \frac{1}{r_d \sigma \sqrt{2\pi}} \exp\left[-\frac{(\ln r_d - \ln r_0)^2}{2\sigma^2}\right] \tag{2.1.29}$$

(A, b, c, σ, and r_0 in these formulas are paramaters of the distributions), then the velocity $\tilde{v}$ can be found (in analytical form or numerically) without any special difficulty using formula (2.1.25) and taking (2.1.27) into account. In particular, if in (2.1.28) we set $b = 2$ and $c = 2/r_m$, then the formulas for $f(r_d)$ and $\tilde{v}$ become

$$f(r_d) = \frac{4r_d^2}{r_m^3} \exp\left(-\frac{2r_d}{r_m}\right), \tag{2.1.30}$$

$$\tilde{v} = v_\infty \left[1 - \frac{64}{(2 + ar_m)^6}\right], \tag{2.1.31}$$

where r_m is the drop radius for which $f(r_d)$ is a maximum (the so-called modal radius). The radius depends on the cloud water content or the precipitation intensity (establishing the precise relationship is one of the main tasks of cloud microphysics).

Lebedev [150] proposed (without presenting, to be sure, any experimental proof) that velocity $\tilde{v}$ be related directly to the cloud water content by the formula

$$\tilde{v} = \tilde{v}_\infty [1 - \exp(-m\delta)], \tag{2.1.32}$$

where $\tilde{v}_\infty$ and m are constants.

In view of the fact that the mean drop radius apparently decreases at greater heights in clouds giving precipitation, in [136] it was proposed to describe $\tilde{v}$ as a function of height using the expression

$$\tilde{v} = \tilde{v}_m \exp\left(-\beta \frac{z - z_l}{z_u - z_l}\right), \tag{2.1.33}$$

where $\tilde{v}_m$ is the maximum value of $\tilde{v}$, reached at the lower cloud boundary z_l, while z_u is the upper cloud boundary, and β is a dimensionless parameter.

If velocity $\tilde{v}$ is known, then the described method can be used to model not only cloudiness, but also precipitation (without resorting to equations describing the microphysical processes). Another approach to the analysis and forecasting of precipitation is possible within the framework of the method of invariants. First, eqns. (2.1.6) and (2.1.7) are solved without taking the fallout of precipitation into account. The cloud then retains all the liquid water formed during the condensation of water vapor.

An analysis of numerous experimental data has indicated that the water content in a cloud depends primarily on temperature. The following temperature dependence of the mean specific water content of a cloud has been established [161, 192]:

$$\delta_0 = 0{,}201 \frac{T}{p} \exp\left[17{,}86\left(1 - \frac{258}{T}\right)\right], \qquad (2.1.34)$$

where δ_0 is in grams per kilogram (g/kg), p is in hectopascals (mb), and T is in kelvins (k).

If δ is the specific water content of a cloud, determined for a time t from the solution of the system (2.1.6) and (2.1.7), then the difference $\rho(\delta - \delta_0)$, summed from the lower cloud boundary to the upper, will represent the amount of precipitation falling out of the cloud during the time t. This is the approach to precipitation calculations that is adopted in the chapters of this monograph where the system of equations is solved without taking into account the fallout of precipitation. Here, the problem of precipitation is generally not even posed at all and, since all the liquid water is assumed to be retained in the cloud, it follows that the cloud's water content, as well as its vertical extent, will as a rule exceed the observed values of these characteristics (often considerably).

2.2. Formation and evolution of stratiform clouds

In order to investigate the cloud-formation process, it is necessary to solve the system of equations (2.1.5)-(2.1.8) for the nonsteady state while retaining the left-hand sides of (2.1.6) and (2.1.7) the local time derivatives of s and Π. Cloud formation will be considered in an air mass moving at a velocity which is independent of height. In this case, eqns. (2.1.6) and (2.1.7) are transformed via a change of variables into eqns. (2.1.13) and (2.1.14), which will now be analyzed.

The first solutions of the system (2.1.13) and (2.1.14) were devised for the simplest assumptions: w and k constant and independent of height (over the entire troposphere). In this case, eqns. (2.1.13) and (2.1.14) have exact solutions in the form of a series of trigonometric functions. Calculations carried out using them have made it possible to establish a number of cloud-formation regularities: *the velocity of motion of cloud boundaries differs substantially from the vertical velocity of the air itself* (there is even a difference in sign); the contributions of vertical motions, turbulent mixing, and condensation heat to the variations in temperature and water content are approximately equivalent; etc.

However, studies of the dynamics of the atmosphere [49] have established that the vertical velocity varies considerably with height in the troposphere. The profile of the vertical velocity (for instance, in a cyclone) is usually as follows: in the absence of mountains, it drops to zero at the Earth's surface; in the lower troposphere it increases with height, reaches a maximum in the middle troposphere (at a height of 3-5 km) and then decreases, approaching zero again close to the tropopause.

In accordance with these properties of the height variation of w, the following formula is obtained:

$$w(z) = 4w_m \frac{z}{H}\left(1 - \frac{z}{H}\right), \tag{2.2.1}$$

where w_m is the maximum value of the vertical velocity, attained at a height $z = H/2$, and H is the height of the tropopause.

Stratiform cloudiness arises, as a rule, in a warm air mass moving over a cooler underlying surface. These conditions apply primarily to Ns-As-Cs cloud systems, and also layer clouds forming in a warm air mass as it moves along with a cyclone. During the motion, the temperature of the air mass generally drops. An analysis of experimental data has indicated that the variations in temperature and specific humidity close to the ground ($z = 0$) are described by formulas like the following:

$$T(0, t) = T_2 + (T_1 - T_2)\exp(-t/\tau), \tag{2.2.2}$$

$$q(0,t) = q_2 + (q_1 - q_2)\exp(-t/\tau), \tag{2.2.3}$$

where T_1 and q_1 are the temperature and specific humidity at a time $t = 0$, when the air mass arrived at a new underlying surface; T_2 and q_2 are these same quantities after considerable time has passed (theoretically as $t \to \infty$); and τ is a constant (with units of time) characterizing the rate of temperature decrease at $z = 0$. Formulas (2.2.2) and (2.2.3) are used as boundary conditions for the functions s and Π at $z = 0$.

The tropopause serves as the second boundary, and there the conditions

$$\frac{\partial T}{\partial z} = 0, \quad \frac{\partial q}{\partial z} = 0, \quad \delta = 0 \quad \text{for } z = H, \tag{2.2.4}$$

hold true, that is, the vertical gradients of T and q approach zero and there is no cloudiness.

An initial condition is also imposed when solving nonsteady-state problems, that is, the distribution of the unknown functions with height at the initial moment ($t = 0$) is specified:

$$T(z, 0) = T_1 - \gamma_0 z; \quad q(z, 0) = 0{,}622 f_1 \frac{E(z, 0)}{p(z)},$$

$$\delta(z, 0) = 0, \quad p(z, 0) = p_0\left(1 - \frac{\gamma_0 z}{T_1}\right)^{g/R\gamma_0}, \tag{2.2.5}$$

where γ_0 is the vertical temperature gradient [lapse rate] for $t = 0$; $E(z, 0)$ is the saturation of water-vapor pressure at a temperature $T(z, 0)$; f_1 is the relative humidity of the air at $t = 0$ (it is the same at all heights) related to q_1 by the formula $q_1 = f_1 q_m \times (T_1; p_1)$; and R is the specific gas constant of air.

The initial condition is imposed specifically in order to exclude any singularities in the distribution of the unknown functions. A suitable choice of the initial condition makes it possible to ascertain in an undistorted form the role of the factors being considered (w k, heat of condensation, etc.) in the cloud-formation process.

All the quantities subsequently vary with time. An exception is the air pressure, which is assumed to remain the same at all heights as it was for $t = 0$ (that is, the cyclone in which the cloudiness forms does not intensify or mature).

The finite-difference method is used to solve the system (2.1.13) and (2.1.14) for the formulated initial and boundary conditions. The derivatives are approximated by central differences, and an explicit calculation scheme is used (with the aid of a computer).

The main results of the calculations [52] will now be given without the finite-difference equations or the time (Δt) and space (Δz) intervals.

For any temperature variation close to the ground, that is, for either a drop $(T_2 < T_1)$ or a rise $(T_2 > T_1)$ in temperature, quite dense cloudiness will form, frequently spreading over the whole troposphere. For a drop in temperature, fog may appear, in addition to cloudiness, and often the fog merges with the clouds.

This happens, due to updrafts $(w > 0)$ and stable stratification (for the calculation, γ_0 was assumed to be 0.6°C/100 m), the air temperature at all heights decreases with time, while the specific moisture content increases $(\partial s/\partial t = -w(\partial s/\partial z)$; since $\partial s/\partial z < 0$, while $w > 0$, it follows that $\partial s/\partial t > 0)$. Turbulent mixing also aids the accumulation of water vapor in the troposphere: vapor arriving from the ground via evaporation does not emerge into the stratosphere, since at the tropopause the condition $\partial q/\partial z = 0$ is imposed.

Figures 2.1 and 2.2 show typical profiles of the specific water content. Figure 2.1 corresponds to a lowering of the temperature near the ground $(T_2 < T_1)$ in a warm mass moving over a colder underlying surface; this is an instance of *warm-front* cloudiness. As is to be expected, with an increase in w_m there is an increase both in the maximum value of the water content δ_m (as a function of height) and in the water content at a specified height. Here,

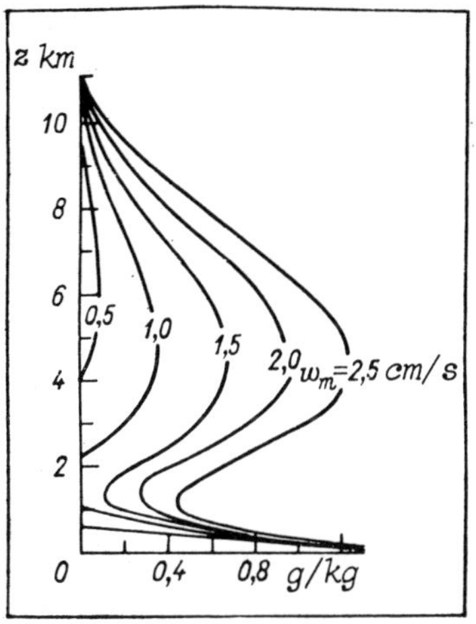

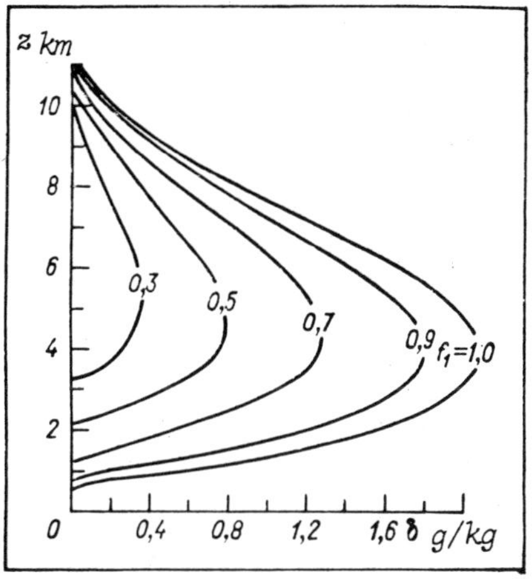

Fig. 2.1. Distribution with height of specific water content of cloud and fog in warm air mass, for various values of w_m and fixed values T_1 = 10°C, T_2 = 5°C, $f_1 = q_1/q_m(T_1)$ = 0.7 (relative humidity at t = 0), q_2/q_1 = 0.9, k = 5 m^2/s, $\tau = t$ = 24 h.

Fig. 2.2. Distribution with height of specific water content of cloud in cold air mass moving over a warmer surface, for various values of f_1 and fixed values T_1 = 10°C, T_2 = 15°C, q_2/q_1 = 0.7, w_m = 2.5 cm/s, k = 5 m^2/s, $\tau = t$ = 24 h.

as w_m increases, the lower boundary of the cloud descends considerably (from 4.2 km for w_m = 0.5 cm/s to 2.2 km for w_m = 1 cm/s), while the upper boundary ascends slightly. The maximum values δ_m, which are practically (with an accuracy to the third decimal place) independent of T_2 (as it varies from 5 to 10°C), are the following:

w_m, cm/s . . .	0.5	1.0	1.5	2.0	2.5	5.0
δ_m, g/kg . . .	0.07	0.36	0.67	0.98	1.28	2.63

For low values of w_m the fog and cloud are separated by a cloudless interlayer; for w_m > 1.5 cm/s the cloudless interlayer disappears and dense cloudiness appears.*)

The initial temperature T_1 and initial relative humidity f_1 of the air mass have considerable effect on the value of δ_m. For instance, for T_2 = 0°C, f_1 = 0.7, q_2/q_1 = 0.9, w_m = 2.5 cm/s, k = 5 m^2/s, and $\tau = t$ = 24 h, the dependence of δ_m on T_1 is

*) Recall that all the liquid water (including precipitation that may fall from the cloud) is retained in the cloud.

T_1, °C	−10	0	10	15	20
δ_m, g/kg . . .	0.38	0.74	1.28	1.60	1.96

For the fixed values T_1 = 10°C, T_2 = 0°C, q_2/q_1 = 0.7, w_m = = 0.5 cm/s, k = 5 m^2/s, τ = 24 h, and t = 36 h, and varying f_1, we have the following:

f_1	0.5	0.6	0.7	0.9	1.0
δ_m, g/kg . .	0.02	0.11	0.22	0.53	0.73

The marked dependence of δ_m on f_1 can be seen in Fig. 2.2, which shows an instance of *cold-front* cloudiness: a temperature rise ($T_2 > T_1$) in a cold mass moving over a warmer underlying surface. In such a mass, fog does not appear, but quite thick clouds form, with a maximum water content in the lower part.

Above it was shown that the air temperature drops as a result of vertical motions. For instance, at a height of 5 km, the temperature drop ΔT during 24 h, for various w_m, is

w_m cm/s	0.5	1.0	1.5	2.5	5.0
ΔT, °C	2.3	3.2	4.2	5.6	7.7

The values of the other parameters are given in the legend to Fig. 2.1.

In addition to the vertical velocity, the heat of condensation and turbulent mixing also greatly influence the water content and the variation of temperature with time. As an example, Table 2.2 presents data on the water content and the temperature drop ΔT at various heights, calculated both taking into account and neglecting the heat of condensation. As is to be expected, the condensation heat slows down the drop in temperature with time, the effect of the condensation being of the same order as the total effect of other factors. For instance, at a height of 5 km, the temperature drops 5.6°C as a result of all factors, while the condensation heat released raises the temperature 7.9−5.6 = 2.3°C.

Table 2.2. Temperature drop ΔT and cloud water content δ, calculated by taking into account (numerator) and neglecting (denominator) the heat of condensation (T_1 = 10°C, T_2 = 0°C, f_1 = = 0.7, q_2/q_1 = 0.9, w_m = 2.5 cm/s, k = 5 m^2/s, τ = t = 24 h)

z, km	0	0.2	0.6	1	2	3	4	5	6	7	8	9
ΔT, °C	$\frac{6.4}{6.4}$	$\frac{5.8}{7.6}$	$\frac{4.2}{5.1}$	$\frac{3.4}{4.0}$	$\frac{2.9}{4.2}$	$\frac{3.5}{5.6}$	$\frac{4.4}{6.9}$	$\frac{5.6}{7.9}$	$\frac{6.6}{8.4}$	$\frac{7.3}{8.4}$	$\frac{7.3}{7.8}$	$\frac{6.5}{6.3}$
δ, g/kg	$\frac{1.36}{1.36}$	$\frac{1.11}{1.72}$	$\frac{0.56}{0.89}$	$\frac{0.43}{0.64}$	$\frac{0.69}{1.01}$	$\frac{1.06}{1.43}$	$\frac{1.25}{1.54}$	$\frac{1.24}{1.41}$	$\frac{1.06}{1.15}$	$\frac{0.80}{0.84}$	$\frac{0.52}{0.54}$	$\frac{0.29}{0.28}$

The higher the temperature T_1 near the ground, the greater is the water vapor condensation (as follows from the foregoing data on the water content), and the smaller is the temperature drop ΔT . For example, at a height of 5 km for various values of T_1 (all the other parameters are the same as in Table 2.2), ΔT has the following values:

T_1, °C	−10	0	10	15	20
ΔT, °C	8.1	7.2	5.6	4.6	3.6

A peculiarity of the temperature variation at various heights should be mentioned. The direct effect of a change in the temperature near the ground (during 24-48 h) is felt only up to a height of 1 or 2 km. For instance, in the above example, the variation in the temperature near the ground during 24 h was -6.3, 0, 6.4, 9.5, and 12.7°C for ground temperatures T_1 of -10, 0, 10, 15 and 20°C, respectively. At a height of 1 km, the values of ΔT are 0.3 and 4.9°C, respectively (as we see, no temperature rise is observed at this height for $T_1 < T_2$, but the effect of the variation in T_1 remains predominant). At a height of 3 km, however, ΔT will be less, the higher the temperature T_1 (close to the ground, the converse is true). This distribution of ΔT with height can be attributed to the effect of the heat condensation. The influence of this factor is also reflected in the dependence of ΔT on f_1 :

f_1	0.3	0.5	0.7	0.9	1.0
ΔT, °C	10.5	7.8	8.6	3.8	3.1

These data were obtained at z = 5 km for the following parameter values: T_1 = 10°C, T_2 = 15°C, q_2/q_1 = 0.7, w_m = 2.5 cm/s k = 5 m^2/s, and t = τ = 24 h.

Turbulent mixing has considerable effect on ΔT and δ. At a height of 5 km, for T_1 = 10°C, T_2 = 15°C, = 0.7, q_2/q_1 = 0.9, w_m = 2.5 cm/s, and $t = \tau$ = 24 h the dependences of ΔT and δ on k are as follows:

k, m^2/s	0	0.5	2.5	5	10	20	40
ΔT, °C	5.7	6.7	5.5	5.6	5.4	4.9	4.0
δ, g/kg.	1.21	1.25	1.21	1,24	1.27	1.32	1.42

A comparison of the calculated results, together with an analysis of the equations, shows that an increase in k of about fourfold produces the same temperature variation as a decrease in the vertical velocity to about half.

Data on the time t^* required for clouds to form will now be given for various values of the parameters entering into the initial equations and boundary conditions (Table 2.3). The clouds form in a moving air mass during a time that fluctuates, depending on the values of the pertinent parameters, between 4 and 6 h (for high w_m, f_1, T_1) and between 50 and 60 h (for low w_m, f_1, T_1). However, t^* as a function of τ and k can be substantially higher or lower than the values in Table 2.3. For example, if τ = 12 h and k = 10 m^2/s, while w_m is twice as great as in Table 2.3, then the time of cloud formation will be only half the time shown in Table 2.3 (this follows from an analysis of the dimensionless parameters used in the problem).

In conclusion, it should be noted that these superficially analyzed quantitative characteristics of cloudiness, obtained via a numerical modeling of the cloud-formation process, point to a great diversity of conditions that can give rise to the formation and evolution of the fields of temperature, humidity, and cloudiness in a moving cyclone. In the middle and upper troposphere updrafts are the initial cause of the drop in temperature with time. In the lower troposphere (up to 1 or 2 km) the ground plays a major role (through transformation effects). However, the contribution of the heat of condensation and the turbulent mixing to the total temperature variation at all levels is great indeed, and comparable to the contribution of vertical motions. Through their influence on the heat of condensation, the air temperature and humidity near the ground have a significant effect on the drop in cloud temperature and water content.

The modeling results show a satisfactory fit with experimental data on the temperature variation in a moving cyclone obtained by various investigators (Kh. P. Pogosyan, A. I. Burtsev, A. A. Pavlovskaya, and others).

Table 2.3. Time t^* of cloud formation (with maximum water content δ_m = 0.2 g/kg) for k = 5 m^2/s and τ = 24 h

T_1=10°C, T_2=0°C, f_1=0.7, q_2/q_1=0.9		T_1=10°C, T_2=0°C, q_2/q_1=0.7, w_m=0.5cm/s		T_2=0°C, f_1=0.7, q_2/q_1=0.9, w_m=2.5cm/s	
w_m cm/s	t^* h	f_1	t^* h	T_1, °C	t^* h
0.5	34.0	0.5	57.0	—15	24.0
1.0	18.0	0.6	44.0	—10	14.4
1.5	12.4	0.7	34.4	—5	11.0
2.0	9.6	0.8	26.0	0	9.6
2.5	7.6	0.9	17.6	5	8.4
3.0	6.0	1.0	9.6	10	7.6
3.5	5.0			15	7.0
4.0	4.2			20	6.6
4.5	3.6				
5.0	3.2				

2.3. Precipitation modeling and the microphysical quantities of a stratus cloud

2.3.1. Droplet clouds

Buikov and Prinach [46] portrayed the dynamic part of the problem (vertical-velocity profile, conditions at the upper and lower boundaries of the troposphere, initial distributions of meteorological quantities, etc.) in the same way as in [52]. However, to the system of equations describing heat and moisture transport they added the equation for the droplet-size distribution function (see Section 3.1).

Allowing for supersaturation introduces only a slight correction (amounting to a few percent) to the cloud water content (see Introduction). Thus, as anticipated, the results of the numerical modeling carried out by Buikov and Pirnach confirmed the regular features in cloud formation that had been established in earlier works [52, 179]. These regular features include the time of beginning of cloud formation, the vertical profile of water content and the height of cloud boundaries, the depndences of these characteristics on vertical velocity, the intensity of turbulent mixing, and the air temperature and humidity near the ground. Some additional cloudiness characteristics calculated in the model of Buikov and Pirnach will now be considered.

Figure 2.3 shows vertical profiles of the specific water content δ, the mean droplet radius $\bar{r}_d$, the number of droplets N per gram of cloudy air, and the supersaturation $\Delta = q - q_m$, relative to a plane water surface. These curves correspond to the following values of the parameters: T_1 = 10°C, T_2 = 5°C, w_m = 2.5 cm/s, k = 5 m^2/s, H = 11 km, f_1 = 0.7, an initial number of condensation nuclei $N_0 = 2.5 \cdot 10^6 \text{ g}^{-1}$, $q_2/q_1 = 0.9$, $\tau = 24$ h, and $\beta = 0$.*)

However, in this case cloudiness appears 5 h 20 min after the initial moment. Eight hours after the moment of formation, the clouds will have a depth of 7 km. The maximum value of the water content (somewhat higher than 0.5 g/kg) is observed in the lower third of the cloud (around the 4 km level). The number of droplets is a maximum close to the lower cloud boundary, and it decreases sharply with the approach to the upper boundary. The mean droplet radius increases with height and reaches a maximum (about 20 μm) near the upper boundary. Throughout most of the cloud, the supersaturation is negligible (less than 10^{-7} g/g), and only in the upper part does it increase, reaching about $8 \cdot 10^{-7}$ g/g (because, by hypothesis, new droplets are formed only at the lower cloud boundary).

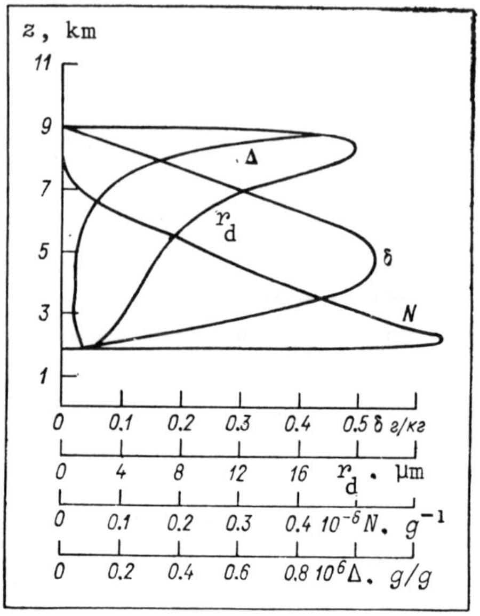

Fig. 2.3. Vertical profiles of specific water content δ, number of droplets N per gram of cloudy air, mean droplet radius $\bar{r}_d$, and supersaturation $\Delta = q = q_m$ relative to plane water surface.

By varying the parameters of the problem, it becomes possible to evaluate the roles of various factors in the formation of the cloud structure. The number of droplets in a cloud depends, first, on the initial number N_0 of condensation nuclei. Modeling has shown that a twofold increase in N_0 does not affect the positions of the cloud boundaries or the water content, but it is accompanied by an increase in the droplet number, and a decrease in the mean droplet radius and the supersaturation, while the form of the height distributions of these parameters remains the same. It was found that the mechanism of cloud-droplet production has considerable effect on the cloudiness characteristics $N, \bar{r}_d, \Delta$ If the droplets are assumed to originate everywhere that supersaturation is reached, the droplet number N will vary substantially less with height (in comparison with the case of droplet production only at the lower cloud boundary), while the mean radius and the

*) The notation here is the same as in Section 2.2.

supersaturation will reach a maximum in the lower part of the cloud. The supersaturation in a cloud varies with time, as we see from the following data (values of parameters are the same as in Fig. 2.3):

t, h.	9.3	13.3	17.3	21.3	24
$10^7\,\Delta$, g/g. .	0.68	0.46	0.35	0.29	0.26

With an increase in the number N_0 of condensation nuclei, the supersaturation Δ decreases:

$10^{-6}\,N_0\,\mathrm{r}^{-1}$	1	2.5	5
$10^7\Delta\,\mathrm{r/r}$	0.61	0.35	0.24

These data pertain to a time 12 h after the cloud originates; the parameters are the same as in Fig. 2.3.

It should be noted that the supersaturation (even its maximum values) does not not excreed 10^{-4} g/kg, that is, it is 2 or 3 orders of magnitude less than the water content of the cloud. Hence, taking the supersaturation into account has almost no effect on the cloud water content.

The discontinuity in specific humidity close to the surface of a droplet can be corrected for, with the aid of the formula

$$\frac{dr_d}{dt}=\frac{D\rho\Delta}{\rho_d(r_d+\beta)},$$

where $\beta = 4D/\alpha_c\bar{v}$ (here α_c is the condensation coefficient, and $\bar{v}$ is the mean velocity of the vapor molecules). From this expression, we see that a variation in β from zero to 2 μm has practically no effect on the water content of the cloud; but the mean radius increases, while the droplet number decreases, by almost 10%.

The calculated droplet-size distribution functions show a satisfactory fit with the gamma distribution

$$f(r_d)=Ar_d^b\exp(-cr_d),$$

where A, b, and c are constants for the given sample. Using the same paramater values as in Fig. 2.3, we obtained the following values of exponent b at various heights z:

z, km . .	2.0	2.4	2.8	3.2	3.6	4.0	4.4	4.8	5.2	5.6	6.0
b	2.32	2.56	2.80	3.12	3.45	3.79	4.12	4.42	4.66	4.84	4.97

Exponent b determines the width of the droplet spectrum (as b increases, the spectrum becomes broader). In the upper part of a cloud, the droplet spectrum is seen to be considerably wider ($b \approx 5$) than in the lower part (where $b \approx 2$-3). At all levels, the exponent b is greater than 2. The broadening of the droplet spectrum in the upper part of the cloud (as compared to the lower part) can be explained as follows: as it ascends, a droplet becomes larger through condensation (in the example considered, droplets originate at the lower boundary of the cloud). An analysis of the spectra for different times indicates that in the middle of the cloud (at a height of 3 or 4 km), the droplet-size distribution is quite stable in time: the exponent b remains practically constant (close to 4). In the upper part of the cloud, on the other hand (at a height of 5 or 6 km), the exponent b decreases with time. This may well be because larger droplets fall out of the upper part, while smaller droplets remain there. However, even in the upper part of the cloud, beginning about 8 h after the cloud originates, the droplet-size distribution function is also essentially invariable with time.

The mechanism of droplet formation has a definite effect on the droplet spectrum. For instance, at a height of 5 km (in the example portrayed in Fig. 2.3), the exponent b varies from 4.4 for droplet origin at the lower boundary, to 3.8 when a droplet originates at any height where supersaturation is attained. In the latter case, the droplet spectrum is more uniform throughout the cloud: the exponent b stays practically constant (3.8) at all heights. It should be noted that when a droplet originates in a region with supersaturation, the additional assumption that droplets originate at the lower cloud boundary has almost no effect on the droplet spectrum.

The droplet spectrum is influenced more markedly by the humidity discontinuity close to the droplet surface (parameter β in the formula for the droplet growth rate). For instance, in the example of Fig. 2.3, the exponent b varies from 4.4 at a height of 5 km, where $\beta = 0$, indicating no humidity discontinuity to 6.8 for $\beta = 1$ µm. A further increase in β from 1 to 2 µm due mainly to the condensation coefficient) is accompanied by only a slight change in the droplet spectrum: the exponent b increases from 6.8 to 7.0 at a height of 5 km, from 2.9 to 3.0 at 2 km, and from 7.4 to 7.7 at 6 km.

For the subsequent growth of droplets due to collisions, an important role is played by the number of droplets that have grown to maximum size, under the influence of condensation. The number of such droplets is 2 or 3 orders of magnitude less than (1/100 to 1/1000 of) the maximum droplet concentration. An analysis of modeling results indicates that the number of large droplets (for instance, 17-18 µm in radius) depends considerably on a number of parameters, in particular on w_m, the temperatures T_1 and T_2, and the number N_0 of condensation nuclei, the parameter β, and the relative humidity. Thus, an increase in w_m from 1.5 to 2.5 cm/s is accompanied by the following variations in the number N of droplets of radius 17-18 μm per gram of air: at a height of 3 km, from 15 to 220; at 5 km, from 160 to 2000; and at 8 km, from 270 to 540. An increase in the turbulence coefficient from 5 to 50 m^2/s is accompanied by a drop in N from 220 to 17 at a height of 3 km. Taking the humidity discontinuity into account (for β = 1 μm) causes N at a height of 2 km to change from 50 to 260.

It would be interesting to compare the results of modeling the main cloud characteristics (water content, temperature, and humidity) using the method employed in [52] with those obtained using the method employed in [46]. Tables 2.4 and 2.5 give the cloud temperatures and water contents calculated by the authors of [46] for a typical example of the formation of dense cloudiness in the troposphere.

Table 2.4. Air temperature (K) at various heights z at t = 24 h

Method	z, km								
	1	2	3	4	5	6	7	8	9
Hydrodynamic	274.5	268.8	262.6	255.9	249.0	242.2	235.6	229.6	224.4
Microphysical	274.7	269.1	263.0	256.3	249.4	242.5	235.8	229.7	224.4

Table 2.5. Variation of specific water content of cloud (g/kg) with height z for t = 24 h

	z, km							
	2	3	3.6	4	4.6	5	6	7
I	0.52	0.93	1.08	1.13	1.16	1.14	1.00	0.77
II	0.49	0.91	1.06	1.12	1.16	1.15	1.02	0.79
III	0.63	1.12	1.26	1.30	1.30	1.26	1.11	0.75
IV	0.55	0.99	1.13	1.19	1.20	1.18	1.03	0.80

The authors of [46] called the method worked out in [179, 181] the "hydrodynamic method". They called the "microphysical method" the one in which, for the thermodynamic part of the problem in [52] (equations of heat and moisture transfer, vertical motions, turbulent mixing, boundary conditions at the ground and the tropopause), additional equations are introduced: one for the droplet-size distribution function, and Maxwell's equation, so that it becomes possible to determine the microphysical characteristics of the cloud, especially the supersaturation Δ. Since the latter is always small in comparison with the specific humidity as well as the specific water content of the cloud, it follows that, taking the supersaturation into account cannot significantly affect the accuracy of the cloud water content calculations, and hence of the air temperature (as was seen).

The data of Tables 2.4 and 2.5 verify this conclusion. Actually, the air temperature (see Table 2.4) at all heights do not differ by more than 0.3-0.4 K. Such a difference lies within the limits of accuracy of the numerical solution of the system of equations. This can easily be shown by comparing the theoretical results (using the same initial conditions and taking the same factors into account) in [52], where the equations were solved using an explicit method, with those in [46], where they were solved using the method of elimination. Such a comparison has shown that because of the numerical integration, errors of 0.4-0.5 K appear; at some heights, the errors are as large as 1.0-1.5 K.

Row I of Table 2.5 shows the results of calculating the cloud water content using the hydrodynamic method. The data in rows II-IV were obtained with the microphysical method. In row II droplet fall was ignored; in row III it was considered; in row IV it was ignored, but droplets were assumed to originate at all levels where supersaturation is reached, whereas in calculating the data of rows II and III, it was assumed that all droplets originate at the lower cloud boundary. It is seen that the values of the cloud water content obtained using the two methods differ even less than the values of air temperature (generally the water content found using the hydrodynamic method lies somewhere in between the values given in rows II and IV; since the mechanism of droplet origin is unknown, the water content found using the first method should be considered the most probable). Taking droplet fall into account (row III) has some effect on the results, but this effect is also allowed for within the framework of the hydrodynamic method.

2.3.2. Mixed clouds

At negative (subfreezing) cloud temperatures, ice is observed in addition to gaseous and liquid water. The probability of the formation of ice crystals increases as the temperature drops (see Chapter 7). Buikov and Pirnach [47] attempted to model the formation and evolution of a mixed cloud. The dynamic part of the problem remains the same as in the case of droplet clouds. In the microsphysical part, in addition to the droplet distribution function f_1, it is necessary to introduce f_2, the size distribution function for ice crystals:

$$\frac{\partial f_2}{\partial t}+(w-v_2)\frac{\partial f_2}{\partial z}-\frac{k}{\rho}\frac{\partial}{\partial z}\rho\frac{\partial f_2}{\partial z}+\frac{\partial}{\partial r_2}(\dot{r}_2 f_2)=I_2'+I_2''. \tag{2.3.1}$$

Here v_2 is the fall velocity of the crystals (cm/s), given by the Krystanov-Schmidt formula:

$$v_2=10^6\rho_2\left(\frac{0{,}787}{r_2^2}+\frac{503}{r_2^{1/2}}\right)^{-1}, \tag{2.3.2}$$

where r_2 is the crystal radius (cm), and ρ_2 is its density (g/cm^3).

The quantity I_2' on the right-hand side of (2.3.1) is the rate of production of the third phase as a result of the freezing of droplets and the activation of crystallization nuclei. If $P(r_1, z)$ is the probability that a droplet of radius r_1 will freeze at a height z, while $F(T)$ is the number of crystallization nuclei activated for a drop in temperature T of 1 K, then the following expression may be written for I_2':

$$I_2'=P(r_1, z)f_1(r_1, z, t)-F(T)\left(\frac{\partial T}{\partial t}+w\frac{\partial T}{\partial z}\right), \tag{2.3.3}$$

where the subtrahend is taken to be zero if supersaturation with respect to ice is not reached or the temperature does not decrease ($dT/dt > 0$) or the radius of a crystallization nucleus does not reach some critical value r_0 .

The following relations are used for $P(r_1, z)$ and $F(T)$:

$$P(r_1, z)=A_1 r_1^3\exp(-B_1 T_s), \tag{2.3.4}$$

$$F(T)=A_2\exp(-B_2 T_s), \tag{2.3.5}$$

where $T_s = T - 273.15$, and A_1, B_1, A_2, and B_2 are empirical multipliers. Quantities P and F are taken to be zero at positive (above-freezing) temperatures ($T_s > 0$).

The quantity I_2'' represents the rate of change of the ice--crystal radius due to collisions between crystals and droplets. Considerations analogous to those pertaining to collisions between large and small droplets lead to the following expression for I_2'':

$$I_2''=-\pi\rho_2\frac{\partial}{\partial r_2}f_2\int_0^R(r_1+r_2)^2(v_2-v_1)\,\Im\,\Delta r f_1\,dr_1, \tag{2.3.6}$$

where $\Delta r=(r_1^3+r_2^3)^{1/3}-r_2$; ; $\Im$ is the collision factor (taken to be unity), and R is the maximum droplet radius.

In addition to the cloud water content δ_1 and the number and size of the droplets, for a mixed cloud the ice content and the number and size of the ice crystals must be calculated. The specific ice content δ_2 of a cloud is defined as the mass of ice crystals per unit mass of cloudy air. If the radius r_2 and the fall velocity v_2 of the ice crystals are known, then, with the assumption that the precipitation is produced through crystal fallout, the precipitation intensity J is easily calculated:

$$J = \frac{4\pi\rho_2}{3} \int_0^\infty v_2 r_2^3 f_2(r_2, 0, t)\, dr_2. \tag{2.3.7}$$

The formation and structure of mixed clouds show marked dependence on the mechanism of ice formation in the atmosphere. Under actual conditions two mechanisms operate simultaneously: the formation of crystals at sublimation nuclei for positive supersaturation with respect to the ice and the freezing of supercooled droplets produced when there is positive supersaturation with respect to water.

Calculations show that for the first mechanism of ice formation, cloudiness appears (for the parameter values indicated above) after 3 h, and from the beginning the cloudiness is everywhere mixed. During the first few hours of cloud development, the rate of water-vapor condensation rises markedly. The precipitation intensity during the first hour or two is close to zero, and then it, too, starts to increase, although not as rapidly as the condensation rate. A state of dynamic equilibrium, where the condensation rate equals the precipitation intensity, is reached after 35 or 40 h, when the cloud is composed entirely of crystals. From the moment dynamic equilibrium is reached, the precipitation intensity remains practically constant with time.

For the same values of the thermohygrometric and dynamic parameters (T_1 = 0°C, T_2 = -5°C, k = 5 m^2/s, τ = 24 h, q_2/q_1 = = 0.9, f_1 = 0.7, w_m = 5 cm/s, w = 0 at a height of 7 km, N_0 = = 2.5·10^6 g^{-1}, A_2 = 0.5·10^{-2} g^{-1}·°C^{-1}, B_2 = 0.2°C^{-1}) as in the first example, but for the second mechanism of ice formation (freezing of supercooled droplets in the presence of supersaturation with respect to water), the cloud development during the first hours will proceed in a similar manner. Then, instead of an equilibrium state as in the first example, a regime will be arrived at in which the condensation rate and precipitation intensity fluctuate with time.

This phenomenon, which was first observed and modeled by Buikov and Pirnach, can be interpreted purely qualitatively (physically) in the following manner: the precipitation intensity increases until all the droplets freeze. Then, for some time, the precipitation intensity continues to increase because of the sublimation of water vapor on ice particles and the fallout of the ice particles. However, as time goes by, the sublimation (over the cloud as a whole) slows down because of a reduction of the water-vapor mass; consequently, the precipitation intensity starts to decrease. This continues until all the ice particles previously formed have fallen out of the cloud as precipitattion. Then the air humidity in the cloud begins to increase, until supersaturation with respect to water is reached and supercooled droplets form. The subsequent formation of ice particles (due to the freezing of droplets) and their growth will once again lead to an increase in precipitation intensity, and the cycle begins again.

Figure 2.4 illustrates the variation in time of the precipitation intensity and condensation rate (over the cloud as a whole). Both quantities are seen to fluctuate considerably as time goes by. On the basis of the vertical profiles of the water content and the ice content of a cloud given in [47], it can be concluded that 14 h after the cloud forms, its main bulk will be of a composite structure, and only in its upper part (about 1 km in thickness) will the cloud be composed wholly of crystals. After 18 h, when the precipitation intensity is a maximum, the cloud will be almost entirely crystalline (only in the layer from 6.6 to 7 km will there be droplets). After 20 h, the precipitation intensity reaches a minimum; the ice content is low (of the order of 10^{-2} g/kg), but at this time droplets start to reappear (in layers at 1-2, 3-4.5, and 6.5-7 km), and their freezing leads to a new rise in the precipitation intensity.

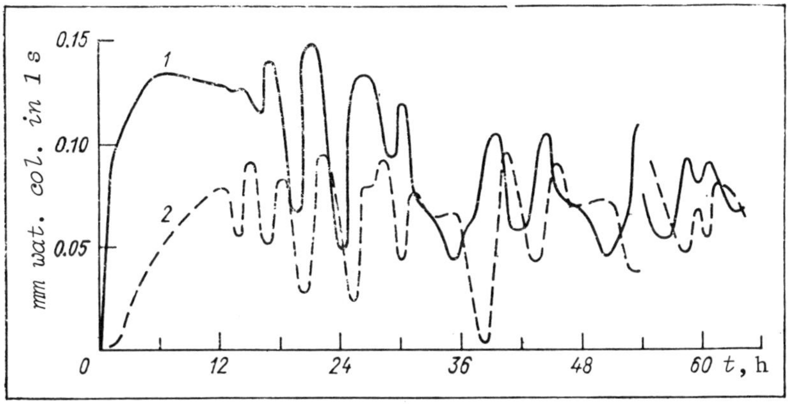

Fig. 2.4. Variation in time of condensation rate (1) (over the cloud as a whole) and precipitation intensity (2).

2.3.3. *Formation of a stratus cloud in the boundary layer of the atmosphere*

A model of cloud formation in the boundary layer of the atmosphere was constructed by Khvorostyanov [315]. In addition to the equations of nonsteady-state transfer of heat and moisture, the following were used as well: the equations of motion and balance of turbulent energy (for horizontal, steady, uniform motion), a number of relations from dimensional analysis, and the equations describing the transfer of long-wave radiation and the thermal conduction of the soil. The vertical velocities w were calculated using a method worked out by the author [176]. In addition to the expression for w_R (the vertical velocity averaged over an isobaric region of radius R), an expression was obtained for w_S, the vertical velocity averaged over a circle of radius R, with its center at the center of the cyclone or anticyclone.

Assuming the turbulent stress to be the same everywhere within the confines of S (that is, the stress does not depend on R), it follows that w_R and w_S are related simply:

$$w_S(z, R) = 2w_R(z, R).$$

With the aid of numerical methods, the vertical profiles of the temperature, vertical velocity, and turbulence coefficient were calculated for the following values of the paramaters: u_g = 10 m/s, T_0 = 268.15 K, γ_0 = 6 K/km, $2\omega_z = 10^{-4}\text{s}^{-1}$, R = 250 km, $f = f_1(0)$ for $z \leqslant 510$ m, and $f = f_2 = 65\%$ for z = 930 m (between levels 510 and 930 m, the relative humidity f is a linear function of the height). The values T_0, γ_0, and $f_1(0)$ pertain to the initial moment t = 0. Since conditions close to those observed for the formation of stratus clouds above a snow cover were modeled, the condition of saturation over ice was specified at the ground.

First the initial vertical profiles of T, w, and k are calculated in the absence of cloudiness, it being assumed that f_1 = 90% for $w < 0$ and f_1 = 75% for $w > 0$.

As a result of radiation heat losses by the Earth's surface (in 4 h it cools down by 7°C), together with turbulent mixing and vertical motions, by which the cooling is transmitted from the ground to the air, a temperature inversion is produced in the lower part of the boundary layer. The magnitude of the inversion will be quite different for $w > 0$ (cyclone), $w = 0$, and $w < 0$ (anticyclone): the temperature differences between the 400 m level (close to the upper boundary of the inversion) and the ground after a time of 4 h, are 1.8, 3.6, and 5.0°C, respectively. In a cyclone, the main variation in T is observed in the lower 30-50 m layer. Since for $w < 0$ the thermal stratification is much more stable than for $w > 0$, the turbulent mixing in a cyclone will be more intensive than in an anticyclone: the maximum values of K reached at a height of 60 or 30 m are 5.0 and 3.3 m^2/s for $w > 0$

and $w < 0$, respectively. A variation in the thermal stratification has practically no effect on the vertical velocity itself (the variation in w does not exceed 10% at any level).

If the relative humidity of the air in the lower layer (up to a height of 510 m) is assumed to be 90% at the initial moment [t = 0], then in a cyclone ($w > 0$) cloudiness will appear at a time t = 1 h 50 min, whereas for $w = 0$ it will appear at t = 4 h. For the next 2 h, for $w > 0$ clouds will form in the layer from 120 to 750 m, while for $w = 0$ they will form between 150 and 480 m (Fig. 2.5).

In or under a cloud, the radiation influx of heat is close to zero (in particular, the effective radiation of the Earth's surface is close to zero). Consequently, here the main roles in shaping the temperature profile will be played by the convective (w), turbulent (k), and condensational heat influxes. Under the influence of these influxes, an inversion observed at t = 1 h 50 min changes by a time t = 3 h 50 min to a drop in temperature from the ground to the upper cloud boundary (780 m), while above the cloud a very slight temperature inversion (amounting to a 0.8 K temperature differential over the layer) remains. It is easily seen that the temperature drop close to the upper boundary of the cloud is attributable to w. Turbulence has no effect here, since in the model being considered k goes to zero even for $w > 0$ at a height of 750 m.

Actually, at the initial moment the air temperature at a height of 750 m is 268−0.6•7.5 = = 263.5 K. At t = 3 h 50 min the air temperature at a height of 750 m will be 260.5 K, that is, it will have dropped by 3 K. On the other hand, a comparison of the temperature-distribution curves for $w > 0$ and $w = 0$ in the absence of cloudiness for t = 4 h indicates that the temperature difference close to the upper limit of the boundary layer (where k = 0) is about 2.5 K. This difference is due exclusively to the effect of w, since the radiation conditions are identical in the two cases.

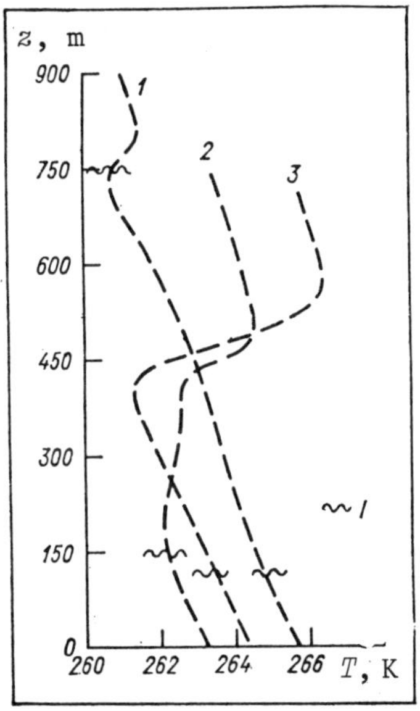

Fig. 2.5. Vertical temperature profiles for $w > 0$ (1), $w = 0$ (2), and $w < 0$ (3) in the boundary layer of the atmosphere.

I = cloud boundaries.

For $w = 0$, under the influence of turbulent mixing, 2 h after the beginning of cloud formation in the layer from the ground up to 150 m the temperature will drop; above this, up to 390 m, the isothermy remains, becoming an inversion cloud in the layer from 150 to 480 m. Here, the air temperature close to the 450 m level will be essentially the same (around 265 K) in the presence or absence of cloudiness. The situation will be completely different, however, if the clouds that formed for $w = 0$ are assumed to drift into an anticyclone region where $w < 0$. The inversion above the clouds intensifies considerably when $w < 0$: the temperature difference after 4 h (when the downward motion has started to have an effect) increases from 2 to 5.2 K, and the thickness of the inversion layer more than doubles. By $t = 10$ h the air temperature for $w < 0$ has dropped throughout the entire layer from the ground to the upper cloud boundary; in 4 h of downdrafts, the water content of the cloud drops by 20-25%, but the cloudiness does not dissipate.

The results presented in [315] are in good agreement with the theories developed by the author [195] concerning the formation of St and Sc clouds.

2.4. Modeling of frontal cloudiness

It is well known that the largest cloud systems are associated with fronts. Fronts are located in troughs; therefore, near the ground and throughout the boundary layer the forces of surface friction and eddy viscosity cause a convergence of air currents, which in turn, gives rise to an ascending motion of the air. The velocity of the ascending motion w increases with increasing height in the boundary layer (up to 1 or 2 km), as is shown by an analysis of the continuity equation; this velocity is assumed to have value w_H at the upper limit H of this layer. Above H, the vertical velocity may either increase (for convergence of the air currents above H) or decrease (for divergence of the wind velocity produced above H because the motion there is inertial and unsteady). However, if it decreases, the vertical velocity will drop to zero ($w = 0$) at some height H^* considerably above level H (level H^* is most frequently close to the tropopause).

It is quite understandable that because of the ascending motion, which does not change direction (but remains upward) throughout most of the troposphere, dense stratiform cloudiness forms in the trough region. The horizontal flow, perpendicular to the trough axis, transports air and clouds along the horizontal. The vertical velocity w_H is of the order of 10^{-3} or 10^{-2} m/s, while the horizontal component u of the wind velocity, perpendicular to the trough axis, is 10^0-10^1 m/s. Consequently, the ratio w_H/u will be of the same order of magnitude (10^{-4}-10^{-2}) as the observed inclination of the frontal surfaces. One more important effect furthering the formation of a frontal surface (zone) must be taken into account, namely, the liberation of latent heat of condensation in an ascending air mass.

Aside from facilitating the appearance of ascending motion, turbulent mixing plays a major role in the transfer of heat and moisture. The great importance of turbulence in the boundary layer of the atmosphere is generally recognized. However, in the frontal zones, the horizontal temperature gradient creates considerable wind shear and generates turbulent fluctuations not just in the boundary layer, but also throughout the entire troposphere. Furthermore, the cloudiness itself intensifies the turbulent mixing as well (the turbulence coefficient in clouds reaches several tens of m^2/s).

Following [206], the main results of the modeling of a frontal zone and the cloudiness forming in it will now be presented. The initial system of equations will be given in Section 3.3 (eqns. (3.3.1)-(3.3.8)). Clearly, the advective, convective, turbulent, and condensational influxes of heat and moisture must all be taken into account. Equation (3.3.3) for the moisture content has an added term allowing for the fall of droplets due to gravity (see eqn. (2.1.26)). For the weighted-mean (over the mass) rate of fall of cloud elements $\tilde{v}$, the following relation, suggested in [136], is used:

$$\tilde{v} = \tilde{v}_m \exp\left(-\beta \frac{z - z_l}{z_u - z_l}\right),$$

where $\tilde{v}_m$ is the maximum value of $\tilde{v}$, reached at the lower cloud boundary z_l; z_u is the upper cloud boundary; and β is a parameter.

The equations of motion retain their inertial terms, but the characteristics of turbulent mixing are found from the equation of balance of eddy energy (3.3.8) and the following equation predicted by similarity theory:

$$\frac{dk}{dz} = \frac{k}{b}\frac{db}{dz} + \sqrt[4]{c}\,\varkappa\sqrt{b}.$$

Here this equation will replace eqn. (3.3.9).

The x axis is normal, while the y axis is parallel, to the trough axis. However, unlike the case considered in Section 3.3, the geostrophic wind in the problem involving a front has two components (Fig. 2.6): one perpendicular (u_g) and the one parallel (v_g) to the front. The first component remains constant upon transition through the front ($u_g^{(1)} = u_g^{(2)}$) and does not vary with the height, since it has the same direction as the horizontal temperature gradient Γ. The second component varies upon transition through the front ($v_g^{(1)} \neq v_g^{(2)}$), and it varies with height as well (under the influence of Γ). For this reason, the last term in eqn. (3.3.5) is written in the form $2\omega_z(v - v_g)$. From the initial system of equations it is seen that the plane problem (all the quantities are independent of y) is being considered here. Moreover, the cloudiness and the frontal zone being modeled close to the time when the cloudiness and the other fields reach an extremal state (in particular, the clouds have reached maximum development along the vertical). Around this time, the local derivatives with respect to time are close to zero, so that they can be neglected in the initial system.

The pressure field in the given problem is assumed to be known. It can be predicted (or modeled) according to some numerical method without the subtler features of the boundary layer of the cloud-formation processes being taken into account.

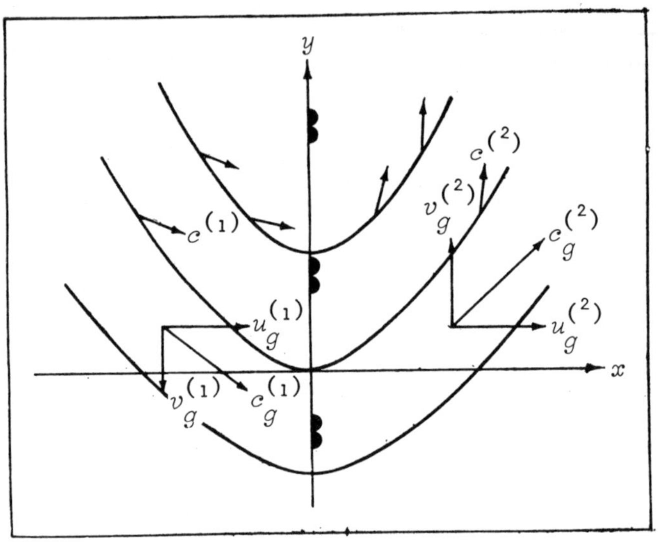

Fig. 2.6. Diagram of a trough. Here $c_g^{(1)}$ and $c^{(1)}$ are, respectively, the geostrophic wind and wind velocity close to the ground in a warm air mass; $c_g^{(1)}$ and $c^{(1)}$ are the analogous quantities in a cold air mass.

The boundary conditions for this problem will now be formulated. Some of the conditions will be of the usual kind. In particular, at the upper boundary of the layer, assumed to be the tropopause, none of the quantities (T, g, w, v) depend on x (the distance from the front), and the energy of turbulence (eddy energy) b approaches zero. At a sufficient distance from the front (theoretically, as $x \to -\infty$), in a warm air mass the temperature is a linear function of the height, that is, $T(-\infty, z) = T_1 - \gamma_1 z$, while the relative humidity f_1 is constant: $q(-\infty, z) = f_1 q_m(z)$. At the ground ($z = 0$) the condition $u = v = \partial b/\partial z = 0$ requires no explanation. It is also evident that $u_g^{(1)} = u_g^{(2)} = u_g = \text{const}$. The component v_g of the geostrophic wind, parallel to the front, depends markedly on x (in Fig. 2.6, it changes sign upon transition through the front). If the pressure field is predicted, then it will be possible, of course, to take specific values of v_g at the grid points and to use these for a numerical solution of the problem. When constructing a model, however, more general results can be obtained if we specify the dependence of $v_g(x)$ on x in analytic form. The following formula is in accordance with experimental findings:

$$v_g(x, 0) = \bar{v}_g + \frac{\Delta v_g}{2} \operatorname{th} \frac{x}{D_v}, \qquad (2.4.1)$$

where $\bar{v}_g = [v_g^{(1)} + v_g^{(2)}]/2$, $\Delta v_g = v_g^{(2)} - v_g^{(1)}$, and $v_g^{(1)}$ and $v_g^{(2)}$ are the components of the geostrophic wind along the y axis at a great distance from the front (theoretically, as $x \to \pm\infty$), in a warm and cold air mass, respectively. In (2.4.1), $\operatorname{th} x = \exp(x) - \exp(-x)/\exp(x) + \exp(-x)$ is the hyperbolic tangent ($\operatorname{th} x \to 1$ as $x \to \infty$, and $\operatorname{th} x \to -1$ as $x \to -\infty$).

The height dependence of $v_g(x,z)$ is described by the familiar formula

$$v_g(x, z) = v_g(x, 0) \frac{T(x, z)}{T(x, 0)} + \frac{gT(x, z)}{2\omega_z} \int_0^z \frac{1}{T^2} \frac{\partial T}{\partial x} dz. \qquad (2.4.2)$$

The air temperature close to the ground can also be assumed to be a known function of x. It can be described using a formula of the same form as (2.4.1):

$$T(x, 0) = \overline{T} - \frac{\Delta T}{2} \operatorname{th} \frac{x}{D_T}, \qquad (2.4.3)$$

where $\overline{T} = (T_1 + T_2)/2$, $\Delta T = T_1 - T_2$: T_1 and T_2 are the air temperatures near the ground a great distance from the front in a warm and cold mass, respectively. On the other hand, it is not necessary to assume (and specify) a known temperature difference. This can be obtained if we take into account the effect of the cloudiness on the influx of solar radiation to the Earth's surface (or in the more general case, on the radiation and heat balance). Physically speaking, it is obvious (and the appropriate calculations and experimental data verify this conclusion) that the appearance of clouds will be accompanied by a drop in the air temperature near the ground for a positive radiation balance and by a temperature rise for a negative balance. Consequently, if in the initial state even $\Delta T = 0$, it follows that subsequently this difference will be different from zero because of the cloudiness. Moreover, if $\Delta T = 0$ close to the ground, then, due to the release of the condensation heat, a temperature difference will be created at higher levels, and a pronounced frontal zone will be produced.

Such fronts are often observed in Central Asia and in the

intertropical convergence zone (ITC). At these fronts the temperature contrast close to the land (water) surface is small ($\Delta T \approx 0$); at higher levels (in the ITC, beginning at the 850 mb surface), ΔT values comparable with the temperature contrasts at fronts in extratropical latitudes are observed.

The variation in the relative humidity f close to the ground is given by the formula

$$f(x, 0) = \bar{f} - \frac{\Delta f}{2} \operatorname{th} \frac{x}{D_f}, \qquad (2.4.4)$$

where $\bar{f} = (f_1 + f_2)/2$, $\Delta f = f_1 - f_2$; and f_1 and f_2 are the relative air humidities far from the front in a warm and cold air mass, respectively.

Parameters D_v, D_T, and D_f in formulas (2.4.1), (2.4.3), and (2.4.4) have units of length and are the horizontal scales of the frontal zone. The lower these scales are, the narrower (along x) will be the zone of transition from $v_g^{(1)}$, T_1, and f_1 to the values of these quantities in the cold mass, and thus the narrower will be the frontal zone itself. The trough in which the front is situated will be deeper, the greater the difference Δv_g. In a well defined (uncamouflaged) trough $v_g^{(2)} > 0$ and $v_g^{(1)} <$ < 0.

Some results of the calculations, for which it can be assumed that $T_1 = 20°C$, $\gamma_1 = 0.7°C/100$ m, $2\omega_z = 1.12 \cdot 10^{-4} s^{-1}$, $\tilde{v}_m = 13$ cm/s, $\beta = 3$, $\Delta f = 0$, $v_g^{(2)} = v_g^{(1)}$ (close to the ground the trough is symmetrical relative to its axis), and $D_v = D_T$ will now be given.

The field of vertical motions, the water content and boundaries of the clouds, and the inclination of the frontal surface are affected the most by such quantities as ΔT, Δv_g, f_1, and D_v.

Figure 2.7 portrays the dependence of the absolute water content δ^* of the cloud on the height above a front near the ground ($x = 0$), for various values of the relative humidity f_1 in warm air. For f_1 from 60 to 70%, only clouds of the middle and upper decks form, with a water content not exceeding 0.1 g/m^3. Fronts with such clouds are observed, for instance, in the summer in Central Asia. For a high relative humidity (80-95%), dense nimbostratus cloudiness forms, with maximum values of the water content (in the middle troposphere) ranging from 0.3 to 0.6 g/m^3. When Δv_g increases and D_v decreases at $x = 0$, the maximum values of the vertical velocity and the cloud water content increase.

Inspection of Fig. 2.8 indicates the variations in the boundaries, thickness, and vertical profile of frontal cloudiness as functions of the distance x from a front near the ground. The densest cloudiness is seen to form close to the front that is near the ground (at $x = 0$ and $x = 100$ km). Further away from the front near the ground, toward cooler air, the lower boundary of the clouds becomes more elevated (from 330 m at $x = 0$ to 1670 m at $x = 200$ km, and then to 3650 m at $x = 300$ km), while the water content decreases (from its maximum value $\delta_m^* = 0.72$ g/m^3 at $x = 0$ to 0.47 g/m^3 at $x = 200$ km, and then to 0.22 g/m^3 at $x = 300$ km).

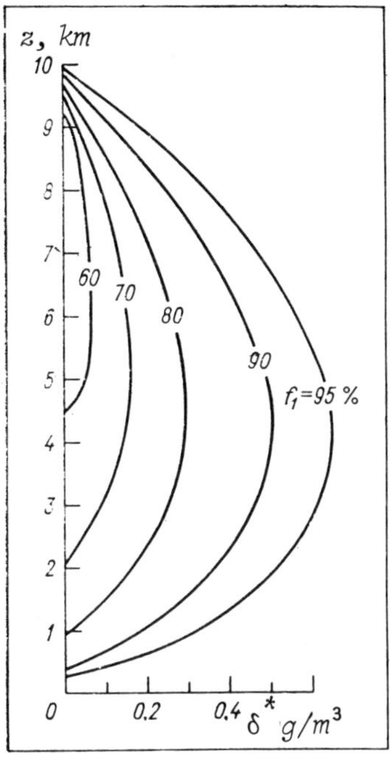

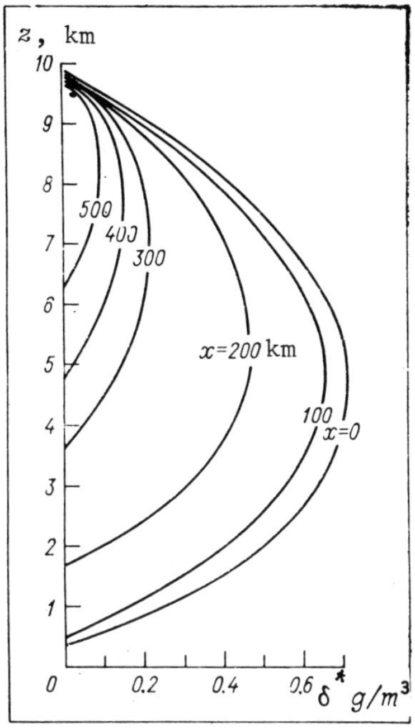

Fig. 2.7. Distribution of absolute water content δ^* with height above front near ground ($x = 0$) for various values of f_1 and fixed values $\Delta T = 10$°C, $\Delta v_g = 20$ m/s, and $D_v = 100$ km.

Fig. 2.8. Distribution of absolute water content δ^* with height at various distances x from front near ground for fixed values $f_1 = 90$%, $D_v = 50$ km, and $\Delta T = 10$°C.

The temperature contrast ΔT between the warm and cold air has the most marked effect on the horizontal extent of the cloud system and the inclination of the frontal surface. Close to the front near the ground the cloudiness characteristics for $\Delta T = 10$°C and $\Delta T = 5$°C are not very different, with height of lower boundary equal to 330 and 450 m, and δ^* equal to 0.72 and 0.67 g/m^3, respectively. However, for $x = 100$ km, and even more so for $x \geqslant 200$ km, significant differences are observed in the cloud parameters for different values of ΔT. For example, at $x = 200$ km the height of the lower boundary and δ^* are, respectively, 1670 m and 0.47 g/m^3 for ΔT = 10°C and 4300 m and 0.25 g/m^3 for ΔT = 5°C.

With an increase in ΔT, the extent of the cloud system in the x direction (normal to the front) increases because of a variation in the inclination of the frontal surface ($\tan \alpha$), According to Fig. 2.8, $\tan \alpha$ has the following values at various distances from a front near the ground:

x, km.	0—100	100—200	200—300
$\tan \alpha$	$0{,}27 \cdot 10^{-2}$	$1{,}20 \cdot 10^{-2}$	$1{,}90 \cdot 10^{-2}$

For ΔT = 5°C the inclination of the frontal surface is about twice as great: $2.65 \cdot 10^{-2}$ between 100 and 200 km, and $3.30 \cdot 10^{-2}$ between 200 and 300 km. These figures agree with experimental data. In the standard monographs and textbooks, on the other hand (beginning with the works of the founders of frontological analysis, J. Bjerknes, T. Bergeron, and others), a frontal surface is so portrayed that its inclination decreases as x increases. However, the data given above suggest that $\tan \alpha$ becomes larger with increasing distance from a front near the ground. This is because the horizontal temperature difference at the front decreases with an increase in x; however, according to the familiar Margules formula, a drop in ΔT is accompanied by an increase in $\tan \alpha$. Consequently, it is not clear why the foregoing regularity was not recognized in earlier works.

Figure 2.9 shows the results of a numerical modeling of frontal cloudiness and of the fields of temperature and vertical motions. The model is seen to reflect the most important features of front structure that have been established by synoptic practice. Since the cloud and front characteristics are known to depend on several factors, by varying these the diverse formation conditions and structural features of frontal cloudiness may be modeled.

The amount of precipitation falling from frontal clouds, will now be evaluated. This was done in [206] by calculating the water reserves of a cloud, neglecting ($\tilde{v}$ = 0) and taking into account ($\tilde{v} \neq 0$) the fallout of cloud droplets for various values of x, ΔT, and f_1. The water reserves of a cloud were found to increase with increasing f_1 and ΔT, and to decrease further away from the ground front. During precipitation, the water reserves of the cloud close to the ground front dropped from 35.82 to 4.49 kg/m^2, for ΔT = 10°C, f_1 = 90%, Δv_g = 20 m/s, and D_v = 50 km. This means that during the time of existence of the cloud, 31.33 kg/m^2 of precipitation falls from a column 1 m^2 in cross section, that is, an amount equivalent to a water layer 31 mm thick. At x = 400 km, only 1.88 mm of precipitation will fall.

Another effect caused by large-scale vertical motions is worthy of mention. As already noted, in the lower troposphere w increases with increasing height; therefore, the temperature w variation with time, $\partial T/\partial t = w(\gamma - \gamma_a)$, will be negative at all

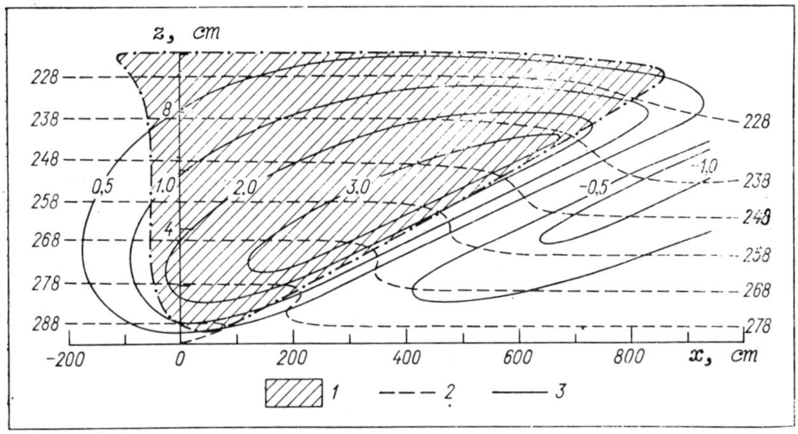

Fig. 2.9. A numerical model of frontal cloudiness, and the fields of temperature and vertical motions, for $\Delta T = 10°C$, $\Delta v_g = 20$ m/s, $f_1 = 80\%$, and $D_v = 150$ km.

(1) cloudiness,(2) isotherms,(K) (3) isolines of vertical velocity (cm/s).

levels (since $\gamma < \gamma_a$) for $w > 0$, but can vary considerably in absolute value. Close to the ground, where w is nearly zero, $\partial T/\partial t$ is small. In the middle troposphere, where w reaches a maximum but the lapse rate γ is no greater than it was close to the ground (for this reason, the difference $|\gamma - \gamma_a|$ in the middle troposphere is not greater than the ground value either), the temperature decrease with time is a maximum.

Such a temperature variation with time makes the stratification of the lower troposphere more unstable with time. Because of this instability, convection appears, accounpanied by the development of convective clouds (Cu, Cb). Studies of this phenomenon by many investigators have shown that convective clouds (thunderclouds, in particular) originate only in regions of lower pressure, with their ascending vertical motions. Frontal cloud systems of the Ns-As type (even at warm fronts) are not, as a rule, uniform formations: inside the clouds convective currents are observed, often manifesting themselves as projections of cloud protruding above the upper boundary of the main mass of frontal clouds. The increased turbulence in clouds (in comparison with the cloudless atmosphere) is also a consequence of the heightened instability of the atmosphere in regions of ascending vertical motions (updrafts).

A graphical example of this effect is the formation of cloudiness in the intertropical convergence zone (ITC), which was studied carefully during the Tropical Experiments of 1972 and 1974

[302, 303]. The ITC is known to be a region of lower pressure and convergence of air currents, as well as a place where the horizontal gradients of temperature and humidity can be large - a system which is more pronounced at the 850 mb surface than at sea level.

In the ITC, therefore, ascending motions are observed, accompanied by the formation of stratiform and convective clouds. An analysis of experimental data - the data in [302, 303, 375, 386], for instance - indicates that cloudiness and precipitation can be interpreted only with the aid of models in which a decisive role is accorded to vertical motions.

Using radar methods, it was established [212] that in frontal zones in the middle latitudes, together with large-scale vertical motions, medium-scale vertical motions of cellular structure are observed as well. The latter extend over regions several tens of kilometers in size along the horizontal (usually 30-60 km at warm fronts) and are accompanied by the formation of convective clouds. The velocities of the updrafts and downdrafts in the central parts of these cells fluctuate between 3-5 and 30-40 cm/s. If only stratiform clouds are present, the field of vertical motions is more uniform. Radar studies have shown that the heaviest snowfalls are observed in regions of descending currents. This fact is in accordance with the interpretation developed in this monograph.

2.5. The lower cloud boundary

In a series of works [139, 298-301], Tkachenko and his co-workers investigated the distributions of temperature, humidity, and water content close to lower cloud boundaries and established the regular features of the variation in height of the lower boundary. These studies were based on the premise of complete entrainment of cloud elements by turbulent air parcels, and their purpose was to put the initial equations of heat and moisture transfer into a form not containing the rate of water-vapor condensation. Therefore, as in the works of the present author, the specific moisture content $s = q + \delta$ and the quantity $\Pi^* = T - L/c_p\delta + \gamma_a z$, related linearly to the quantities s and $\Pi = T - L/c_p + \gamma_a z$ introduced by the present author, are linked by the equation

$$\Pi^* = \Pi - \frac{L}{c_p} s.$$

Thus, the quantities s and Π^* like s and Π, are invariant to phase transitions of water.

From the equation for heat balance at the lower cloud boundary z_K, it follows that a difference (or discontinuity) in the vertical gradients of Π and Π^* for $z = z_K$ can exist only because of a difference (or discontinuity) in the vertical radiation fluxes at this level. In 1967, 25 aircraft soundings of the atmosphere over the Ukraine were made at times of stratiform cloudiness (a total of 181 measurements of the temperature and humidity were made on horizontal platforms during each flight). An analysis of these soundings showed that the vertical gradients of Π and Π^*, like the gradient of s, do not have a discontinuity at the lower cloud boundary: the coefficient of correlation between the observed difference $\Pi(z) - \Pi(z_K)$ and the same difference determined for level z using an equation of linear regression, was found to be 0.94. These data provide still another proof of the absence of any radiation influx in the layer under the cloud and throughout most of the cloud itself.

The vertical profiles of Π^* and s, from the ground up to the bottom z_2 of the layer within the confines of which the upper cloud boundary fluctuates, are described, according to experimental data, by the expressions

$$\Pi^*(z) = \Theta_0(1 + \beta_1 z - \beta_2 z^2), \qquad (2.5.1)$$

$$s(z) = q_0 \exp(-\alpha_1 z + \alpha_2 z^2), \qquad (2.5.2)$$

where Θ_0 and q_0 are the potential temperature and specific humidity at $z = 0$. The mean values of the parameters (according to the data from the above-mentioned 25 soundings) are the following: $\beta_1 = 1.163 \cdot 10^{-3}\ \text{m}^{-1}$, $\beta_2 = 10.0 \cdot 10^{-9}\ \text{m}^{-2}$, $\alpha_1 = 4.4 \cdot 10^{-4}\ \text{m}^{-1}$, $\alpha_2 = 1.6 \cdot 10^{-7}\ \text{m}^{-2}$.

The other quantities have the following mean values: height of lower cloud boundary, 390 m (with fluctuations in individual instances from 150 to 700 m); height of upper cloud boundary; 690 m (fluctuations from 360 to 1150 m); z_2 = 660 m (fluctuations from 360 to 1000 m); and height of cloud summit 750 m (fluctuations from 460 to 1230 m).

With the aid of the Clausius-Clapeyron equation and the barometric formula, Tkachenko [298] obtained an expression for the saturation specific humidity q_m. Then, by equating it to the specific moisture content (2.5.2), he obtained, for the height of the lower cloud boundary, the formula

$$(1 - a_1) z_d + a_2 z_d^2 = h, \qquad (2.5.3)$$

where

$$a_1 = \frac{\alpha_1 + m\beta_1}{F_0}, \quad a_2 = \frac{\alpha_2 + m\beta_2}{F_0},$$

$$h = -\frac{\ln f_0}{F_0}.$$

Here $m = nT_0$ (where $n = L/R_v T_0^2$, and T_0 is the air temperature at $z = 0$), and f_0 is the relative humidity at $z = 0$. Finally,

$$F_0 = n(\gamma_a - B), \quad B = \frac{g}{nR_c\bar{T}},$$

where $\bar{T}$ is the mean temperature of the air column from zero up to z_K. The author does not evaluate the accuracy of calculating z_K using the formula (2.5.3).

The rate of variation of the height of the lower cloud boundary z_K was considered in [300, 301]. Starting from the condition that at the lower cloud boundary the specific moisture content equals the specific humidity of saturation, the following equation for determining z_K is obtained:

$$\varphi_s(z_d, t) = \exp\{m[\varphi_\Theta(z_d, t) - 1] - [z_d - h(0)]F_0\}, \qquad (2.5.4)$$

where $\varphi_s = s(z, t)/s(0,0)$, $\varphi_\Theta = \Pi^*(z, t)/\Pi^*(0,0)$; all the other notation is the same as in (2.5.3).

Functions $\varphi_s(z, t)$ and $\varphi_\Theta(z, t)$ are found by solving equations having the form

$$\frac{\partial\varphi}{\partial t} + \tilde{w}\frac{\partial\varphi}{\partial z} = k\frac{\partial^2\varphi}{\partial z^2}, \qquad (2.5.5)$$

where $\tilde{w} = w - \partial k/\partial z$ (here k is the turbulence coefficient).

The nonperiodic variation of z_K was studied in [139]. At the initial moment, φ_s is an exponentially decreasing function, and φ_Θ is a linear function of the height. The boundary conditions are the following: at the ground, φ_s and φ_Θ are linear functions of time; while at the upper cloud boundary z_u, the turbulent flows of heat and moisture approach zero. The height of the upper cloud boundary is assumed to be known; it varies linearly with time.

Two models are used for the vertical velocity w:

$$w(z) = w_\infty[1 - e^{-\lambda z}(\sin\lambda z + \cos\lambda z)], \qquad (2.5.6)$$

$$w(z) = 4w_m\frac{z}{z_B}\left(1 - \frac{z}{z_B}\right). \qquad (2.5.7)$$

The first of these was proved by Dyubyuk [103], and the second by the authors of [52].

The main results of the numerical solution of the problem can be summarized as follows:

(a) The variations in the ground-level values of the relative humidity f_0 and the temperature T_0 have a decisive effect on the variation of z_K with time. The parameter h, the height of the lower cloud boundary for constant s and Π^* over the height, is a function of f_0 and T_0.

(b) A variation in the height of the upper cloud boundary has practically no effect on z_K.

(c) Equally negligible is the effect on z_K of the vertical velocity: a transition from $w_m = -1$ cm/s to $w_m = 1$ cm/s is accompanied by a more rapid decrease in z_K only during the first moments; afterwards, the rate of change of z_K is practically the same, no matter what the value of w_m.

In cases where h decreases with time or remains constant, the lower boundary z_K decreases with time more rapidly than h decreases. On the other hand, h increases, then during the first moments z_K decreases, reaches a minimum, and then begins to increase, but more gradually than h increases.

It should be noted that the lack of any relation between the variations of z_K and the height of the upper cloud boundary z_u is due to the boundary condition placed on z_u by the authors of [139]. Since heat and moisture do not penetrate the upper cloud boundary (moreover, the height of the upper boundary z_u at the initial moment is taken to be the level where this condition is imposed, and why z_u has to vary with time remains unclear), it is physically obvious that when the problem is posed this way, no such relation need exist.

The vertical velocity has no significant effect on the variation of z_K, apparently because the velocity w goes to zero at level z_u (the authors point out that when w is specified according to Dyubyuk's model, the right-hand side of the formula (2.5.6) must be supplemented by the factor $1 - (z/z_u)^\nu$, where $\nu = 20$. The vertical velocity naturally cannot affect the variation in z_u, and hence does not affect the relation between the time variations of z_K and z_u either. In a more recent work [299], Tkachenko and Kolezhuk retained in the equations of heat and moisture transfer the advective influx as well as the convective and turbulent terms. Moreover, while formulating the boundary condition at the ground, they assumed the temperature and specific humidity to be periodic functions of the time and the horizontal distance. Clearly, z_K is also subject to periodic fluctuations in time and space, especially under the influence of the fluctuations of h which are caused, in turn, by the periodic variations in temperature and humidity near the ground. The fluctuations of z_K will have smaller amplitudes than the fluctuations of h, especially over a portion of the Earth's surface that is uniform along the horizontal (in the absence of advection).

2.6. The role of turbulence and radiation in the formation of stratus clouds in the Arctic

Herman and Goody [364] studied the formation of stratus clouds in the Arctic Basin. They noted that the amount of cloudiness in the Arctic increases from 40-60% in winter to 70-90% in midsummer, in a comparatively short time (about 4 weeks). Arctic stratus clouds frequently consist of two layers, each 300-500 m thick, with water content from 0.05 to 0.01 g/kg. The field of the air temperature close to the coasts of America and Eurasia is nonuniform, whereas the temperature in the Central Arctic is practically constant along the horizontal.

In [364], the system of equations was written in the form

$$u_0 \frac{\partial \Pi}{\partial x} = \frac{\Pi}{c_p \rho T} \varepsilon_{rad} - \frac{\partial}{\partial z} (\overline{w'\Pi'}) \tag{2.6.1}$$

$$u_0 \frac{\partial s}{\partial x} = - \frac{\partial}{\partial z} (\overline{w's'}) - \tilde{v} \frac{\partial \delta}{\partial z}, \tag{2.6.2}$$

where u_0 is the speed of the geostrophic wind, with respect to which the advective influxes of heat and moisture are linearized; Π is the equivalent potential temperature; ε_{rad} is the radiative influx of heat; s is the specific moisture content; δ is the cloud water content; and $\tilde{v}$ is the mass-weighted mean falling speed of the droplets.

The system (2.6.1) and (2.6.2) is supplemented by the equations of steady-state motion, also linearized:

$$u_0 \frac{\partial u}{\partial x} = 2\omega_z v - \frac{\partial}{\partial z} (\overline{w'u'}) \tag{2.6.3}$$

$$u_0 \frac{\partial v}{\partial x} = -2\omega_z u - \frac{\partial}{\partial z} (\overline{w'v'}). \tag{2.6.3}$$

in which u and v are the projections of the deviation of the wind from the geostrophic. In the following, the derivatives with respect to the variable x will be replaced by the derivatives with respect to time, with the aid of the relation $u_0 \partial/\partial_x[\ \ ] = \partial/\partial t[\ \ ]$.

The system (2.6.1)-(2.6.4) shows that Herman and Goody are solving the problem of the formation of clouds and fog over a nonuniform underlying surface, under the influence of advective and turbulent influxes of heat and moisture. The radiative influx of heat was taken into account using a method worked out previously by Guirache and Goody for the clouds of Venus. This method takes into account absorption and scattering by cloud droplets, as well as absorption by water vapor (in a band from 0.7 to 6.3 μm), of both solar and thermal (infrared) radiation. Herman and Goody assume that with regard to radiation, their approach has a sound physical basis, and the only unknowns are the values of the parameters entering into the problem.

The turbulence was allowed for in a very simple manner; in the ground layer the scale of the eddies and the turbulence coefficient were taken to be linear functions of the height: above the ground layer, these quantities were assumed to be constant. For unstable stratification, the hypothesis of convective acommodation was invoked: the lapse rate was assumed to be γ_a, and the momentum and specific moisture content were assumed to be constant with height because of intensive mixing. The cloud water content was found from the condition of reaching saturation.

It should be noted that in eqn. (2.6.1), vertical air movements must generally be taken into account (settling of the air mass). The variation of the ordered vertical velocity w in a layer from the ground up to a height of 3 km was taken to be linear: $w = Az$, where in the Arctic $A = 1.4 \cdot 10^{-7}$ s^{-1}, which gives $w \approx$ ≈ -0.04 cm/s at $z = 3$ km.

Calculations were made for two cases: (1) for stable (inversion type) stratification of an air mass flowing over a melting ice surface ($T_0 = 273$ K, air temperature 4 K above the temperature of the ice surface); (2) for unstable stratification ($T_0 = 273$ K, air 3 K cooler than the ice surface). In both cases, the relative humidity of the air throughout the boundary layer at the initial moment is taken to be 90%, the wind distribution is described by an Ekman spiral for a turbulence coefficient of 2.5 m^2/s, the albedo of the ice is 40%, and the roughness parameter $z_0 = 1$ mm.

Herman and Goody evaluated quantitatively the roles of the various factors participating in the formation of clouds and fog. In six models the role of radiation was evaluated, and in five models the role of turbulence. Assuming radiation to be absent (either completely or else in some of its components, such as radiation of water vapor or of droplets), they found that clouds and fog still form in the boundary layer, but that no layering of the clouds is observed. The greatest effect of radiation, in the opinion of Herman and Goody, manifests itself in the formation of cloud layers, although the role of turbulent diffusion is greater than that of radiation: estimates show that in a layer about 1 km thick, the average situation becomes rearranged in about 16 min because of turbulence; but it takes about a day for the radiation to reaarrange the average situation to the same extent.

For a stratification that is initially stable, the condensation of water vapor is completed in 34 h at a height of 500 m. Then, for 7 days (the time over which the system of equations is integrated), the lower cloud boundary remains at practically the same height, whereas the upper boundary gradually rises and after 5 days becomes stationary. It takes about 66 h for the cloud to stratify. At the moment of stabilization, the boundaries of the lower layer are at heights of 500 and 1050 m, while those of the upper layer are at 1450 and 1700 m. The water content in the upper layer reaches a maximum close to the upper boundary (0.37 g/m^3); the lower layer is more tenuous (with maximum water content only about 0.05 g/m^3).

For stratification that is initially unstable, clouds form much earlier than in the first case (in about 8 h). The clouds form into layers in 50 h. However, unlike the first case, a steady distribution of the water content is arrived at in the lower layer; the upper boundary of the upper layer continues to rise slowly, for as long as 5 or 6 days.

After 4 or 5 days, the initial state comes to an end; the cloud characteristics in the first and second case become practically the same.

The physical mechanism by which a cloud forms layers will now be considered. During the initial period, cooling is caused mainly by the effective radiation of the Earth's surface. Air is cooled, bringing water vapor to the saturation state and leading to cloud formation, mainly as a result of turbulent mixing. Absorption of radiation by air slows down this cooling somewhat, but does not prevent it. As soon as a cloud forms, the heat loss via radiation is mainly from the upper boundary of the cloud (comprising a layer about 100 m thick). Cooling spreads inside the cloud because of turbulent exchange, at the same time decreasing very slightly because of absorption by the air of infrared radiation.

The absorption of solar radiation by cloud droplets compensates considerably the loss of heat via radiation in the infrared range, especially in the upper part of the cloud. Since the long-wave radiation is close to zero inside the cloud, there the droplets will be heated by the solar radiation so much that they evaporate. In the lower part of the cloud, where the infrared radiation is also very low, the flux of solar radiation becomes greatly attenuated, so that the absorption of radiation does not lead to droplet evaporation. Consequently, the formation of cloud layers in this model is caused by the absorption of solar radiation by cloud droplets, which, in the middle of the cloud may well evaporate (for a certain ratio between the absorption coefficients for infrared and solar radiation).

CHAPTER 3

MODELS OF FOG FORMATION

3.1. Radiation fogs

The author of [190] was the first to attempt to explain the observed features of fog structure. In that paper, an analysis of the equations of heat influx and moisture transfer was used to explain, for the first time, the drop in temperature close to the ground and the creation of a lifted temperature inversion in a fog (high-inversion fog). The ideas of [190] were developed in [162], where the equation for the heat conduction of the soil was added to the system of equations describing heat and moisture transport. Radiation effects were taken into account in the boundary conditions (at the ground before the fog forms and at the upper boundary of the fog after the fog forms).

Melkaya [211] calculated the characteristics of the ground layer and the formation of fog in it, assuming steady motion uniform along the horizontal. As in the earlier works [162, 190], the concept of entrainment of droplets by turbulent air parcels was used, and the following functions were introduced:

$$\Pi = T + \frac{L}{c_p} q + \gamma_a z, \quad s = q + \delta, \tag{3.1.1}$$

whereby the initial system of equations becomes

$$\frac{d}{dz} \alpha_H k \frac{d\Pi}{dz} = 0, \quad \frac{d}{dz} \alpha_H k \frac{ds}{dz} = 0. \tag{3.1.2}$$

However, the turbulence coefficient k entering into (3.1.2), which in [162] was assumed to be known and was approximated by an exponential height function [175, 185], was in [211] obtained from equations and relations established in boundary-layer physics on the basis of similarity theory and dimensional analysis [222]. Consequently, the system (3.1.1.) and (3.1.2) must be supplemented by the equation describing uniform horizontal motion,

$$\frac{d}{dz} k \frac{du}{dz} = 0 \tag{3.1.3}$$

and the relations

$$k\psi = cb^{3/2}/l, \quad \psi = \left(\frac{du}{dz}\right)^2 - B\frac{d\Pi}{dz}, \tag{3.1.4}$$

$$k = lb^{1/2}, \quad l = -\tilde{\varkappa}\psi/(d\psi/dz), \tag{3.1.5}$$

$$B = \frac{\alpha_H\left[\frac{g}{T} + g\left(\frac{R}{R_v}\right)^2 \frac{E'(T)}{p}\right]}{1 + \frac{L}{c_p}\frac{R}{R_v}\frac{E'(T)}{p}}, \quad \alpha_H = k_H/k. \tag{3.1.6}$$

Here u is the wind velocity; b is the kinetic energy of the turbulent fluctuations [eddy kinetic energy]; l is the mixing length; $\varkappa = 0.4$ is the Kármán constant; $\tilde{\varkappa} = 2c^{1/4}\varkappa$; $c = 0.046$; k_H and k are the turbulence coefficients for heat (and water vapor) and momentum, respectively: and R and R_v are the specific gas constants of air and water vapor.

The unknown functions entering into the system (3.1.1)-(3.1.5) satisfy the following boundary conditions:

(a) at the ground ($z \to 0$)

$$T = T_0, \quad q = q_m(T_0), \quad \delta = 0,$$

$$u = 0, \quad k = k_0 = \varkappa v_* z_0, \quad k\,(du/dz) = v_*^2;$$

(b) at the upper boundary of the fog ($z \to h$)

$$k_H \frac{d\Pi}{dz}\bigg|_{z \to h-0} = k_H \frac{d\Pi}{dz}\bigg|_{z \to h+0} - \frac{B^*}{\rho c_p}\bigg|_{z \to h+0},$$

$$k_H \frac{ds}{dz}\bigg|_{z \to h-0} = k_H \frac{dq}{dz}\bigg|_{z \to h+0}, \quad \delta\,\big|_{z \to h-0} \to 0.$$

The first of the latter pair of equations is the heat-balance equation (B^* is the effective radiation of the upper fog boundary), and the second is the equation of moisture balance. Now, if we introduce the dimensionless quantities

$$\tilde{z} = z/L_*, \quad \tilde{u} = \varkappa u/v_*, \quad \tilde{\Pi} = \alpha_H(\Pi - \Pi_0)/\Pi_*, \tag{3.1.7}$$

$$\tilde{s} = \alpha_H(s - s_0)/s_*, \quad \tilde{k} = k/\varkappa v_* L_*,$$

then the equations for the three principal variables $\tilde{u}$, $\tilde{s}$, and Π assume the same form:

$$\tilde{k}\left(d\tilde{u}/d\tilde{z}\right)=1,\quad \tilde{k}\left(d\tilde{s}/d\tilde{z}\right)=1,\quad \tilde{k}\left(d\tilde{\Pi}/d\tilde{z}\right)=1. \qquad (3.1.8)$$

The solution of these equations has the form

$$\tilde{u}=\tilde{s}=\tilde{\Pi}=\frac{2}{\Phi}+\ln\left|\frac{1-\Phi}{1+\Phi}\right|+2\operatorname{arctg}\Phi+\text{const}, \qquad (3.1.9)$$

where Φ is a new independent variable, related to z as follows:

$$\tilde{z}=\frac{2}{\Phi}-\frac{2}{3}\Phi^3-\frac{4}{3}. \qquad (3.1.10)$$

In relations (3.1.7), the following notation was employed: s_0 and Π_0 are the values of functions s and Π for $z = 0$, and

$$s_*=\frac{k_H}{\varkappa v_*}\frac{dq}{dz}\bigg|_{h+0};\qquad \Pi_*=\frac{1}{\varkappa v_*}\left[k_H\frac{d\Pi}{dz}-\frac{B^*}{\rho c_p}\right]_{h+0}.$$

Finally, $L_* = v_*^2/\varkappa^2 B\Pi_*$ represents the scales of quantities s and Π and the height z (the dynamic velocity v_* serves as the scale of the velocity). The turbulence coefficient $\tilde{k}$ is related as follows to variable Φ :

$$\tilde{k}=1-\Phi^4. \qquad (3.1.11)$$

From formulas (3.1.3)-(3.1.5) it is easy to obtain the dimensionless mixing length $\tilde{l}$ and energy $\tilde{b}$:

$$\tilde{l}=-2\tilde{\psi}/\left(d\tilde{\psi}/d\tilde{z}\right),\quad \tilde{b}=\left(\tilde{k}/\tilde{l}\right)^2,\quad \tilde{\psi}=\left(d\tilde{u}/d\tilde{z}\right)^2-d\tilde{\Pi}/d\tilde{z}. \qquad (3.1.12)$$

In the problem being considered, the "external" parameters are z_0, v_*, s_*, Π_*, T_0, and $q_0 = q_m(T_0)$, that is, the water vapor close to the ground (the temperature of which is T_0) is saturated.

Figure 3.1. shows the calculated characteristics of the ground layer (u, T, k, l, b), as well as the fog water content δ. The curves were plotted for the following parameter values: $z_0 = 0.02$ m, $T_0 = 283$ K, $v_* = 0.1$ m/s, $\alpha_H = 1$, $\Pi_* = 0.684$ K, and $s_* = 1.6\cdot10^{-6}$ g/g. It is seen that the air temperature in the layer from the ground up to a height of 14 m, within which fog is observed in the given example, increases with height by more than 2 K. Such a temperature profile is predetermined by the condition at the upper fog boundary: since the effective radiation B^* at the upper boundary is positive ($B^* > 0$), it follows from the condition $\Pi_* > 0$ that for $z = h + 0$ the derivative of the total heat content, equal to the sum of the sensible heat and the latent heat, must be positive: $d\Pi/dz > 0$. From the expression for Π_* we then get

$$\frac{k_H}{\varkappa v_*}\frac{dT}{dz}\bigg|_{h+0} = \Pi_* + \frac{B^*}{\rho c_p} - \frac{L}{c_p}s_* - \gamma_a\frac{k_H}{\varkappa v_*}\bigg|_{h+0}.$$

If we assume that $s_* = 1.6\cdot 10^{-6}$ g/g and $v_* = 0.1$ m/s, and if, according to Fig. 3.1, k_H close to level h is about 0.2 m^2/s,

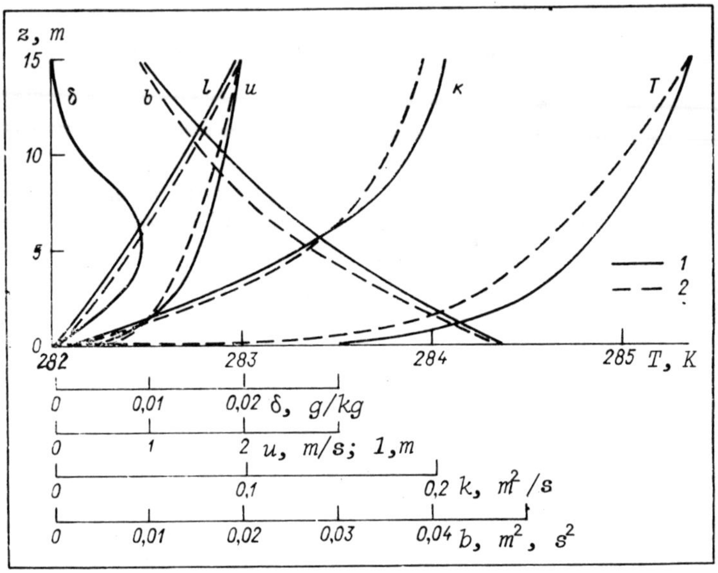

Fig. 3.1. Distribution of meteorological quantities with height in fog (1) and in dry air (2) for the same values of the external parameters.

then the last two terms in this equation will be equal to 0.004 and 0.05 K, respectively. Since $\Pi_* = 0.647$ K, it follows that $dT/dz|_{z\to h+0} > 0$, that is, close to the upper fog boundary the derivatives both of the temperature and of Π with respect to z are positive.

The wind velocity u and the turbulence coefficient k increase with height over the entire layer in which fog is observed (the increase in k is practically linear). The dashed curves in Fig. 3.1 show the distributions of T, u, k, l, and b in the case where the effects of water-vapor condensation are neglected (the air is considered to be dry). The air temperature is seen to be affected most by water-vapor condensation (it is several tenths of a degree higher at all levels in the fog than in the absence of fog). The water content of the fog at first increases with height, reaching a maximum at a height of 5 m, and then decreases. In the given example, a fog containing little water is modeled: even the maximum water content only amounts to 0.01 g/kg. Naturally, the temperature rise (in comparison with dry air) due to the condensation heat will be comparatively small (and proportional to δ).

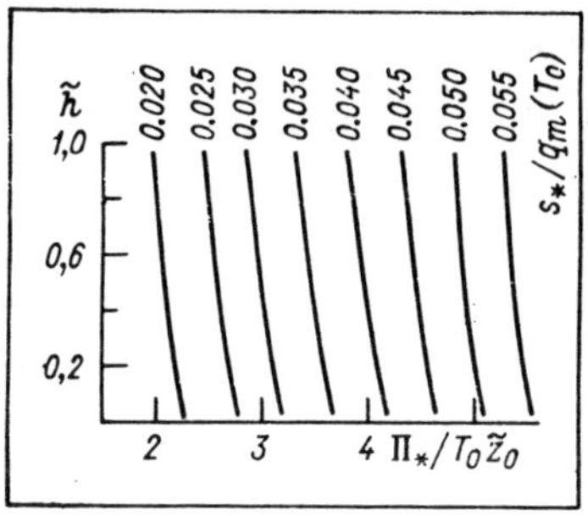

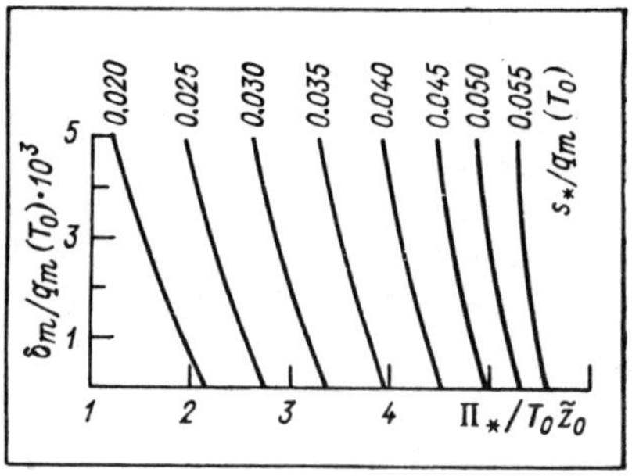

Fig. 3.2. Dimensionless fog depth $\tilde{h} = h/L_*$ as a function of dimensionless quantities $\Pi_*/T_0 z_0$ and $s_*/q_m(T_0)$. Here $\tilde{z}_0 = z_0/L_*$.

Fig. 3.3. Ratio $\delta_m/q_m(T_0)$ as a function of $\Pi_*/T_0\tilde{z}_0$ and $s_*/q_m(T_0)$. Here δ_m is the maximum water content (as a function of height).

Fog has a negligible effect on the purely dynamic characteristics of the ground layer (u, k, l, b). This conclusion was arrived at earlier by Kuznetsov [148] and others, who evaluated the effects of water-vapor condensation in clouds. Therefore, as a first approximation, the inverse effect of fog and stratiform clouds on the dynamic regime of the atmosphere need not be taken into account.

An analysis of the solution and the calculation results portrayed in Figs. 3.2 and 3.3 indicates that the fog characteristics (water content and depth) are extremely sensitive to the external parameters of the problem. Moreover, the range of variation of these parameters within which fog formation is possible turns out to be quite narrow. For instance, for a specified value of $s_*/q_m(T_0)$, say 0.035, fog may appear for a variation in parameter $\Pi_*/T_0\tilde{z}_0$ (see Fig. 3.3) just from 3.5 to 4.0, while the water content of the fog varies more than fivefold. Parameter $s_*/qm(T_0)$ has an equally marked effect on $\tilde{h}$ and δ_m: for $\Pi_*/T_0 z_0 = 4$, a change in $s_*/q_m(T_0)$ from 0.035 to 0.045 causes δ_m to vary more than tenfold. Such a pronounced dependence of the fog characteristics on the fluxes of sensible and latent heat, and also on the temperature T_0, can be explained physically as follows: since the fog water content is about an order of magnitude less than $q_m(T)$, a slight variation of the vapor content (due to flow through the upper boundary or to the influence of temperature) causes sizable changes in the water content, and hence also the depth, of the fog.

The effect which the radiative influx of heat directly to the air has on the formation and evolution of fog was discussed in [438, 439]. In the first of these papers the turbulence coefficient was specified as a known function of height, whereas in the second it was obtained from the equations describing the boundary-layer dynamics, plus Blackadar's formula for the scale of eddies.

Berlyand and Kanchan [21], developing the earlier studies of Berlyand and his co-workers generalized in [20], formulated a theory of the formation of radiation fogs that takes into account the turbulent and radiative influxes of heat; they also investigated the influence of fog on the propagation of pollutants.

Zakharova [108, 109] included in her inital system of equations the equations for functions Π and s, which take into account the radiative influx of heat and the influx of moisture due to the fall of droplets in the gravitational field, respectively. The fall velocity of the droplets is described by the Stokes formula, and the droplet-size distribution by the Khrgian-Mazin formula. For such a distribution, the mean (weighted over the mass) fall velocity of the droplets $\tilde{v}$ is related as follows to the mode radius r_m of the droplets:

$$\tilde{v}(z, t) = \frac{7}{3} \frac{g\rho_\text{к}}{\eta} r_m^2 (z, t),$$

where ρ_d is the droplet density, η is the dynamic viscosity, and g is the free-fall acceleration.

The radiative influx of heat in the long-wave region is described by taking into account the results presented in generalized form by Feigel'son [311).

Since the turbulence coefficient is also assumed to be an unknown quantity, the following must be added to the system of equations: the equation of nonsteady-state motion (horizontal and uniform), the equation for the eddy energy, and a number of relations suggested by dimensional analysis (for the scale of the eddies, the rate of dissipation of eddy energy, and the turbulence coefficient). Finally, to the system of equations are also added the equation of heat conduction of the soil and the equation giving the number of droplets N per unit volume.

The boundary conditions at the ground ($z = 0$) presuppose: (a) an absence of droplets ($N(t,0) = 0$, $s(t,0) = s(0,0)$; (b) equality of the air and soil temperatures; and (c) validity of the heat-balance equation:

$$c_\text{п} \Delta\zeta \frac{\partial\tau}{\partial t} = \rho c_p k_H \frac{\partial\Theta}{\partial z}\Big|_{z+0} + L\rho k_H \frac{\partial q}{\partial z}\Big|_{z+0} +$$

$$+ c_s k_s \frac{\partial\tau}{\partial\zeta}\Big|_{\zeta=\Delta\zeta} - B^*(t, 0),$$

where τ is the soil temperature, c_s is the volume specific heat of the soil, k_s is the temperature conductivity of the soil, $\Delta\zeta$ is the thickness of the soil layer within which sudden temperature variations are observed, and $B^*(t, 0)$ is the effective radiation of the Earth's surface. The components of the wind velocity and the vertical flux of eddy energy go to zero at $z = 0$, while the profile of the mixing length (eddy scale) becomes linear as the ground is approached $(l \to \varkappa z,\ l_0 \to \varkappa z_0)$.

As the upper boundary z_∞ of the layer in which the fog-formation process is being modeled, the following conditions obtain: (a) fog is absent ($\delta = 0$, $N = 0$); (b) the temperature and the specific humidity do not vary with time ($\Theta(t, z_\infty) = \Theta(0, z_\infty)$, $q(t, z_\infty) = q(0, z_\infty)$): and (c) the wind velocity reaches the geostrophic value ($u = u_g$, $v = 0$), while the eddy energy goes to zero.

At the lower boundary ζ^* of the soil layer, the temperature is constant in time: $\tau(\tau, \zeta^*) = \tau(0, \zeta^*)$.

At the initial moment ($t = 0$), the temperature distribution was described by a linear function of height ($T(0, z) = T(0, 0) - - \gamma z$), and the specific humidity by an exponential function ($q(0, z) = q(0, 0) \cdot 10^{-az-bz^2}$); at $t = 0$, fog is absent ($\delta(0, z) = = 0$, $N(0, z) = 0$).

The problem as formulated was solved using numerical methods with the aid of a computer, for the following values of the parameters: temperature $T(0, 0) = 278$ K, relative humidity $f(0, 0) = 88\%$, velocity of geostrophic wind $u_g = 6$ m/s, $k_s = 0.86 \cdot 10^{-6}$ m^2/s, $z_0 = 2.5$ cm, ratios $\alpha_H = k_H/k = 0.8$, $\alpha_b = k_b/k = 0.73$ (where k_b, k_H and k are the mixing coefficients for the eddy energy, heat, and momentum, respectively) and $c_0^4 = c_1^{4/3} = 0.046$ (where c_0 and c_1 are the proprotionality factors in the formulas $k = c_0 l b$ and $\varepsilon = c_1 b^3/l$, l being the eddy scale, b the eddy energy, and ε the rate of dissipation of eddy energy).

The interval in time is taken to be 2.5 min, the depth interval is 0.05 m, and the height interval is irregular, varying from 0.2 m in the bottom meter layer to 100 m in the layer from 500 to 3000 m (78 levels in all).

Some calculations of the results will now be presented. In the author's earlier works [162, 190], it was established that prior to the formation of fog, a rapid cooling of the ground and of the air close to the ground takes place: in an hour the ground temperature drops 2.4°C and the air temperature at a height of 0.2 m drops 1.0°C; at a height of 31 m, the air temperature drops 0.5°C. It takes an hour for the temperature inversion to spread to a height of 37 m, after which the rate of cooling of the soil and air decreases. However, since the air temperature continues to drop and spreads over a thicker layer, fog forms; after 3 h the fog is 25 m deep. The deeper the fog becomes, the lower will be the effective radiation of the Earth's surface.

Because of the influx of heat to the Earth's surface from the atmosphere and from the soil, the temperature inversion close to the ground gradually starts to disappear. Turbulent heat exchange ensures that in the absence of any radiation effects on the ground temperature, the temperature in the fog begins to drop with height (under the influence of turbulence, the temperature distribution approaches dry-adiabatic in unsaturated air and moist-adiabatic in saturated moist air). Figures 3.4 and 3.5 give the calculated profiles of the fog temperature and water content for various times.

Figure 3.4 illustrates the features just mentioned: after about 3 or 4 h the depth of the fog has increased so much (in 3 h 30 min it is already 91 m) that the effective radiation of the Earth's surface practically stops having any cooling effect on the ground under the fog; and air temperature decreases with height over most of the layer in which fog has formed; and the lapse rate γ after 4 h becomes close to the moist-adiabatic value γ_a', and after 5 h, somewhat higher than γ_a'. Because the temperature at the upper fog boundary decreases slowly with time (see Fig. 3.4), while the temperature in the upper part of the fog drops much more rapidly through mixing, therefore close to the upper fog boundary a sharp temperature inversion is produced. This inversion is located in the fog, and not above it. These data are convincing proof that

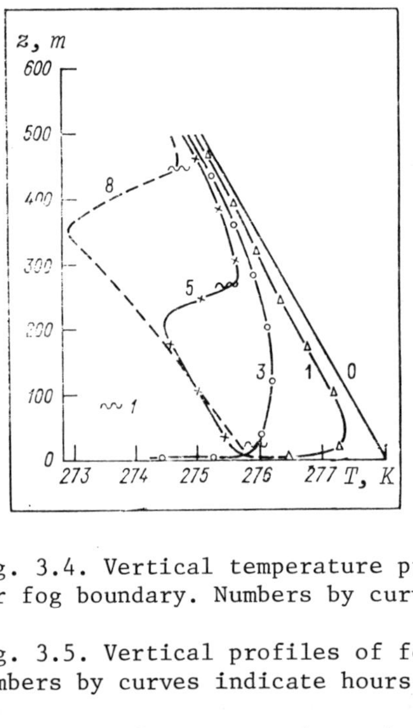

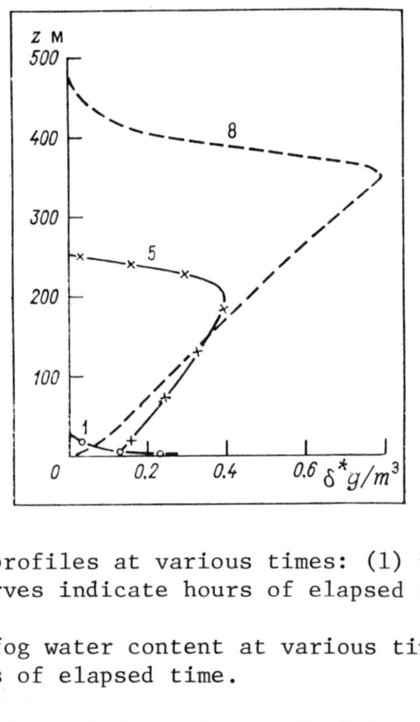

Fig. 3.4. Vertical temperature profiles at various times: (1) upper fog boundary. Numbers by curves indicate hours of elapsed time.

Fig. 3.5. Vertical profiles of fog water content at various times. Numbers by curves indicate hours of elapsed time.

the dynamic factor, namely turbulent mixing, plays a decisive part in producing the temperature inversion in the upper part of the fog. The radiative influx of heat in the atmosphere causes a general lowering of the temperature, practically the same at all levels where a substantial temperature variation is observed (this is borne out by the graphs of Zakharova's papers [108, 109].

This model also clarifies the role of the gravitational settling of droplets. For instance, if the effect of droplet fall is taken into account, then after 8 h the fog will be about 300 m deep, while the water content of the fog in the layer from 2 to 149 m will vary with height from 0.15 to 0.24 g/m^3; if the effect is not taken into account, the fog will be somewhat more than 400 m deep by t = 8 h, while the water content in the layer from 2 to 149 m will vary with height from 0.04 to 0.35 g/m^3.

Buikov and Khvorost'yanov [48] constructed a model of the formation of radiation fog by taking into account the effect of varying degrees of moistening of the ground as well as the part played by the radiative influx of heat directly to the air. The initial system of equations includes equations of steady, horizontally uniform motion in the boundary layer of the atmosphere; the energy-balance equation, together with relations from dimensional analysis and similarity theory; equations of radiation transfer; equations of heat influx (in the atmosphere and soil); the equation of water-vapor influx; and the equation for the droplet-size distribution function $f(r_d)$:

$$\frac{\partial f}{\partial t} - v_r \frac{\partial f}{\partial z} = \frac{\partial}{\partial z} k \frac{\partial f}{\partial z} - \frac{\partial}{\partial r} (\dot{r}_d f),$$

where Maxwell's formula was used for the rate of droplet growth $\dot{r}_d$, and the Stokes formula ($v_r \sim r_d^2$) for the droplet fall velocity v_r. The coefficients of turbulent mixing for momentum, heat, water vapor, and eddy energy were all taken to be the same.

Of all the results obtained in [48], only one will be given here. The droplet-size distribution function arrived at by the numerical modeling of the system of equations is close to the gamma distribution

$$f(r_d) = A r_d^b \exp(-c r_d),$$

where exponent b varies over a wide range (for one specified time it varies from 3 to 19 as a function of height; for another it varies from 5 to 41), but in all cases b is considerably greater than two.*)

*) Note that in the equation for $f(r_d)$, the effect of condensation on the variation in droplet size is taken into account, but the effect of coalescence is not, and neither is the deviation of the droplet fall velocity from the Stokes value (a deviation that is sizable for large droplets). The condensation coefficient α_c is traditionally assumed to be 0.04, but the experimental data in [228] indicate that α_c is closer to unity.

3.2. Evaporation fogs

Evaporation fogs form during the autumn-winter period, when highly cooled air streams out from a continent or an ice field over an unfrozen water surface. Such fogs are frequently produced over arctic seas, and also over inland water bodies (at night during summer; in winter, over unfrozen rivers and lakes both in the daytime and at night). Before it reaches the water surface, air of continental origin has an inversion-type stratification. Under the influence of the warmer water surface and turbulent mixing, the temperature inversion in the layer adjacent to the water disappears, and the stratification becomes highly unstable. Other characteristics of the layer next to the water (wind velocity; humidity; fluxes of heat, water vapor, and momentum; intensity of turbulent mixing) also change. Because of evaporation, the moisture content of the air in the water-adjacent layer increases, and, under certain conditions, produces a quite stable fog of considerable depth.

According to experimental data [66], evaporation fogs are produced when there is a significant difference ΔT (10-15°C or more) between the water temperature and the air temperature (before the air reaches the water surface). However, this difference is not the sole criterion for the appearance of fog. Also important are the air humidity, wind velocity, the thermal stratification of the air mass (before it reaches the water surface), and the surface roughness of the land (ice) and water. For instance, with a decrease in the relative humidity f, the minimum difference ΔT for which fog forms will increase:

f, % . . .	90—85	85—80	80	80—75	75—70
ΔT, °C . .	8.6	9.7	10.3	12.6	14.5

An increase in wind velocity is accompanied by an intensification of the turbulent mixing, which spreads over a layer of greater thickness; thus the minimum difference ΔT also increases for a higher wind velocity u. For example, for f = 85... 80%, the difference T has the following values for various u (according to observational data obtained in the laboratory):

u,m/s.	0.22	0.35	0.45	0.52	0.66
ΔT, °C.	8.5	9.3	9.8	10.3	10.8

Timofeev [292] was the first to construct a theoretical

model of the formation of radiation fog. He solved the equations of heat and water-vapor transport by approximating the vertical profiles of the wind velocity u and the turbulence coefficient k by step functions. In [201] a theory of the formation of evaporation fogs was worked out, taking into account recent advances in the physics of the ground layer and in cloud physics. Following [201], the theory of fog formation, together with the results of calculating the main characteristics of evaporation fogs, will now be presented.

Assume that an air mass passing through a land-water or interface is incident upon an open (unfrozen) sea surface ice-water or a surface of some other water body and moves over it for a long time. The motion is assumed to be steady and uniform in a direction normal to the wind velocity. Evaporation fogs form in a water-adjacent layer with a thickness of several tens of meters or 100-200 m. Consequently, in the equations of motion the Corolis force can be neglected, and the wind direction can be assumed to be constant with height. Now, if the invariants Π and s are introduced, the initial system of equations (taking all the foregoing remarks into account) may be written as

$$u\frac{\partial\Pi}{\partial x}=\frac{\partial}{\partial z}k\frac{\partial\Pi}{\partial z}, \tag{3.2.1}$$

$$u\frac{\partial s}{\partial x}=\frac{\partial}{\partial z}k\frac{\partial s}{\partial z}. \tag{3.2.2}$$

For a fog in which the water vapor is in the saturated state, the system (3.2.1), (3.2.2) is supplemented by the relation

$$q=q_m=0.622\frac{E(T)}{p}. \tag{3.2.3}$$

The pressure was calculated with the aid of the fundamental equation of statics, with the pressure at water level being taken as 1000 mb.

When an air mass streams over a surface with other properties, an internal boundary layer (transformation layer) is produced above this surface. The distribution of meteorological quantities and turbulence characteristics in this layer will differ from the distribution above the "old" surface. The thickness of the transformation layer increases with the distance from the interface between the surfaces. Very large superadiabatic temperature gradients are observed in the layer, so that the intensity of the turbulent mixing there increases. However, above this layer the turbulence coefficient remains close to its value near the interface. Such a pattern of the mixing is also verified by experimental data.

For the turbulence coefficient k in the internal boundary layer, a formula obtained by Monin and Obukhov [222] on the basis of similarity theory and dimensional analysis may be used:

$$k = \frac{\varkappa u_* L_*}{f'(\zeta)}, \tag{3.2.4}$$

where u_* is the dynamic velocity, L_* is the Monin-Obukhov scale, $\zeta = z/L_*$ is the dimensionless height (parameter of hydrostatic stability, $f'(\zeta) = d\ /d\zeta$, and $\varkappa$ is Kàrmãn's constant ($\varkappa \approx 0.40$).

As an analytical expression approximating function $f(\zeta)$, the formula

$$f(\zeta) = \begin{cases} \ln\zeta + 10\zeta & \text{for } \zeta > 0, \\ \ln|\zeta| & \text{for } -0{,}07 < \zeta \leqslant 0, \\ 1.2\zeta^{-1/3} + 0{,}25 & \text{for } \zeta \leqslant -0.07 \end{cases} \tag{3.2.5}$$

may be used. With regard to another parameter in eqns. (3.2.1) and (3.2.2), namely the wind velocity u, its variation above the water surface will be due mainly to the roughness parameter. The effect of stratification on the wind profile, on the other hand, is comparatively slight. Therefore, the distribution of the wind velocity with height over the water surface was described using the logarithmic formula (3.2.6). The dynamic velocity u_*, related to the wind velocity u_1 at some level z_1 by the expression $u_* = \varkappa u_1/\ln\,(z_1 + z_0)/z$, also varies only as a result of a change in z_0, since u_1 is independent (by assumption) of the horizontal distance x.

As an air mass moves forward over a water surface, the distribution of the air temperature with height, as well as the other meteorological quantities and the turbulence characteristics, will vary accordingly. The lapse rate γ is very large and positive close to the water surface; with increasing height γ decreases, and at a certain altitude it passes through zero, after which it takes on negative values (the inversion-type temperature distribution remains). In the face of such a temperature distribution, there will be variations in the scale L_* and in the temperature scale T_*. In order to take this effect into account, a method is employed whereby, if information is available on the meteorological quantities at many levels (an overdetermined system of equations), scales L_* and T_* can be calculated using least squares. In the present case, information about the lapse rate at various distances from the shore was taken from calculated data. Scale L_* was determined by successive approximations, with the aid of the relation $L_* = u_*^2/(\varkappa\beta T_*)$, where $\beta = g/T$.

The influence of the thermal stratification on parameters L_* and T_*, and also on the turbulence coefficient k, was taken into account only in the internal boundary layer, where the temperature drops with height. In the part of the layer where the temperature distribution with height was still of the inversion type, the turbulence coefficient was assumed to have the values observed for it at the corresponding heights for $x = 0$ (over the shoreline).

The initial and boundary conditions will now be formulated. At the moment the air mass reaches the water surface (at $x = 0$), the distributions of the wind velocity and air temperature with height are described by the logarithmic expressions

$$u(0, z) = u_1 \frac{\ln \frac{z + z_0}{z_0}}{\ln \frac{z_1 + z_0}{z_0}}, \tag{3.2.6}$$

$$T(0, z) = T_0 + T_* \ln \frac{z + z_0}{z_0} - \gamma_a z \tag{3.2.7}$$

where z_0 is the roughness parameter of the land (ice) surface, and T_0 is the air temperature at $z = 0$.

The distribution of the specific humidity at $x = 0$ was taken to be

$$q(0, z) = f_0 q_m(0, z), \tag{3.2.8}$$

where f_0 is the relative humidity of the air, which at $x = 0$ is assumed to be independent of the height.

The lower boundary ($z = 0$) is the water surface, the temperature T_w of which is close to 0°C under the given conditions:

$$T(x, 0) = T_w = 0\,°\text{C}. \tag{3.2.9}$$

The water vapor in the immediate vicinity of the water surface is in the saturated state:

$$q(x, 0) = q_m(x, 0). \tag{3.2.10}$$

At great heights (z_∞), an air mass moving over a water surface retains its initial properties:

$$T(x, z_\infty) = T(0, z_\infty).$$

$$u(x, z_\infty) = u(0, z_\infty),$$

$$q(x, z_\infty) = q(0, z_\infty),$$

$$k(x, z_\infty) = k(0, z_\infty). \tag{3.2.11}$$

The system of equations (3.2.1)-(3.2.4), with initial conditions (3.2.6)-(3.2.8) and boundary conditions (3.2.9)-(3.2.11), was solved by the factorization method, using a varying height interval and a constant interval along the horizontal. For the calculations, an implicity six-point difference technique, which ensured calculational reliability, was employed. The region of fog formation is defined by the condition $s > q_m$. Its water content is found using the relation

$$\delta = s - q_m. \tag{3.2.12}$$

The problem as formulated may be solved by a computer.

Theoretical results. The formation of fog and its intensity (water content) are affected considerably by certain quantities entering into the initial equations and the initial and boundary conditions of the problem. These are the difference ΔT between the water temperature T_w = 0°C and the air temperature $T_{0,5}$ at a height 0.5 m at the moment when the air mass passes over the shore-line ($\Delta T = T_w - T_{0 \cdot 5}$), the relative humidity of the air f_0 over the land (at x = 0), the wind velocity u_1 at a height z_1 = 1 m; the roughness parameter z_0, and the lapse rate (nature of the stratification) at x = 0. Variations in these quantities can give rise to a great diversity of conditions for the formation of evaporation fogs. The calculation results presented below make possible a quantitative evaluation of the effect of various parameters on the fog-formation process, on the depth and water content of the fog, etc.

Table 3.1. Minimum values of ΔT(°C) corresponding to the appearance of fog at a height z = 2 m, for u_1 = 3 m/s and various f_0

f_0, %	x, km						
	1	5	10	20	30	40	50
30	16.30	11.85	8.50	7.10	6.50	6.10	6.00
40	15.60	9.90	6.82	6.15	5.70	5.45	5.20
60	13.70	7.20	4.80	4.35	4.10	3.90	3.80
80	9.40	5.45	3.30	3.13	3.04	2.94	2.86
90	5.75	4.08	3.00	2.73	2.60	2.45	2.30
95	4.50	3.34	2.75	2.40	2.18	2.10	2.02

Table 3.1 gives the minimum temperature differences ΔT required to obtain the specified relative humidity f_0 for fog formation over the sea at various distances x from the shoreline (for fixed u_1 and z). A value of δ^* = 0.01 g/m^3 was assumed to be the minimum water content for which fog becomes a perceptible phenomenon. For a given x (say, 20 km), the difference ΔT will be smaller, the greater the value of f_0 (thus for f_0 = 60%, ΔT = 4.35°C; for f_0 = 95%, ΔT = 2.40°C). For f_0 = const, the difference ΔT decreases

with increasing x (for instance, if f_0 = 80%, then ΔT= 5.45°C for x = 5 km and ΔT = 2.86°C for x = 50 km).

If, for a specified f_0, the difference ΔT exceeds the value in Table 3.1, then a thick fog will form ($\delta^* > 0.01$ g/m). Figure 3.6 illustrates the dependence of the fog water content on ΔT and f_0 for specified values of u_1, x, and z: the water content is seen to rise with increasing ΔT and f_0.

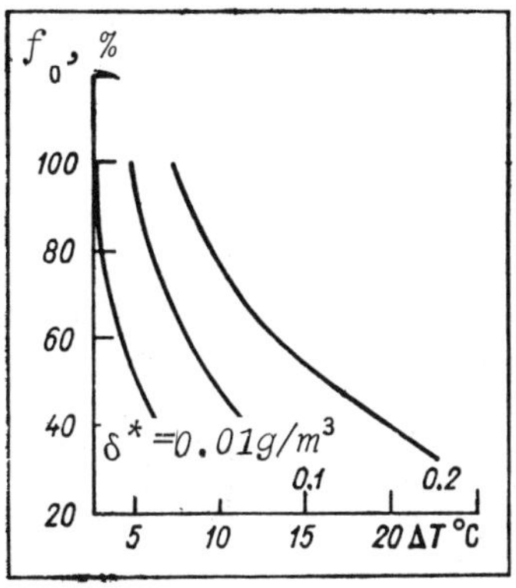

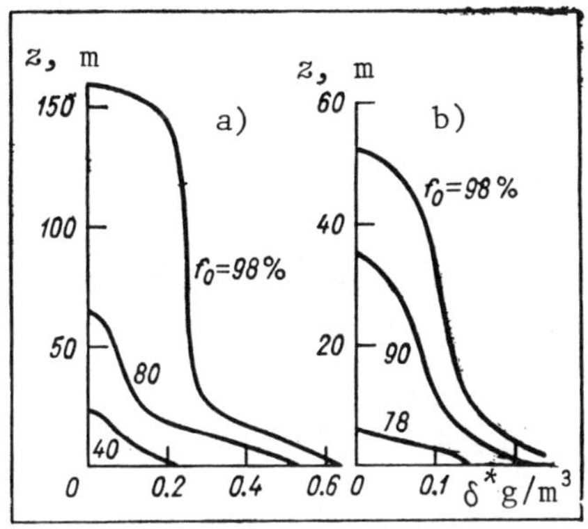

Fig. 3.6. Fog water content as a function of ΔT and f_0, for x = = 10 km, z = 2 m, and u_1 = 3 m/s.

Fig. 3.7. Vertical profiles of fog water content for x = 10 km, u_1 = 3 m/s. (a) ΔT = 20°C; (b) ΔT = 10°C.

x, km	z, m					
	0,5	2	8	30	60	100
10	0.50	0.38	0.26	0.10		
20	0.51	0.38	0.31	0.14		
30	0.51	0.39	0.33	0.17	0.10	
40	0.50	0.40	0.35	0.18	0.11	0.07
50	0.48	0.38	0.36	0.20	0.12	0.10

Table 3.2. Distribution of fog water content (g/m^3) with height for ΔT = 20°C, f_0 = 60%, and u_1 = 3 m/s.

Figure 3.7 shows how ΔT and f_0 affect the vertical profiles of the fog water content. The highest water contents in the fog are observed close to the sea surface, and there is a steady decrease with increasing height. The depth of the fog (the level where δ^* = 0) is seen to depend markedly on ΔT: other conditions being the same, the depth of the fog is about three times as great for $\Delta T = 20°C$ as for $\Delta T = 10°C$.

Since further away from the shore, the thickness of the internal boundary layer increases, the depth of the fog increases as well (Fig. 3.8, Table 3.2): for a fixed f_0, the fog depth is practically a linear function of the distance x.

All the foregoing calculations were carried out on the assumption that the wind velocity u_1 at a height $z_1 = 1$ m is 3 m/s. Figure 3.9 illustrates the dependence of the fog depth on the wind velocity u_1. According to this figure, the depth of the fog, other conditions being equal, increases comparatively slowly (almost linearly) with increasing wind velocity.

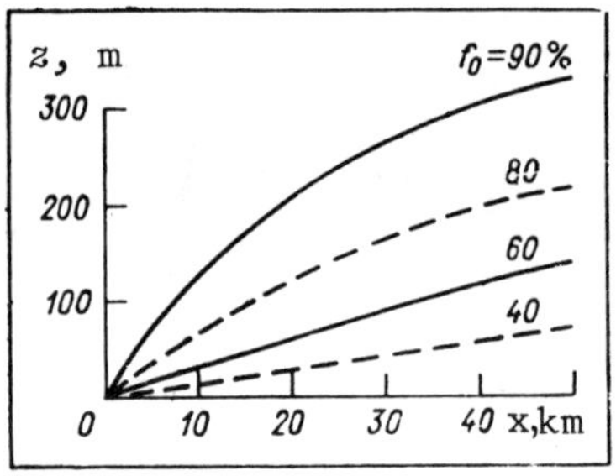

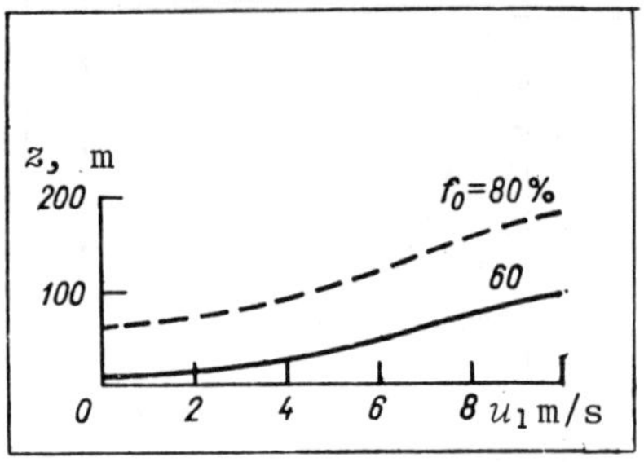

Fig. 3.8. Fog depth at various distances from the shore, for $T = 20°C$, $u_1 = 3$ m/s.

Fig. 3.9. Fog depth as function of wind velocity u_1, for $\Delta T = 20°C$ $x = 10$ km.

$\Delta_z T$, °C	ΔT, °C	f_0, %			
		60	80	90	95
0.2	10.0		0.10	0.28	0.30
	20.0	0.13	0.22	0.30	0.30
0.4	10.0		0.10	0.25	0.28
	20.0	0.11	0.21	0.27	0.31
0.7	10.0		0.10	0.15	0.20
	20.0	0.09	0.15	0.18	0.22

Table 3.3. Fog water content (g/m^3) at a height of 4 m, for $x = 1$ km and $u_1 = 3$ m/s.

The initial air-temperature distribution over the height has some effect on the water content of a fog (Table 3.3). Table 3.3 shows the fog water content as a function of the air-temperature difference $\Delta_z T = T_2 - T_{0.5}$ at heights of 2 m (T_2) and 0.5 m ($T_{0.5}$)

at the moment the air mass reaches the water surface.*) Other conditions being equal, the water content of the fog decreases somewhat with an increasing difference $\Delta_z T$ (that is, with an enhancement of the temperature inversion). For instance, for ΔT = 10°C and f_0 = 90%, the fog water content is 0.28 g/m^3 for $\Delta_z T$ = 0.2°C, 0.25 g/m^3 for $\Delta_z T$ = 0.4°C, and 0.15 g/m^3 for $\Delta_z T$ = 0.7°C.

Table 3.4. Air-temperature distribution (°C) with height at various distances from shore, for ΔT = 20°C, f_0 = 60%, and u_1 = 3 m/s.

x, km	z, m					
	0,5	2	8	16	30	60
0	−20.00	−19.80	−19.60	−19.50	−19.41	−19.31
10	−8.92	−14.18	−17.50	−18.44	−19.37	−19.30
20	−5.15	−7.63	−12.14	−15.03	−18.24	−19.11
30	−4.48	−6.95	−10.21	−12.02	−13.25	−15.30
40	−4.26	−6.58	−9.80	−10.98	−12.78	−14.56
50	−4.15	−6.28	−9.38	−10.15	−12.20	−13.88

x, km	z, m				
	100	200	300	400	500
0	−19.24	−19.14	−19.08	−19.04	−19.00
10	−19.23	−19.12	−19.06	−19.01	−19.00
20	−19.21	−19.09	−19.02	−18.97	−18.96
30	−18.80	−18.98	−18.96	−18.93	−18.90
40	−17.90	−18.89	−18.89	−18.86	−18.81
50	−15.94	−17.08	−18.80	−18.79	−18.77

In conclusion, in Table 3.4 we present calculated vertical profiles of the air temperature at various distances from the shore. As mentioned above, as the distance x increases, the air temperature close to the sea surface rises (at a height of 0.5 m, for instance, it rises from -20°C for x = 0 to -5.15°C for x = = 20 km, and then to -4.15°C for x = 50 km). Beause of turbulent mixing and condensation heat, the temperature increase also spreads to higher levels. According to the data of Table 3.4, at a distance of 10 km from the shore, the thickness of the layer within which the temperature drops with height (that is, where the inversion created when the air mass reached the water surface disappears) is about 30 m; further away from the shore, the thickness of this layer has the following values: about 100 m at a distance of 20 km, about 200 m at 30 km, about 300 m at 40 km, and about 400 m at a

*) When calculating the data of all the tables and figures, except Table 3.3, the difference $\Delta_z T$ was assumed to be 0.2°C. The roughness parameter z_0 is everywhere equal to 2 cm.

distance of 50 km. Above these levels, the inversion-type temperature is retained, although here, too, there is a gradual rise in temperature (at a height of 400 m, for instance, the temperature increases from -19.04°C at $x = 0$ to -18.79°C at $x = 50$ km).

3.3. Advection fogs

Advection fogs are produced when warm air flows over a cooler underlying surface. For a certain ratio between the temperature T_2 and T_1 and the specific humidities q_2 and q_1, the motion of the air mass may give rise to a fog.

Advection fogs occur very frequently. For instance, according to the data of [260], in the central Ukraine, advection fogs account for 59% of all the foggy days, and radiation fogs for only 34%. The other 7% are frontal fogs, and advection plays a significant role in these as well (from 1951 to 1960 there were a total of 831 days with fog in Kiev). Advection fogs also last the longest: in 30 and 9% of the cases, respectively, they persisted for more than 12 and 24 h (there have been instances where fog persisted for more than 100 h without a break). The most common continuous durations of advection fogs are between 7 and 12 h. For comparison, we note that radiation fogs are observed continuously for less than 3 and 6 h, respectively, in 51 and 75% of the cases; whereas for all types of fogs together, durations of up to 2 h are noted in 39% of the cases, durations of 2-4 h in 23%, and durations of 4 to 6 h in 14% of the cases.

Most advection fogs are produced by heat advection caused by the cooling of an air mass (the increase in moisture content plays a role). On the other hand in the literature (for instance, in [66]), it is sometimes stated that fogs may be produced by the advection of cold, as a consequence of evaporation from a well moistened land surface and an increase in the moisture content of the air mass. However, such fogs should preferably be called evaporation fogs.

Theoretical models of advection fogs were worked out in [227]. The conditions for fog formation, on the basis of synoptic and aerological data, were discussed in [145, 304].

Since an advection fog forms, as a rule, in a layer a few hundred meters thick, it is necessary to invoke the fundamental relations describing the structure of the boundary and ground layers of the atmosphere.

In connection with the idea of the complete entrainment of fog droplets by turbulent air parcels (following [261]) some data on the size distribution of fog droplets [$f(r_d)$ being the probability density] will now be presented:

r_d, µm . . .	3	4	5	6	7	8	9	10	11
$f(r_d)$, % . .	35.9	26.4	14.7	10.4	5.9	2.0	1.5	1.0	2.6

According to these data, the mean value $\bar{r}_d$ is 3.5 µm. In 97.4% of all cases, the fog-droplet radius is less than 11 µm, and droplets of this size will all be entrained by turbulent air parcels.

The initial system of equations is written as follows [202]:

(a) the equations of heat and moisture transfer, taking into account the effects of water-vapor condensation,

$$u \frac{\partial \Pi}{\partial x} + w \frac{\partial \Pi}{\partial z} = \frac{\partial}{\partial z} k \frac{\partial \Pi}{\partial z}, \tag{3.3.1}$$

$$\Pi = T + Lq/c_p + \gamma_a z, \tag{3.3.2}$$

$$u \frac{\partial s}{\partial x} + w \frac{\partial s}{\partial z} = \frac{\partial}{\partial z} k \frac{\partial s}{\partial z}, \tag{3.3.3}$$

$$s = q + \delta, \tag{3.3.4}$$

where q is the specific humidity, δ is the specific water content of the fog, k is the turbulence coefficient, $c_p\Pi$ is the total potential heat content, comprising the sensible and latent heat, and L is the latent heat of condensation;

(b) the equations of motion and continuity,

$$u \frac{\partial u}{\partial x} + w \frac{\partial u}{\partial z} = \frac{\partial}{\partial z} k \frac{\partial u}{\partial z} + 2\omega_z v, \tag{3.3.5}$$

$$u \frac{\partial v}{\partial x} + w \frac{\partial v}{\partial z} = \frac{\partial}{\partial z} k \frac{\partial v}{\partial z} - 2\omega_z (u - u_g), \tag{3.3.6}$$

$$\frac{\partial \rho u}{\partial x} + \frac{\partial \rho w}{\partial z} = 0, \tag{3.3.7}$$

where u, v, w are, respectively, the x, y, z components of the velocity (the x axis lies along the horizontal temperature gradient Γ; the z axis is directed vertically upward), $2\omega_z$ is the Coriolis parameter, and u_g is the velocity of the geostrophic wind, directed along Γ;

(c) the equation of balance of eddy energy b,

$$u\frac{\partial b}{\partial x}+w\frac{\partial b}{\partial z}=k\left[\left(\frac{\partial u}{\partial z}\right)^2+\left(\frac{\partial v}{\partial z}\right)^2-\frac{g}{T}\frac{\partial\theta}{\partial z}\right]+ \\ +\alpha_b\frac{\partial}{\partial z}k\frac{\partial b}{\partial z}-\frac{cb^2}{k}, \qquad (3.3.8)$$

where g is the free-fall acceleration, $\alpha_b \approx 0.73$, and $c \approx 0.046$;

(d) the expressions for the mixing length l and the turbulence coefficient k,

$$l=-\varkappa\frac{b/k}{\frac{\partial}{\partial z}\left(\frac{b}{k}\right)}\left(1+\frac{1{,}48\cdot 10^4}{\mathrm{Ro}}\frac{z}{z_0}\right)^{-1}, \quad k=l\sqrt{b}, \qquad (3.3.9)$$

where $\varkappa = 0.40$, $\mathrm{Ro} = u_g/(2\omega_z z_0)$ is the Rossby number, and z_0 is the roughness parameter.

As the system (3.3.1)-(3.3.9) implies, it is the steady motion of an air mass whose properties do not change in the y direction, that is, along a normal to the horizontal temperature gradient $\mathbf{\Gamma}$ (where $\Gamma = -\partial T/\partial x$) that is being considered. Symbolically expressed, these conditions are $\partial u/\partial y = \partial v/\partial y = \partial w/\partial y = \partial T/\partial y = 0$. The properties of the air mass in the x and z directions, on the other hand, will vary because of the advective $[u(\partial f/\partial x)]$, convective $[w(\partial f/\partial z)]$, and turbulent influxes of heat, moisture, and momentum, and because of phase transitions of water. The distribution of the velocity with height is affected by the Coriolis force.

The unknown quantities satisfy the following conditions:

(a) at the ground ($z = 0$),

$$u=v=w=0, \quad \partial b/\partial z=0, \quad l=\varkappa z_0, \qquad (3.3.10)$$

$$T(x, 0)=T_1+(T_2-T_1)\exp(-x/D), \qquad (3.3.11)$$

$$s(x, 0)=s_1+(s_2-s_1)\exp(-x/D), \qquad (3.3.12)$$

$$s_2=f_2 q_m(T_2), \quad s_1=f_1 q_m(T_1), \qquad (3.3.13)$$

where D is a parameter with units of length, characterizing the rate of decrease of the temperature and moisture content in the x direction; T_2 and T_1 are the air temperatures close to the ground, respectively, at the interface ($x = 0$) and at a great distance away from it ($x \gg D$); s_1 and s_2 are the specific moisture contents for these same values of x; and f_2 and f_1 are the relative humidities of the air at $x = 0$ and $x \gg D$, respectively.

(b) at the upper limit of the boundary layer ($z = H$),

$$u = u_g, \quad v = 0, \quad \partial b/\partial z = 0, \tag{3.3.14}$$

$$s(x, H) = s(0, H), \quad T(x, H) = T(0, H); \tag{3.3.15}$$

(c) at the interface of the underlying surfaces ($x = 0$),

$$T(0, z) = T_2 + T_* \ln \frac{z + z_0}{z_0} - \gamma_a z, \tag{3.3.16}$$

$$q(0, z) = f_2 q_m(0, z), \tag{3.3.17}$$

where T_* is the temperature scale, γ_a is the dry-adiabatic gradient, and $q_m(0, z)$ is the saturation specific humidity for temperature $T(0, z)$ and a pressure $p(0, z)$ determined with the aid of the barometric formula for $p(x, 0) = 1000$ mb. The initial (for $x = 0$) vertical profiles of u, v, k, b, and l were taken to be those obtained from a solution of the system of equations (3.3.5), (3.3.6), (3.3.8), and (3.3.9) in the case of uniform conditions along the horizontal, when the nonlinear terms in these equations go to zero.

The formulated problem was solved by numerical methods with the aid of a computer.

Theoretical results. In the initial equations and boundary conditions, quantities that affect the fog-formation process and the physical characteristics of the fog appear. These quantities include the air temperature (T_2 and T_1) and the relative humidity (f_2 and f_1) close to the "old" or "new" underlying surfaces, as well as the velocity of the geostrophic wind u_g, the roughness parameter z_0, and the width of the transition zone D. It is quite reasonable to assume that in a fog the relative humidity is close to 100%.

For most of the above-mentioned calculations we assumed that $u_g = 12$ m/s, $z_0 = 0.02$ m, $D = 25$ km, and $T_* = 0.1°$C, and also that at very great distances from the interface ($x \gg D$) the air near the ground is close to saturation ($f_1 \approx 100\%$).*)

Some results of the numerical solution of the problem will now be considered. According to Fig. 3.10, which shows the fog water content δ^* as a function of the temperature difference $\Delta T = T_2 - T_1$ and the temperature T_2, fog does not appear for $f_2 = 80\%$, provided that $\Delta T < 5°$C. With an increase in ΔT and T_2, the fog water content rises, reaching, for example for $\Delta T = 8°$C, a value close to 0.2 g/m^3 for $T_2 \approx 1°$C and close to 0.3 g/m^3 for $T_2 \approx 17°$C.

Figure 3.11 allows us to observe the transformation of the water-content field of the fog as the air mass streams out above a cooler underlying surface. Because of cooling effects, fog will form closer to the interface (at $x = 0$), the higher the relative

*) The Coriolis parameter was taken for a latitude $\varphi = 60°$.

humidity of the air (f_2) reaching the cooler underlying surface (if f_2 = 70%, fog forms at $x \geqslant 44$ km; if f_2 = 95%, fog forms at $x \geqslant 19$ km).

The further the air mass advances over the cool surface, the higher the cooling will spread, because of turbulent heat exchange, and this the higher the upper fog boundary will be as well: for f_2, values of 70 and 80%, the depth of the fog will increase monotonically with increasing x. However, if f_2 is higher (90, 95, or 98%), then a qualitatively new phenomenon is observed: along with

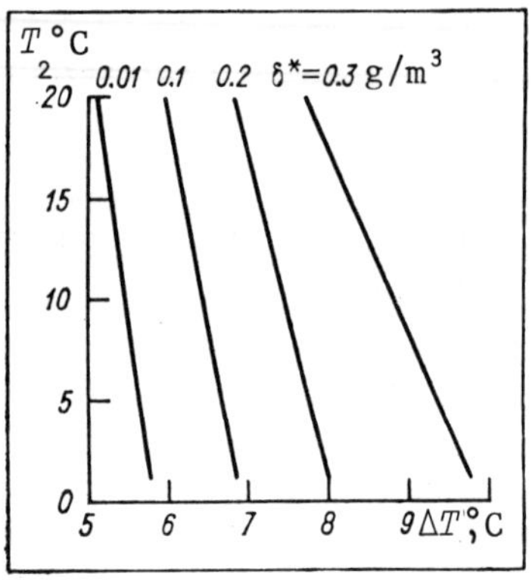

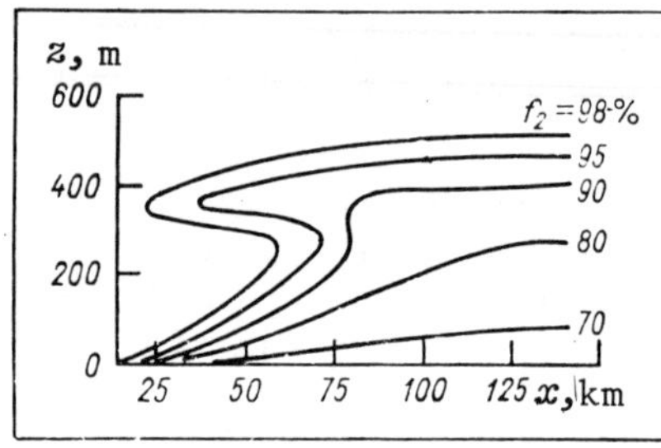

Fig. 3.10. Fog water content as a function of $\Delta T = T_2 - T_1$ and T_2, for f_2 = 80%, z = 2 m, and x = 50 km.

Fig. 3.11. Depth of fog and height of cloud boundaries for fixed values ΔT = 8°C and T_2 = 10°C and various values of f_2.

the fog, in a certain region of x (for instance, from 36 to 73 km, if f_2 = 95%), clouds form as well, the lower cloud boundary being lower with an increase in x and the upper cloud boundary being higher; at some value $x = x^*$ (for instance, x^* = 60 km for f_2 = 98%), the upper fog boundary and the lower cloud boundary merge, so that there is a discontinuous variation in the fog depth at x^*. For $x > x^*$, the upper fog boundary remains at practically the same height.

It is to be expected that with an increase in the difference ΔT the air mass becomes cooled more and more, while the depth and water content of the fog (for given x, T_2, and f_2) both increases (see Tables 3.5 and 3.6). According to Table 3.5, the fog becomes deeper mainly because of f_2 rather than ΔT: for ΔT = 10°C a variation in f_2 from 70 to 98% is accompanied by a deepening of the fog from 80 to 500 m; at the same time, for f_2 = 90% a variation in ΔT from 6 to 12°C is attended by a deepening of the fog just from 365 to 430 m. The fog water content, as Table 3.6 indicates, varies equally strongly with both f_2 and ΔT: for ΔT = = const a rise in f_2 from 70 to 98%, and for f_2 = const an increase in ΔT from 6 to 12°C, will be accompanied by an approximately tenfold increase in the fog water content.

Table 3.5. Depth of fog (m) for T_2 = 10°C and x = 100 km

f_2,%	ΔT, °C			
	6	8	10	12
70	35	60	80	95
80	190	205	230	260
90	365	385	400	430
95	430	445	465	490
98	480	490	500	515

Table 3.6. Fog water content (g/m^3) at height z = 2 m for T_2 = = 10°C and x = 50 km

f_2,%	ΔT, °C					
	6	7	8	9	10	12
70	0.0	0.0	0.031	0.036	0,048	0.062
80	0.054	0.155	0.246	0.311	0,380	0.451
90	0.061	0.164	0.251	0.328	0,422	0.505
95	0.070	0.174	0.258	0.340	0,462	0.586
98	0.081	0.188	0.280	0.394	0,521	0.620

Table 3.7 gives data on the vertical profiles of the fog water content. Since the air mass is cooled most near the ground, it is precisely there that the maximum water contents of the fog are observed. Close to the ground (up to heights of 2–4 m) the water content also exhibits its greatest variation with height: in the layer from 0.5 to 2 m it decreases to apprpximately 1/4 for f_2 = = 70% and to approximately 2/3 for f_2 = 95%. For the most part, a fog is quite uniform over the height: for instance, for f_2 = 98% the water content in the layer from 2 to 100 m varies from 0.281 to 0.228 g/m^3, that is, it decreases by only 0.053 g/m^3. In the uppermost part, the fog water content decreases sharply with height (for f_2 = 95%, the water content at a height of 100 m is 0.12 g/m^3; at a height of about 120 m, according to Fig. 3.11, one already has reached the upper boundary of the fog, where the water content is of course zero). Figure 3.11 shows that for high values of f_2 (95 and 98%), above x = 50 km cloudiness is observed along with fog. At a height of 400 m the cloud water contents are 0.014 and 0.083 g/m^3, respectively, for f_2 = 95% and f_2 = 98% (for the latter value of f_2, cloudiness with a water content of 0.02 g/m^3 is

Table 3.7. Distribution of fog water content (g/m^3) with height for x = 50 km, T_2 = 10°C, and ΔT = 8°C

f_2 %	z, m								
	0.5	1	2	4	10	20	50	100	150
70	0.121	0.070	0.031	0.027	0.011	0	0	0	0
80	0.340	0.278	0.246	0.235	0.220	0.208	0.011	0	0
90	0.363	0.295	0.251	0.246	0.239	0.230	0.218	0	0
95	0.380	0.312	0.258	0.254	0.248	0.237	0.226	0.120	0
98	0.510	0.422	0.281	0.277	0.260	0.254	0.247	0.228	0.01

observed at a height of 300 m).

Figure 3.12 depicts the transformation of the vertical profiles of the temperature and turbulence coefficient as an air mass streams over a cooler underlying surface. There is an interesting detail that has hitherto escaped notice. The temperature distribution with height at the interface (x = 0) is described by formula (3.3.16). The last term in this formula ($-\gamma_a z$) is usually neglected, on the assumption that it is small compared to the second term. Actually, however, already at a height of several tens of meters, the quantity $\gamma_a z$ is quite comparable with the logarithmic term. Moreover, if this term is retained, it becomes possible to describe the most important property of the vertical temperature profile, namely the transition from the temperature inversion ($\gamma < 0$) observed in the ground layer to the normal distribution ($\gamma > 0$) typical of the upper (Ekman) part of the boundary layer and free atmosphere. On the other hand, if the term $\gamma_a z$ is dropped, formula (3.3.16) describes a temperature profile with only one sign of γ (for $T_* > 0$, an inversion, with $\gamma < 0$).

The initial temperature profile (for x = 0) is shown in Fig. 3.12 (solid curve 1): from a height of about 20 m, the temperature inversion changes into a drop. As evaluations have shown, for $x \geqslant 25$ km the initial temperature profile (in particular, the depth of the inversion and the temperature differential in it) does not greatly affect the distribution of T with height.

As Fig. 3.12 shows, in a moving air mass, first (at distances of 25 and 50 km) a temperature inversion that intensifies with height is formed. At greater distances (75 and 100 km) the inversion is lifted slightly, the temperature differential in it decreases, and the temperature distribution approaches isothermal (for x = 100 km, the temperatures at the ground and at a height of about 400 m practically do not differ from $T_1 = T_2 - \Delta T$ = 2°C, the temperature that, in the given example, is observed at the ground far away from the interface).

The distribution with height of the turbulence coefficient (see Fig. 3.12) in the initial state (x = 0) is typical for the boundary layer: it increases sharply with height in the ground

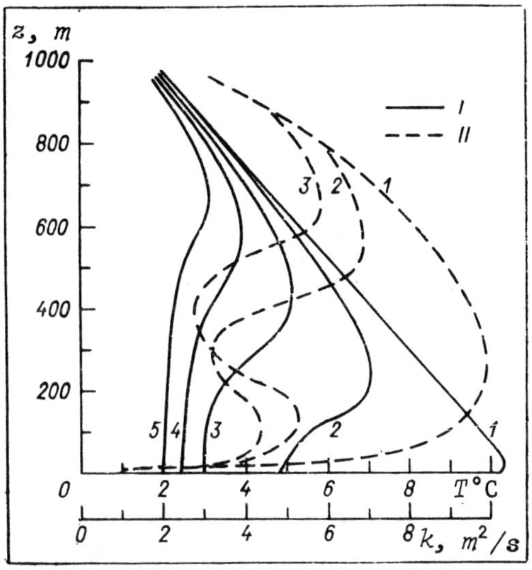

Fig. 3.12. Vertical profiles of temperature (I) and turbulence coefficient (II) for ΔT = 8°C, T_2 = 10°C, f_2 = 90%, and various values of x:

curve ...	1	2	3	4	5
x, km ...	0	25	50	75	100

layer, reaches a maximum at a height of about 250 m, and then decreases slowly. However, for other values of x, the vertical profiles of k, as long as moist saturated air is being considered and fog has formed in the boundary layer, will differ considerably from the profiles that were established by the theory for dry (or moist unsaturated) air. After the maximum is reached (at a height of 100–150 m), the turbulence coefficient starts to decrease, going through a minimum at a height of 300 or 400 m (just above the upper boundary of the temperature inversion); then it starts to increase again, goes through a new maximum, and finally begins to decrease with increasing height, as is typical for the upper part of the boundary layer. This pattern indicates that in the presence of fog and clouds, the regularities of the boundary-layer structure may differ qualitatively from those that were established without allowing for the effects of phase transitions of water.

In conclusion, let us consider how the parameters that were considered above to be constant (specified) affect the characteristics of fog and clouds. The most important of these are u_g and z_0. Table 3.8 gives data on the fog depth as a function of the geostrophic-wind velocity u_g and ΔT. With an increase in u_g, the intensity of turbulent mixing becomes higher, and, accordingly, the fog becomes deeper (for instance, for ΔT = 10°C the depth of the fog increases from 365 to 490 m, if u_g varies from 8 to 25 m/s).

Since we have been interested in fog-formation conditions, everywhere in the foregoing we assumed that for $x \gg D$ the water vapor close to the ground is saturated (f_1 = 100%). But this may not be the case. If f_1 < 100%, low cloudiness forms over a cool underlying surface. Table 3.9 gives the results of modeling such cloudiness, and these data show that for a fixed value of f_2,

Table 3.8. Depth of fog (m) for x = 100 km, f_2 = 90%, T_2 = 10°C, f_1 = 100%, and z_0 = 0.02 m

ΔT,°C	u_g, m/s				
	8	10	12	15	25
6	340	350	365	395	440
8	360	370	385	420	470
10	365	380	400	440	490
12	390	410	430	470	530

the rise in f_1 is accompanied by a comparatively gradual lifting of the upper boundary and by a much more rapid lowering of the lower cloud boundary.

Table 3.9. Heights of upper (numerator) and lower (denominator) cloud boundaries (m) for x = 100 km, ΔT = 8°C, T_2 = 10°C, u_g = = 12 m/s, z_0 = 0.02

f_2 %	f_1 %					
	70	80	90	95	98	100
70				45/30	55/25	60/0
80				140/80	170/45	205/0
90			350/150	370/40	380/20	385/0
95	340/300	380/210	415/120	430/35	440/20	445/0
98	410/250	440/160	464/95	485/50	490/25	490/0

3.4. On urban fogs

A city shapes it own meteorological regime, and in addition it has a significant and many-faceted effect on the meteorological regime of its immediate surroundings. The cloud of pollutants as a by-product of industry, transport, and heating systems that forms over a city can be several hundred meters deep (and is clearly visible from an aircraft approaching a city or from a satellite). The cloud of pollutants reduces the effective radiation of the Earth's surface, and during the daytime it cuts down the influx of solar radiation by 5-10%. The modification of the radiation balance, in conjunction with turbulent mixing, is the main reason that the temperature rises in the city more than in its surroundings.

The difference $\Delta T = T_{city} - T_{surr}$ between the air temperatures over a city and over its surroundings varies over a wide range. Data from observations carried out on eight different dates between 1970 and 1974 in Leningrad and at five points located 80-100 km away from it, were used to plot the distribution functions of difference ΔT (the samples included 17,366 winter values of ΔT and 14,720 summer values). The mean values of ΔT are a maximum at night (1.5-1.6°C in winter and 2.5-2.7°C in summer) and a minimum in the daytime (1.0°C in winter and 0.4-0.5°C in summer). The probability density reaches maximum values of 33.9 and 25.6%, respectively, in the intervals from 0 to 1°C in winter and from 1 to 2°C in summer. The extreme values of ΔT lie between 16 and -9°C in winter and between 11 and -11°C in summer. Table 3.10 gives

Table 3.10. Integral distribution function $F(\Delta T \geqslant X)$, in %, of difference $\Delta T = T_{city} - T_{surr}$

Season	X, °C									
	12	11	10	9	8	7	6	5	4	3
Winter	0.1	0.2	0.3	0.6	1.1	2.1	3.3	5.5	9.1	15.0
Summer	0.0	0.0	0.0	0.1	0.3	0.8	2.0	4.2	8.4	16.7

Season	X, °C								
	2	1	0	−1	−2	−3	−4	−5	−6
Winter	25.5	48.9	82.9	94.9	97.4	98.6	99.4	99.7	99.9
Summer	32.4	58.0	80.0	91.9	97.1	98.8	99.5	99.8	99.9

data on the integral distribution function $F(\Delta T \geqslant X)$ of the difference ΔT. From these data we see that a city is warmer than its surroundings ($\Delta T \geqslant 0$°C) 82.9% of the time in winter and 80% of the time in summer. In approximately 20% of the cases, however, a city is seen to be cooler than its surroundings.

An analysis of these and other data indicates that it is geophysical factors (in particular, the reduction of the effective radiation of the Earth's surface due to the presence of the pollutant cloud) which play a decisive part in the creation of the urban "heat island", rather than the heat emanating from workplaces, dwellings, and transport. Otherwise, the difference ΔT would always be positive and greater in the daytime than at night, since industrial facilities, and especially transport, work and emit heat during the day more than at night.

The urban temperature rise also influences other phenomena in and around the city, such as the conditions of fog formation. It has been traditionally assumed that fogs are much more frequent in the city than in its surroundings (see, for instance, Smit's monograph [276], which presents data of Landsberg, according to which fogs in the city are 30% more frequent than in rural regions in summer, and 100% more frequent in winter). The reason given is that condensation nuclei are more prevalent in urban areas than in the country.

However, recent studies [196, 320] have shown that despite the above-mentioned opinions, in large cities the probability that fogs will form is not greater than in the country, but in fact is considerably less (1/2-1/3 as great). Table 3.11 gives some observational results pertaining to fogs in Leningrad and in places 20-80 km away from it.

Table 3.11. Number of days with fog

Observation period	Leningrad	Voeikovo	Sosnovo	Belogorka	Volkhov
1965—1969	80	194	142	191	206
1970—1974	74	265	206	238	129

The table shows that during both the first and second five-year periods, fogs were observed in Leningrad only 1/3-1/2 as often as in its surroundings. This is also borne out by data on the total duration of fog t^* during the winters of 1970-1974 and the probability P of fog observation:

Place	Leningrad	Voeikovo	Sosnovo	Belogorka
t^*,h	115	319	443	265
P,%	1.1	3.0	4.1	2.4

Data pertaining to many years show that fogs form in Moscow 1/2-2/5 as often as in its surroundings:

Place	Moscow (GMO)	Moscow (VDNKh)	Klin	Dmitrov	Zagorsk	Kashira
Mean annual number of foggy days	20	26	36	37	46	49

Since the above data on the frequency of occurrence of fogs directly contradicts the opinion given previously, some data on the meteorological visibility will also be cited. According to data for winter during the period 1970-1974, the frequency *P* of a visibility of less than 1 km (S_M = 1 km) is the following:

Place	Leningrad	Voeikovo	Sosnovo	Belogorka
P %	1.4	2.8	4.7	3.5

Clearly, these data also verify that conditions favoring a visibility of less than 1 km are 1/2-1/3 as probable in Leningrad as in its environs.

How can a phenomenon so paradoxical vis-à-vis traditional concepts be explained?

Pollution of the urban atmosphere, as we know, reduces visibility. However, the pollutants increase substantially only the frequency of a thin haze with a visibility of 6-10 km (Table 3.12). The likelihood of a visibility S_M = 6-10 km in Leningrad is about twice as great as in its environs. Compared with Sosnovo and Belogorka, a visibility S_M = 2-6 km is also approximately twice as likely in Leningrad. But S_M = 1-2 km (thick smoke) is somewhat less likely in the city than in the surroundings, whereas $S_M \leqslant 1$ km (fog), as noted above, is 1/2-1/3 as likely in Leningrad as in the other places.

Table 3.12. Frequency (%) of different visibility intervals (3608 samplings were made at each point) for winter periods 1970-1974

Place	S_m, km				
	$\leqslant 1$	1–2	2–6	6–10	>10
Leningrad	1.4	2.1	22.9	41.8	31.8
Voeikovo	2.8	2.5	19.4	21.2	54.1
Sosnovo	4.7	3.4	8.5	19.0	64.4
Belogorka	3.5	2.7	10.3	21.7	61.7

The reference to condensation nuclei, which allegedly ensure a more frequent formation of fog in the city, cannot be taken as valid. Many condensation nuclei are available in the atmosphere, not just over cities but everywhere else as well (even over the oceans); only a small portion of these nuclei are droplets formed [245, 273]. Moreover, when smoke is present ($S_M < 10$ km), the nuclei are already surrounded by water [265]; the higher the relative humidity of the air, the smaller the portion of the condensation nucleus in the total mass of the droplet. If the droplet radius increases to 1 μm (the tiniest fog droplet), then even so-called gigantic condensation nuclei (radius 0.1 μm) will constitute only 0.001 of the total mass of the droplet.

The saturation vapor pressure for such droplets, regardless of the nature of the condensation nuclei, is almost identical to the saturation pressure over a plane water surface. The conditions for the existence and growth of such droplets are determined by the variation of the relative humidity of the air, in particular, under the influence of the temperature. Assume that the meteorological situation is such that the relative humidity in the surroundings of the city,

$$f_{surr} = \frac{e_{surr}}{E(T_{surr})} \cdot 100$$

is close to 100%, so that $e_{surr} \approx E(T_{surr})$. At the same time, the relative humidity in the city,

$$f_{city} = \frac{e_{city}}{E(T_{city})} \cdot 100,$$

will be, as a rule, less than 100%. Actually, the vapor pressure in the city e_{city} will be close to e_{surr}, because of horizontal mixing: $e_{city} \approx e_{surr} = E(T_{surr})$. Consequently, at the moment when fog starts to form in the surroundings of the city, the relative humidity of the air over the city will be

$$f_{city} = \frac{E(T_{surr})}{E(T_{surr} + \Delta T)} \cdot 100.$$

Values of f_{city} (in %) for various T_{surr} and for two values of ΔT are the following:

T_{surr}, °C	—20	—10	0	10
$\Delta T = 1$ °C	92	92	93	94
$\Delta T = 2$ °C	84	85	87	88

Since $f_{city} < 100\%$, urban fog is absent, whereas at the same time fog forms in the surroundings in this same situation. For fog to originate in the city, the air temperature in the surroundings must drop considerably below the dew point. Then in the surroundings the fog will be thick (with a high water content), while in the city it will be moderate or thin. Thus, thin and moderate fogs will be observed in cities more frequently than in the surroundings, while thick fogs will be observed less often. It is primarily thin fogs which occur in the city. An analysis of experimental data carried out by Tsarev [320] verifies this conclusion.

CHAPTER 4

ATMOSPHERIC MOISTURE CONTENT AND CLOUD WATER RESERVES

Of great theoretical and practical interest are the moisture content of a vertical air column

$$Q_s = \int_0^H \rho s \, dz,$$

the water reserves of a cloud

$$Q_\delta = \int_{z_l}^{z_u} \rho \delta \, dz,$$

and the vapor content in a vertical air column

$$Q_q = \int_0^H \rho q \, dz.$$

In these formulas, s, δ, and q are, respectively, the specific values of the moisture content of the atmosphere, the cloud water content, and the air humidity; ρ is the air density; H is the height of the upper boundary of the atmosphere (this can be assumed to be the height of the tropopause, since the water-vapor content above the latter is low); and z_l and z_u are the heights of the lower and upper cloud boundaries.

4.1. Modeling fields of atmospheric moisture content and cloud water reserves

Quantities Q_s and Q_δ are determined with the aid of the equation for the specific moisture content s (see Section 2.1)

$$\frac{\partial s}{\partial t} + u \frac{\partial s}{\partial x} + v \frac{\partial s}{\partial y} + w \frac{\partial s}{\partial z} = \frac{\partial}{\partial z} k \frac{\partial s}{\partial z} + \frac{\partial \tilde{v} (s - q_m)}{\partial z}, \qquad (4.1.1)$$

where $\tilde{v}$ is the mass-weighted mean of the fall velocity of the cloud elements; and q_m is the saturation specific humidity.

We could, of course, assume a steady-state (quasi-stationary) distribution of the humidity if there is little or no advective influx of moisture. However, the following circumstance, suggested by synoptic practice, should be noted. On the left-hand side of eqn. (4.1.1), all the terms are of the same order of magnitude [192]. On the other hand, terms $\partial s/\partial t$ and $u(\partial s/\partial x) + v(\partial s/\partial y)$ have opposite signs in most cases. Actually, synoptic experience indicates that for the advection of moister (as a rule, warmer) air, when the term $u(\partial s/\partial x) + v(\partial s/\partial y)$ is less than zero, the moisture content of the air at a fixed point in space increases with time ($\partial s/\partial t > 0$). For the advection of dryer (as a rule, cooler) air, the signs are opposite:

$$u\frac{\partial s}{\partial x} + v\frac{\partial s}{\partial y} > 0, \quad \frac{\partial s}{\partial t} < 0.$$

Consequently, the nonsteady-state and advective terms on the left-hand side of (4.1.1), being quantities with the same order of magnitude but with opposite signs, will, when added, give a sum that is about an order of magnitude lower than the other terms in eqn. (4.1.1). For this reason, as a first approximation eqn. (4.1.1) can in general (not just in the case of a steady state and slight advection) be written as

$$\frac{d}{dz}k\frac{ds}{dz} - w\frac{ds}{dz} + \frac{d\bar{v}\,(s - q_m)}{dz} = 0. \qquad (4.1.2)$$

This equation is supplemented by the equation for the heat content of the air

$$\frac{d}{dz}k\frac{d\Pi}{dz} - w\frac{d\Pi}{dz} - \gamma_a\left(w - \frac{dk}{dz}\right) = 0. \qquad (4.1.3)$$

Here,

$$\Pi = T + \frac{L}{c_p}q, \qquad (4.1.4)$$

$$s = q + \delta. \qquad (4.1.5)$$

The vertical velocity w and the turbulence coefficient k entering into system (4.1.2)-(4.1.5) are given [136, 185] by formulas like

$$w = w_m\left(1 + \frac{1}{n}\right)^n (1 + n)\frac{z}{H}\left(1 - \frac{z}{H}\right)^n, \qquad (4.1.6)$$

$$k = k_\infty[1 + mz_0 - \exp(-mz)], \qquad (4.1.7)$$

where w_m is the maximum value of the vertical velocity (as a function of height), which is attained at a height $z_m = H/(1 + n)$; H is the height of the tropopause; n is a parameter defining the maximum and the slope of the profile of w; z_0 is the roughness parameter; k_∞ is the value of k above the ground layer; and m is a parameter depending on the stability of the ground layer (and related simply to the vertical scale L_* of Monin and Obukhov by $m = 1/L_*$).

The weighted-mean fall velocity $\tilde{v}$ of the cloud elements, which enters into (4.1.2), is also in general a function of the height. The theoretical results presented below were obtained by assuming that $\tilde{v}$ is independent of height.

If the dimensionless quantities

$$z' = \frac{z}{H}, \quad w' = \frac{w}{w_m}, \quad k' = \frac{k}{k_\infty}, \quad \eta = \frac{w_m H}{k_\infty}, \quad \zeta = \frac{\tilde{v} H}{k_\infty},$$

are introduced, eqn. (4.1.2) can be put into the form

$$\frac{d}{dz'} k' \frac{ds}{dz'} - \eta w' \frac{ds}{dz'} + \zeta \frac{d(s - q_m)}{dz'} = 0. \qquad (4.1.8)$$

The boundary conditions are

$$q(0) = f_0 q_m(0), \quad q(H) \approx 0, \quad \delta(0) = \delta(H) = 0,$$

$$T(0) = T_0, \quad p(0) = p_0, \quad T(H) = T_H, \qquad (4.1.9)$$

where f_0 is the relative humidity of the air close to the ground, and $q_m(0) = q_m(T_0, p_0)$ is the saturation specific humidity for $z = 0$.

The temperature distribution with height in the first version was approximated by the linear function $T(z) = T_0 - \gamma z$; in the second case it was found by solving eqn. (4.1.3). The air pressure was in all cases found from the hydrostatic equation $-dp = g\rho\, dz$ for $p_0 = 1000$ mb and $g = 9.80615$ m/s^2, while the air density was found from the equation of state $p = R\rho T$ for $R = 287$ m^2/(s K).

Equations (4.1.3) and (4.1.8) were solved numerically using the Runge-Kutta method, with the aid of a computer, for an integration interval $\Delta z' = \Delta z/H = 0.002$ and parameter values $H = 10{,}000$ m, $m = 0.1$ m^{-1}, and $z_0 = 0.01$ m.

The parameters determining the cloud water content and the cloud depth, as well as the positions of the cloud boundaries, are η, ζ, and f_0. With an increase in η, that is, with a rise in w_m or a drop in k_∞, there will be sizable increases in the maximum (over the height) water content δ_m and cloud depth $z_u - z_1$ mainly because of a dropping of the lower cloud boundary. The ratio $\delta_m/q_m(0)$ has the following values for $T_0 = 288$ K, $\gamma = 0.65$ K/100 m, $f_0 = 0.9$, $n = 1$, and $\zeta = 0$:

η	−2	0	2	5	10
$\delta_m/q_m(0)$	0.108	0.286	0.458	0.632	0.749

Cloudiness may be observed for either ascending ($\eta > 0$) or descending ($\eta < 0$) vertical motions (updrafts or downdrafts). However, for $\eta < 0$ cloudiness persists only if the parameter $\zeta = \tilde{v}H/k_\infty$ is small, that is, if $\tilde{v}$ is low (small droplets) or k_∞ is high. Developed turbulent mixing (high k_∞) also furthers the persistence of cloudiness for downward motions. Under real conditions, this is frequently the cloudiness observed (primarily during the cold half of the year) in anticyclones and ridges. However, such cloudiness also forms in regions with updrafts (cyclones and troughs). The sign of may change, because of either a rearrangement of the baric field or movement of the clouds (for instance, from a cyclone region to the outskirts of an anticyclone).

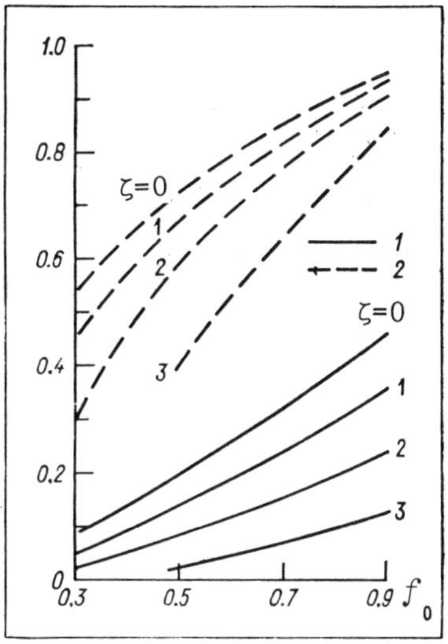

Fig. 4.1. Dimensionless ratios δ_m/q_m (1) and $(z_u - z_l)/H$ (2) as functions of f_0 and ζ for $\eta = 2$, $n = 0.5$, $T_0 = 15°C$, and $\gamma = 0.65$ K/100 m.

As could be expected from physical considerations, with a rise in parameter $\zeta = \tilde{v}H/k_\infty$ (that is, with an increase in the mean fall velocity $\tilde{v}$ of the droplets or a decrease in k_∞), there will be a substantial reduction of δ_m and of the cloud depth. For $T_0 = 288$ K, $\gamma = 0.65$ K/100 m, $f_0 = 0.9$, $n = 1$, and $\eta = 2$, the ratio $\delta_m/q_m(0)$ has the values:

ζ	0	1	2	3	4
$\delta_m/q_m(0)$	0.458	0.358	0.248	0.139	0.042

Figure 4.1 shows to what extent the relative humidity near the ground f_0, as well as the parameter ζ, affect the cloud depth $z_u - z_l$ and the maximum water content δ_m. Both δ_m and $z_u - z_l$ are seen to increase with increasing f_0; with a rise in the value of ζ, on the other hand, these quantities decrease. Exponent n in formula (4.1.6) has some effect on the cloud characteristics, the height of the level where the water content is a maximum being affected most (albeit only for high η and ζ). For example, for $\eta = 5$ and $\zeta = 4$, the height of this level drops from 5040 to 2100 m

as n varies from 0.25 to 9, while for $\eta = 10$ and $\zeta = 7$ it drops from 5620 to 1420 m.

Further data on the cloud water reserves Q_δ, calculated for a given (linear) temperature profile, will now be presented. For fixed values $\zeta = 0$, $f_0 = 0.9$, $T_0 = 288$ K, and $n = 1$, the cloud water reserves increase with increasing η:

η	—2	0	2	5	10
Q_δ, kg/m^2. . .	4.84	13.72	22.31	30.78	36.30

For a fixed value $\eta = 2$ and the same values of f_0, T_0, and n, the water reserves of the clouds decrease with a rise in ζ:

ζ.	0	1	2	3	4
Q_δ, kg/m^2 . . .	22.31	17.12	11.40	5.79	1.28

In a more recent work [138], the moisture content of the atmosphere and the cloud water reserves were modeled on the basis of the system of equations (4.1.2)-(4.1.5), that is, the temperature distribution was not specified, but rather determined from this system. Equations (4.1.2) and (4.1.3) were supplemented by the terms $k_y(d^2\Pi/dy^2)$ and $k_y(d^2s/dy^2)$, respectively, representing the influxes of heat and moisture under the influence of horizontal large-scale exchange (here k_y is the coefficient of large-scale turbulent exchange in the y direction, tangent to the meridian). In addition to the boundary conditions (4.1.9), the following conditions with respect to y were imposed at the equator ($\varphi = 0$):

$$T = T_e(z), \quad q = q_e(z) \tag{4.1.10}$$

and at the pole

$$T = T_p(z), \quad q = q_p(z). \tag{4.1.11}$$

Some of the results of modeling the monthly average fields of the moisture content of the atmosphere Q_g and the cloud water reserves Q_δ over the Atlantic and Pacific Oceans will now be evaluated. Generally, the horizontal large-scale exchange along both the meridian and the parallel should be taken into account. However, in the results presented below, only the exchange in the y direction was taken into account, on the assumption that close to the meridians passing through central oceanic regions, the influence of large-scale mixing along the parallel (x direction) will be negligible: close to these meridians the monthly average values of temperature and humidity reach extreme (along the parallel) values, making the influxes of heat and moisture due to exchange in the x direction close to zero. It was also assumed that all the condensed moisture remains in the cloud ($\tilde{v} = 0$); in this way the water reserves of the clouds and the precipitation that could fall from them were evaluated. The exponent in formula (4.1.6) was taken to be unity.

The initial data figuring in the boundary conditions (4.1.9)-(4.1.11) were taken to be the mean monthly (for July and January) climatic values of the temperature and humidity at meridians 20°W (Atlantic Ocean) and 180° (Pacific Ocean). Q_s and Q_δ were calculated for various k_∞ (from 5 to 100 m^2/s) and k_y (from 10^3 to 10^8 m^2/s), in order to evaluate the sensitivity of these quantities to variations of the turbulent mixing in the vertical and horizontal directions. In [138] the values of Q_s and Q_δ are given for all latitudes (from 5 to 60-80°, every 5°) over the Atlantic and Pacific in January and July. As an example, Table 4.1 shows Q_s and Q_δ over the Pacific for $k_\infty = 25$ m^2/s and $k_y = 10^5$ m^2/s.

Table 4.1. Values of Q_s and Q_δ (kg/m^2) over Pacific Ocean (Northern Hemisphere)

Parameter	Latitude, °						
	5	15	25	35	45	55	65
			January				
Q_s	64	58	41	30	18	7.8	2.3
Q_δ	4.8	4.7	3.2	4.1	3.3	0.56	0.04
			July				
Q_s	62	62	54	48	42	30	15
Q_δ	4.7	5.1	4.8	4.4	4.9	3.2	1.3

It is seen that both Q_s and Q_δ decrease with increasing latitude, reflecting the influence of the temperature T_0 near the ground on these quantities. Since the maximum of T_0, the so-called thermal equator, is close to 20°N in July, it follows that Q_δ remains unchanged. This, in turn, is explained by the effect of the relative humidity f_0, which in the equatorial region decreases with increasing latitude. At a given latitude the July values of Q_s and Q_δ are higher than the January values, especially in the middle and high latitudes. Although in January f_0 is also higher than in July (and, with the rise in f_0, the moisture content and water reserves increase as well), still the influence of T_0 (lower in January than in July) is predominant.

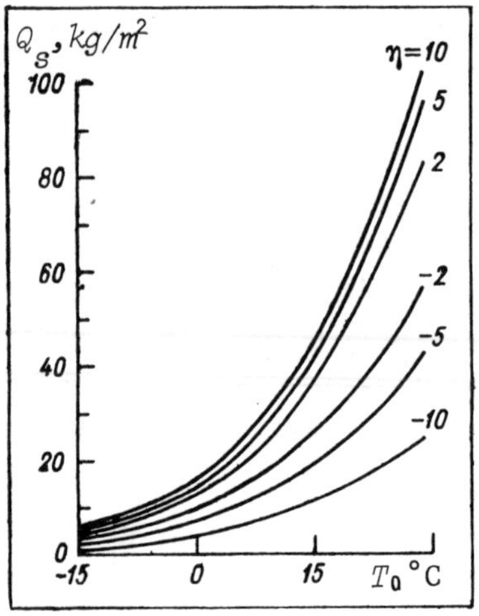

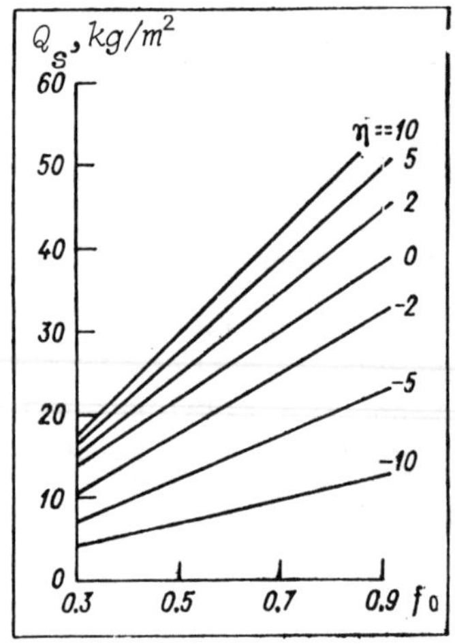

Fig. 4.2. Q_s as a function of T_0 and η for f_0 = 0.7 and ζ = 0.

Fig. 4.3. Q_s as a function of f_0 and η for T_0 = 15°C and ζ= 0.

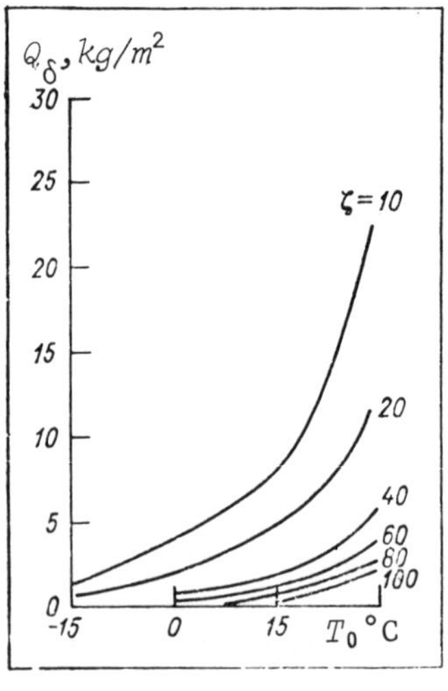

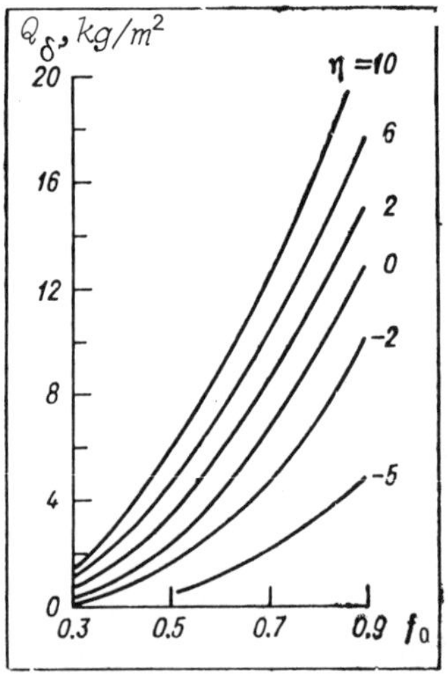

Fig. 4.4. Q_δ as a function of T_0 and ζ for η = 5 and f_0 = 0.7.

Fig. 4.5. Q_δ as a function of f_0 and η for T_0 = 15°C and ζ = 0.

The distribution of Q_s and Q_δ along the meridian depends on the intensity of the large-scale turbulent mixing. Physically, it is clear that an enhancement of the mixing will tend to equalize Q_s and Q_δ (to reduce the difference and ratio of these quantities at two latitudes). As an example, the values of the difference $\Delta Q_s = Q_s(10°) - Q_s(70°)$ and the ratio $Q_s(10°)/Q_s(70°)$ for fixed $k_\infty = 25$ m²/s and various k_y (here $Q_s(10°)$ and $Q_s(70°)$ are the moisture contents of a vertical column at latitudes of 10 and 70°, respectively) are given below:

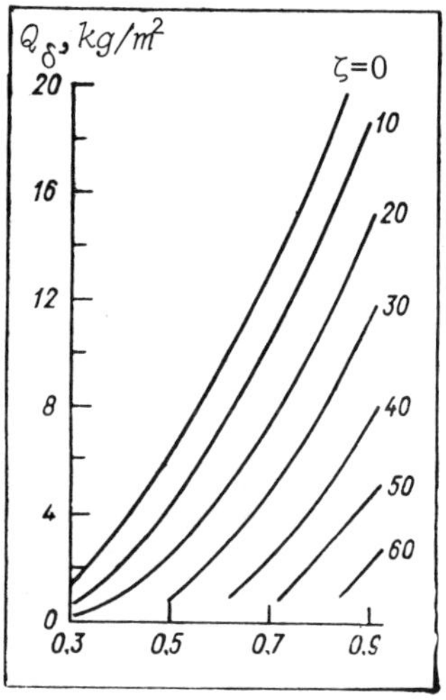

Fig. 4.6. Q_δ as a function of f_0 and ζ for $T_0 = 15°$C and $\eta = 10$.

k_y, m²/s	10^3	10^4	10^5	10^6	10^7	10^8
ΔQ_s, kg/m².	63	63	63	58	44	36
r	38	38	38	27	10	6

It is evident that for $k_y > 10^5$ m²/s, ΔQ_s and r decrease with increasing k_y.

An intensification of the turbulent mixing along the vertical (a rise in k) will be accompanied by a comparatively small increase in Q_s at low altitudes (for instance, if $k_y = 10^7$ m²/s, then at a latitude of 15°, Q_s increases from 41 kg/m² for $k_\infty = 5$ m²/s to 53 kg/m² for $k_\infty = 100$ m²/s), and a decrease at high latitudes (at a latitude of 70°, Q_s decreases from 6.2 kg/m² for $k_\infty = 5$ m²/s to 3.0 kg/m² for $k_\infty = 100$ m²/s). The cloud water reserves Q_δ will depend similarly on k_y and k_∞.

The results of modeling Q_s and Q_δ were used to construct graphs relating these quantities to the parameters exerting the greatest effect on the moisture content of the atmosphere and the cloud water reserves: T_0, f_0, η, and ζ (Figs. 4.2–4.6). As was shown above and as follows from an inspection of Figs. 4.2–4.6, the moisture content of a vertical column and the water reserves of the clouds increase with a rise in the temperature T_0, the relative humidity f_0, and the dimensionless parameter η, and with a decrease of parameter ζ. The variations with T_0, η, and ζ are considerably nonlinear, while the dependences of Q_s and Q_δ on f_0 are close to linear.

4.2. The water-vapor content in an air column and cloud water reserves according to data of radio and aircraft sounding

To date, a considerable volume of observational material has been collected on the moisture content of the atmosphere and the water reserves of clouds. This information has come mainly from network (large-scale) sounding of the atmosphere with radiosondes and aircraft. An analysis of the data shows that the measured Q_s and Q_δ are of the same order of magnitude as the calculated values of these characteristics. Moreover, observations verify the relations predicted by theory, in particular the dependence of Q_s and Q_δ on the temperature near the ground. Unfortunately, however, it does not appear to be possible to use observational data to check other theoretical relations in view of the lack of information about the vertical velocity, turbulence, fall velocity of droplets, etc., during the measurement period.

During recent decades, several studies have been made in which radiosonde and aircraft-sounding data were used to determine the water-vapor content Q_q in a vertical air column. If the hydrostatic equation $-dp = g\rho dz$ is employed, the formula for this quantity can be rewritten as

$$Q_q = \int_0^H \rho q\, dz = \frac{1}{g} \int_{p_0}^{p_H} q\, dp,$$

where $p_0 = p_H$ are the pressures at $z = 0$ and $z = H$. Now, if $g = 9.80615$ m/s^2 is used and the integral is expressed as a sum, the latter formula can be put into the form

$$Q_q = 0{,}0102 \sum_{k=1}^{N} \bar{q}_k \Delta p_k.$$

Here, $\bar{q}_k$ is the mean value of the specific humidity (g/kg) in the layer between p_{k-1} and p_k; the difference $\Delta p_k = p_{k-1} - p_k$ (mb); N is the number of layers into which the atmosphere is divided from the Earth's surface up to height H; and Q_q is in kg/m^2.

In [281], vapor content Q_q was determined for 35 points in the Soviet Union from radiosonde data obtained between 1961 and 1970. Q_q was calculated on the basis of specific-humidity values at the ground and at six standard isobaric surfaces (850, 700, 500, 400, 300, and 200 mb). The statistical characteristics presented below were obtained using 200-400 values of Q_q at each point and during each month.

According to the theory (see Section 4.1), the water-vapor

content in a vertical column fluctuates markedly in both time and space. The dependence of Q_q on the temperature and relative humidity of the air is thus clearly discerned. As an example, Table 4.2 gives the mean monthly values of Q_q at several places in the USSR.

Table 4.2. Mean monthly values of Q_q (kg/m^2)

Place	I	IV	VII	X
Murmansk	4.43	6.74	18.60	10.52
Leningrad	5.75	10.30	21.90	13.80
Moscow	5.76	11.24	25.64	13.70
Rostov-on-Don	7.93	13.38	26.89	15.72
Tiflis	10.85	17.88	36.40	20.59
Ashkhabad	10.37	19.79	25.36	17.33
Vladivostok	3.08	9.34	39.34	12.70

Because of the annual temperature variation, in every place listed the July values of Q_q are the highest and the January values are the lowest. The dependence of Q_q on the latitude was similarly determined: Q_q increases toward lower latitudes. Ashkhabad is an exception, since the values of Q_q there are lower in July and October than at Tiflis (even though the temperature is higher); Vladivostok, where in January, April, and October Q_q is considerably lower than at places further north, is another exception. At Ashkhabad (and throughout Soviet Central Asia) the relative humidity is low in summer and autumn, while at Vladivostok (and throughout the Maritime Territory) the temperature and relative humidity are low in winter (compared, for instance, with Transcaucasia).

The lowest mean monthly values of Q_q for January are observed in the vicinity of Yakutsk (less than 1.5 kg/m^2); for July the lowest values are in the Central Arctic (less than 15 kg/m^2). The highest January values of Q_q are found in the southern part of Soviet Central Asia (about 10 kg/m^2), and the highest July values in Transcaucasia (about 36 kg/m^2) and the southern part of the Maritime Territory (38-40 kg/m^2).

The vapor content Q_q at each point varies over a wide range, because of fluctuations in the temperature and relative humidity, not just annually, but also over short intervals of time. This is borne out by the rms deviations of Q_q, which are of approximately the same order of magnitude as the mean values of Q_q.

Figures 4.7 and 4.8 compare the calculated vapor contents in a vertical column for various values of T_0, f_0, and η [207] with experimental values of Q_q from a handbook [281]. These figures give all the data for all points and for every season; separate graphs are plotted for slight cloudiness and for overcast skies. Here, Q_{1000}^{200} is the vapor content in a vertical column between the 1000

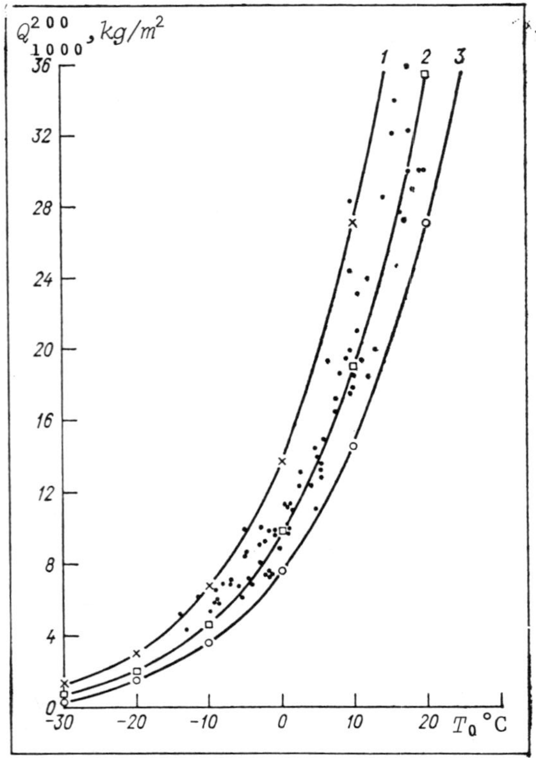

Fig. 4.7. Relation between T_0 and Q^{200}_{1000} for an overcast sky (cloudiness of 8–10). Calculated curves: (1) η=0.5; f_0=70%; (2) η=0.5, f_0=50%; (3) η=0.1, f_0=60%.

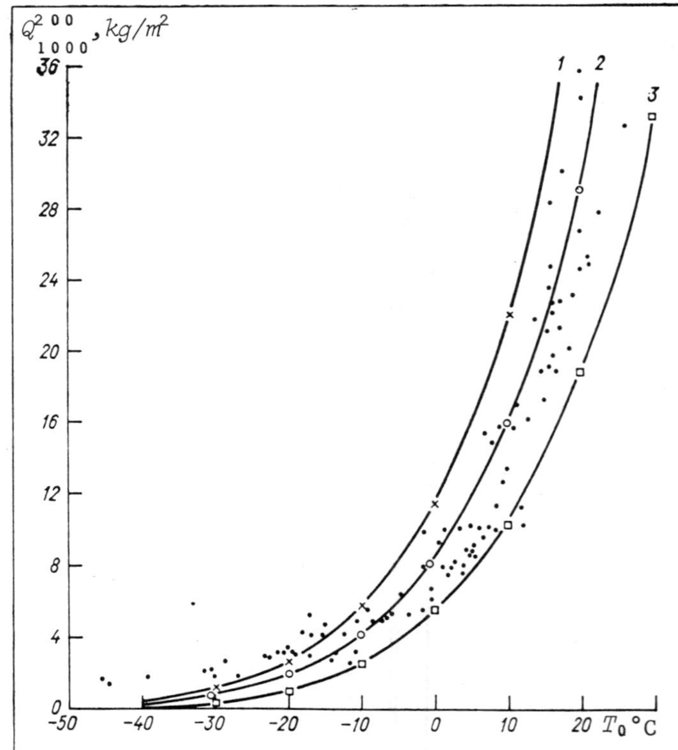

Fig. 4.8. Relation between T_0 and Q^{200}_{1000} for a slightly cloudy sky (cloudiness of 0–2). Calculated curves: (1) η=−1, f_0=70%; (2) η=−1, f_0=50%; (3) η=−5, f_0=60%.

and 200 mb levels; dots correspond to observational data, and curves to calculations. Inspection of these figures shows that in every geographical region, the nature of Q_q as a function of T_0 and η agrees with the predictions of theory. For a cloudiness 8 or 10 (see Fig. 3.7), all the experimental points lie between the calculated curves corresponding to positive values of the parameter η, whereas for slightly cloudy weather (see Fig. 4.8), they lie between the curves corresponding to negative η. Consequently, as is to be expected from physical considerations as well, for overcast skies updrafts will prevail ($\eta > 0$, $w > 0$), while for slight cloudiness downdrafts prevail ($\eta < 0$, $w < 0$).

Snopkov [277] calculated the water-vapor content in a vertical column using radiosonde data from the central part of the Atlantic (the area measured 133 X 113 nautical miles and was centered at 16°30' N, 33°30' W), collected from March to September 1970 (total number of soundings 110). In addition to the value of Q_q for a column 10 km high, the vapor content was calculated for columns from 0 to 2 km (Q_{0-2}) and 2 to 6 km (Q_{2-6}). Figure 4.9 shows ratios Q_{0-2}/Q_q and Q_{2-6}/Q_q as functions of Q_q. It is seen that with an increase in Q_q, the first of these ratios decreases while the second increases. For Q_q values from 55 to 60 kg/m^2, about 55% of the water-vapor mass is concentrated in the layer from 0 to 2 km, and about 40% in the layer from 2 to 6 km. According to observational data for both 1970 and 1972, a quite close statistical connection was established between the water-vapor pressure e_0 close to sea level and the vapor content in the layer from 0 to 10 km. The correlation coefficient r (according to 147 soundings in 1970) between e_0 and Q_q was 0.81, with a probable error of ± 0.018 in determining this coefficient.

On the basis of an analysis of numerous radiosonde data collected over the oceans, Timofeev [293, 294] studied the vertical distribution of the absolute humidity a and the vapor content in a vertical column $Q_q(z)$ for various amounts of cloudiness. A total of 1187 radiosoundings (from ships) were used to establish the relationship between the water-vapor pressure e_0 close to the sea surface, the cloudiness n, and the vapor content Q_q over the whole vertical column:

$$Q_q = (1.16 + 0.034n)\, e_0^{1.075}, \tag{4.2.1}$$

where Q_q is in kg/m^2, e_0 is in mb, and n is in points on the cloudiness scale.

For a fixed n, the coefficient of the correlation between $\log Q_q$ and $\log e_0$ for an ice-free surface of the World Ocean ranges from 0.90 to 0.94. The relative standard deviation of Q_q from the value calculated using (4.2.1) is 14%

Sounding data were used to obtain the mean values of the vapor content in a layer from 0 to 5 km over the Pacific for February and May (Table 4.3). With decreasing latitude, in both hemispheres the water-vapor content in the lower half of the troposphere is seen to increase. The annual variation in Q_q also shows up; in February (winter) the vapor contents at all latitudes of the Northern Hemisphere are lower than in May (spring); in the Southern Hemisphere the vapor content is higher in February (summer) than in May (autumn). These data reflect the dependence of Q_q on the air temperature.

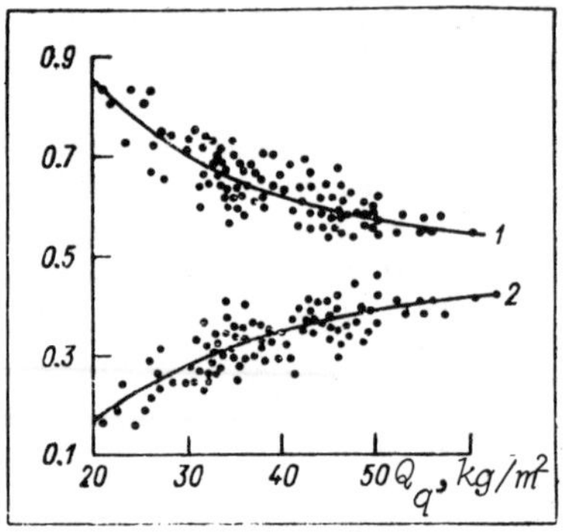

Fig. 4.9. Ratios Q_{0-2}/Q_q (1) and Q_{2-6}/Q_q (2) as functions of Q_q.

Ushakova [308] used data from network aircraft soundings of the atmosphere to evaluate the water reserves of clouds over the USSR. In accordance with the predictions of theory, these data revealed a pronounced dependence of Q_q on the air temperature T_0. For instance, the lowest water reserves are observed in winter, and the highest in summer.

Table 4.3. Mean vapor content (kg/m^2) in layer from 0 to 5 km over the Pacific

Latitude	II	V	Latitude	II	V
60—50° N	6.3	10.6	0—10° S	42.1	40.3
50—40	11.0	14.8	10—20	40.3	34.1
40—30	16.0	21.6	20—30	34.1	26.6
30—20	23.0	30.4	30—40	24.8	18.3
20—10	32.5	38.8	40—50	17.8	12.9
10—0	38.3	42.2	50—60	12.6	8.6

In the European USSR in winter, the water reserves of nimbostratus clouds vary from 0.44-0.46 kg/m^2 in the west and on the Black Sea coast to 0.17-0.20 kg/m^2 in the east. The water reserves of Ns in winter are sizable (around 0.35 kg/m^2) in the mountains of Soviet Central Asia (comparatively high temperature and relative humidity). In the Asian USSR, the water reserves of Ns show a typical drop from south to north and from west to east (from 0.19 kg/m^2 at Omsk and Novosibirsk to 0.10-0.14 kg/m^2 at Krasnoyarsk and Yakutsk).

In summer the water reserves of Ns increase (relative to the winter) to 0.60-0.65 kg/m^2 in the western part of the European USSR, to 0.9 kg/m^2 on the Black Sea coast, to 0.58 kg/m^2 near Omsk and Novosibirsk, and to 0.17 kg/m^2 near Yakutsk. The spring and summer water reserves of Ns in the Maritime Territory are considerable (0.40-0.60 kg/m^2). Over the arctic seas (Kara and

Laptev), the Ns water reserves range from 0.10 kg/m^2 in autumn to 0.12-0.15 kg/m^2 in summer.

The water reserves of the stratocumulus clouds vary from 0.20 kg/m in summer to 0.03-0.04 kg/m^2 in winter (the minimal values, around 0.01-0.02 kg/m^2, are observed at Yakutsk). On the Black Sea coast and in the mountains of Soviet Central Asia, the water reserves of Sc in summer and spring reach 0.22 kg/m^2.

It should be noted that the cloud water reserves, according to the data of [308], are comparatively low. In order to reconcile these figures with the calculated results (see Figs. 4.4-4.6), it must be assumed that in the clouds investigated the fall velocity of the cloud elements is considerable (parameter ζ is high). Such an assumption with regard to Ns is quite reasonable, since precipitation falls from these clouds. The data for stratocumulus clouds, on the other hand, once more point to a predominance of downdrafts in them (in Fig. 4.5 these are the curves corresponding to $\eta < 0$; since with decreasing η the water reserves diminish, it follows that for $\eta < 0$ the values of Q_δ are very low).

4.3. The atmospheric moisture content and cloud water reserves according to radiometric data

During the last 10 or 15 years, radiometric equipment, ground--based as well as airborne and satellite-mounted, has come to be heavily used to determine the moisture content of the atmosphere and the cloud water reserves. The physics of the determinations of Q_s and Q_δ, as well as of the characteristics of other terrestrial objects, with the aid of this equipment has been described in monographs [32, 142]. Radiometers are used to measure the fluxes of thermal radio-frequency self-radiation of the atmosphere, clouds, precipitation, and underlying surface in the microwave (so-called superhigh, or SHF) range (wavelengths from 1 m to 1 mm). Since these fluxes vary with the water content, temperature, and phase of the clouds and precipitation, and with the air temperature and humidity, it becomes possible to determine the moisture content of the atmosphere and the water reserves and temperature of the clouds. Without discussion of the derivation of the theoretical formulas relating Q_q and Q_δ to the radiation fluxes (which were characterized by introducing the radio-brightness temperature), the main results obtained using these methods will now be presented.

Mel'nikov and others [302] used data on the thermal radio--frequency self-radiation of the atmosphere at wavelengths close to the resonance line of monomeric H_2O vapor ($\lambda \approx 1.35$ cm) to investigate the distribution of the vapor content Q_q in a vertical column over the tropical Atlantic from June to September 1972. The measurements of Q_q were made aboard R/V *Akademik Kurchatov* during the main periods of meteorological observation in the form of individual series (lasting 7-10 min), each of which included 5 or 6 reckonings of Q_q. The rms error in determining Q_q (instrumental and systematic errors) did not exceed 10%.

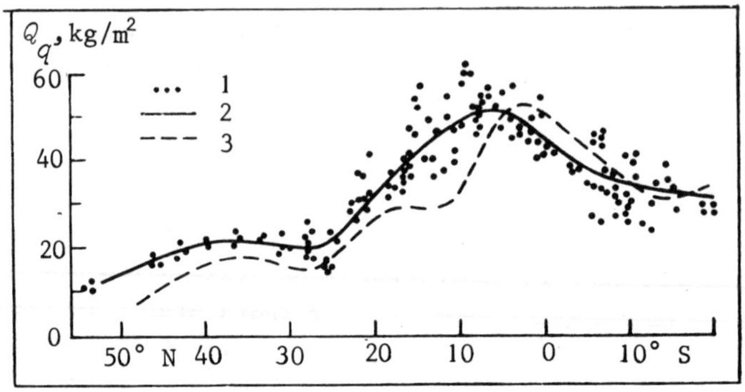

Fig. 4.10. Meridional profile of Q_q over the Atlantic: (1) data of measurements from 17 June to 31 August 1972; (2) average of these data; (3) average of measurements made from February to May 1969.

Figure 4.10 depicts the meridional profile of the vapor content in a vertical column over the Atlantic at latitudes from 53° N to 22° S, according to observations carried out from 17 June to 31 August 1972 and from February to May 1969. In the subtropics (45–25° N), the vapor content in a vertical column is seen to be comparatively low (22-23 kg/m^2). The dryest air (Q_q = 15-20 kg/m^2) is observed between 26 and 30° N, that is, in the vicinity of the Azores anticyclone. The highest values of Q_q are found in the intertropical convergence zone (ITC), which in June-August 1972 lay between 7 and 9° N and in February-May 1969 lay between 3 and 4° N. The mean value of Q_q in the ITC was 50 kg/m^2, its maximum was 65 kg/m^2, and its minimum was 32 kg/m^2. The standard deviations σ_Q of Q_q were comparable for the various latitude zones of the ocean. In the zone ranging from 40° N to 20° S, the values of σ_Q varie from 5.4 to 7.0 kg/m^2.

The vapor content in a vertical column fluctuates quite sharply from one time to another, with the largest fluctuation amplitudes (maximum of spectral density Q_q) corresponding to periods from 24 to 40 h. It should be noted that Q_q is closely connected with ITC: with an approach to the ITC or an accentuation of it, Q_q increases, as a rule. For instance, on 12-13 July 1972, Q_q was measured in the ITC over a width of about 2.5°, and the vapor content in a vertical column was found to exceed 60 kg/m^2 at various times. On 10 August 1972, the vessel was located 1-1.5° south of the ITC: the sky was slightly cloudy and Q_q was only 40 kg/m^2. On the very next day, there were thick clouds over a vast region (from 4 to 10° N and from 15 to 22° W), accompanied by heavy rain with thunder and lightning (90.5 mm of rain fell in 24 h), and Q_q increased sharply (to 55 or 60 kg/m^2). In [302], Mel'nikov and others pointed out that the fluctuations of Q_q are considerably larger than the fluctuations of the absolute humidity close to the sea surface. For example, in the ITC the maximum values of Q_q are mo e than twice as great as the minimum values of Q_q, whereas the ratio of the maximum absolute humidity to the minimum does not, as a rule, exceed 1.1-1.3.

In order to measure the cloud water reserves over the tropical Atlantic, Galaktionov et al. [302] made use of an assembly of five radiometers, with the aid of which Q_δ was found by soundings from the vessel at wavelengths of 0.8 and 1.6 cm for $Q_\delta < 1$ kg/m^2 and at wavelengths of 1.6 and 2.5 (3.4) cm for $Q_q > 1$ kg/m^2. According to these data, the water reserves of the middle and upper cloud levels varied from 0.05 to 0.6 kg/m^2, and in the case of thick convective clouds reached 15 kg/m^2.

On the TROPEKS-74 voyages, Q_q and Q_δ were determined from the difference between the radio-brightness temperatures of a cloudy and cloudless sky at wavelengths of 0.8 and 1.37 cm. According to [303], when the vessel intersected the ITC, the cloud water reserves reached 4-8 kg/m^2; the cloud formations in this case had a typical scale of a few km (which agrees with data from radar observations of clouds).

Toong and Staelin [432] used radiometric measurements at five frequencies in the microwave range to investigate fluctuations of Q_q and Q_δ during the passage of a cold front through Boston on 2-3 August 1968 (Table 4.4). In a direction perpendicular to the front, Q_q is seen to vary from 22 to 44 kg/m^2, while Q_δ varies from 0.2 to 14 kg/m^2.

Table 4.4. Variations of Q_q and Q_δ during passage of cold front

Time,h	Q_q,kg/m^2	Q_δ,kg/m^2	Precipitation
0—1	33	0.2	
1—3.2	38.5	0.2	
3.2—3.5	44	1.8	Slight
3.5—4	44	14	Slight
4—4.8	44	1.8	
4.8—5	38.5	4	Slight
5—7	38.5	0.6	
7—10.5	22	0.2	

The authors of [82] determined the cloud water reserves with the aid of two radiometers, which picked up atmospheric radiation at wavelengths of 0.8 and 1.35 cm, together with a Doppler radar operating in the decimeter range. The water reserves of clouds from which 1.5-18 mm/h of precipitation fell varied from 0.36 to 2.9 kg/m^2, with the rain accounting for 15-43% of the total water reserves (clouds plus precipitation).

In late March 1975, measurements with airborne radiometers picking up atmospheric radiation from the upper hemisphere at 0.8, 1.35, and 2.25 cm established [5] that in the vicinity of a

Central Asian front the values of Q_q ranged from 8 to 10 kg/m^2 in cold air and there was no cloudiness. As the front was approached, Q_q increased to 16 kg/m^2 and cumulus clouds with water reserves of 0.15-0.6 kg/m^2 appeared. As the aircraft moved away from the front, the clouds gradually decayed and Q_q and Q_δ decreased.

Using radiometers recording the atmospheric radiation at 0.34, 0.8, and 1.5 cm, the effective cloud temperature, as well as Q_q and Q_δ [217], can be found. Measurements made with aircraft flying over the Caspian in 1976 indicated that clouds 0.6-0.7 km deep had water reserves not exceeding 0.20-0.23 kg/m^2 (T_{cl} = 5°C), for a mean water content of about 0.3 g/m^3. According to measurements made in this same region in March-April 1978, the water reserves of stratiform clouds vary smoothly from place to place and are comparatively small $Q_\delta \leqslant$ 0.10-0.25 kg/m^2); on the other hand, the water reserves of convective clouds not giving precipitation vary over a wide range, reaching 3-6 kg/m^2.

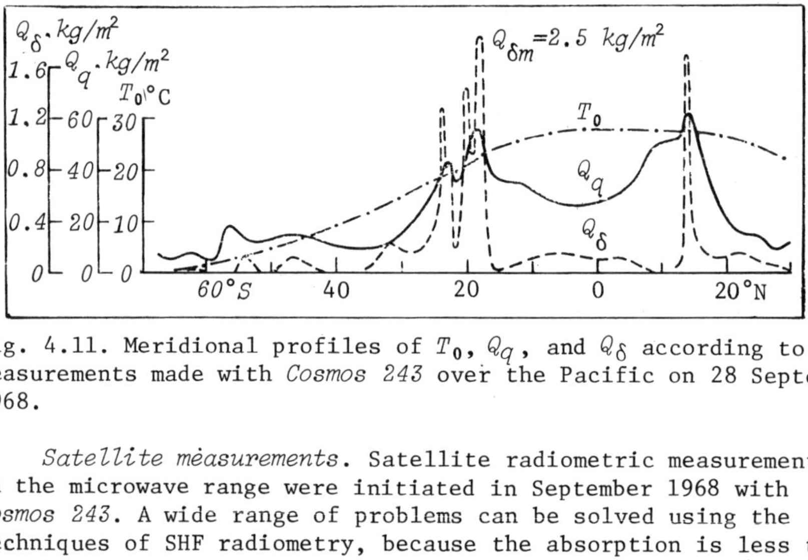

Fig. 4.11. Meridional profiles of T_0, Q_q, and Q_δ according to measurements made with *Cosmos 243* over the Pacific on 28 September 1968.

Satellite measurements. Satellite radiometric measurements in the microwave range were initiated in September 1968 with *Cosmos 243*. A wide range of problems can be solved using the techniques of SHF radiometry, because the absorption is less than that in the infrared region, and because this absorption depends on the water content, temperature, and phase of the clouds and precipitation, as well as on the humidity and temperature of the air. A comparative analysis of the fluxes of thermal self-radiation of clouds and of the atmosphere, measured with *Cosmos 243* at 0.8, 1.36, 3.4, and 8.5 cm, have enabled a determination of the vapor content Q_q in a vertical column, the cloud water reserves Q_δ, and the temperature T_0 of the ocean surface along the satellite orbit over the ocean. As an example, Fig. 4.11 shows curves of these quantities over the Pacific. The parameter values were averaged over a 400-500 km^2 area. The cloud water reserves at latitudes $\varphi \geqslant 40°$ S did not exceed 0.2 kg/m^2. Between 17 and 25° S and close to 15° N, where cumulonimbus clouds were observed,

Q_δ was considerably higher. In the ITC, in some cases the water reserves of cumulonimbus clouds exceeded 3-5 kg/m^2 [7], with several (up to 5 or 6) peaks of Q_δ being observed. A comparison of radiometric data with television and infrared cloud photos indicated that a value of $Q_\delta \approx 0.2$ kg/m^2 usually corresponds to the boundaries of unbroken cloudiness. The average size of a cloud system with $Q_\delta \geqslant 0.2$ kg/m^2 is in the ITC from 1000 to 1500 km.

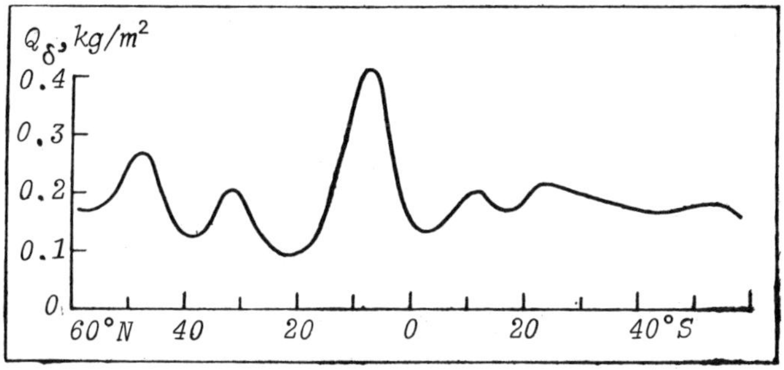

Fig. 4.12. Average meridional profile of cloud water reserves Q_δ for the period 23-27 September 1968.

Radiometric data obtained with *Cosmos 243* in front regions (53 cases in all) were considered in [5]. In every case, Q_δ increased as the front was approached. The mean value of Q_δ near a front ranged from 0.7 to 0.8 kg/m^2, and the maximum values were 1.5-2 kg/m^2. The extent of the cloudiness fields where $Q_\delta \geqslant 0.2$ kg/m^2 varied from 100 to 1300 km, while for $Q_\delta \geqslant 0.5$ kg/m^2 the range was from 40 to 500 km (here the probability of precipitation was high).

The water reserves of frontal clouds fluctuate considerably [94]. In the southern part of the ITC at a cold front (32-55° N) on 24 September 1968, cumulonimbus clouds with $Q_\delta > 0.8$ kg/m^2 prevailed. In other regions, Q_δ varied from 0.1 to 0.5 kg/m^2. The widths of individual peaks of Q_δ ranged from 100 to 150 km. In the vicinity of a relatively immobile front (in a baric field with small gradients), the values of Q_δ did not exceed 0.6 kg/m^2; in the vicinity of a rapidly moving front, Q_δ reached 1 kg/m^2 (for the same high variability); close to the cyclone center Q_δ increased to 1.5 kg/m^2.

Figure 4.12 shows radiometric data obtained with *Cosmos 243* over the oceans and averaged over the observation period 23-27 September 1968 for five-degree latitude bands. The principal maximum of Q_δ is observed close to 10° N, where the ITC was situated during this period. Where the trade winds blow (15-30° N) the cloud water reserves are low ($Q_\delta \approx 0.1$ kg/m^2); the same is true (values around 0.15 kg/m^2) in the Southern Hemisphere. In the Northern Hemisphere, two second maxima of Q_δ are observed: between 30-35° N and between 45-50° N, where Q_δ is only about 0.2 and 0.3 kg/m^2.

The standard deviations σ_Q of the cloud water reserves are of the same order of magnitude as $Q_\delta(\sigma_Q/Q_\delta \approx 1)$.

Radiometric measurements were also carried out with other Soviet satellites (in particular, with *Cosmos 384* and *Meteor 18*).

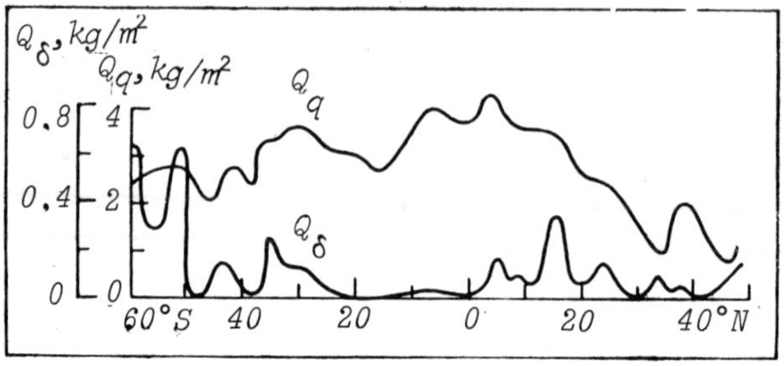

Fig. 4.13. Meridional profiles of vapor content Q_q and cloud water reserves Q_δ, according to measurements made with *Nimbus 5* on 9 March 1973 over the Pacific.

An extensive program of radiometric studies of the Earth at six wavelengths was implemented on the basis of observations with the American satellite *Nimbus 5*, launched in December 1972 into a circular solar-synchronous orbit at a height of 1100 km. The characteristics of the cloudiness and of other terrestrial objects were averaged over an area of 220 x 220 km; the error in determining Q_δ was about 0.10 kg/m^2, and the error in Q_q was about 2.5 kg/m^2. As an example, Fig. 4.13 shows the meridional profiles of the water content Q_q and the cloud water reserves Q_δ, obtained from measurements at 0.96 and 1.35 cm with *Nimbus 5* [426]. The Q_δ and Q_δ profiles from these data are smoother than the profiles in Fig. 4.11, because the instruments on *Nimbus 5* had a lower resolution (about 200 km) than the instruments aboard *Cosmos 243* (about 20 km). The maximum values of Q_q and Q_δ, 55-60 and 1.6-2.0 kg/m^2, respectively, were recorded with *Nimbus 5* over the ITC and tropical cyclones.

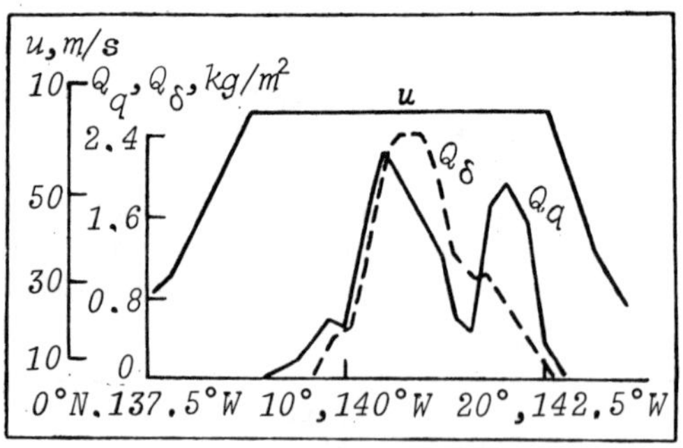

Fig. 4.14. Profiles of vapor content Q_q , water reserves Q_δ, and wind velocity u for passage of a satellite through tropical cyclone June, according to radiometric data obtained with *Nimbus 6* on 19 November 1975 over the Pacific.

An attempt was made in [418] to take into account the effect of the wind velocity on the radiometric determinations of the atmospheric characteristics. Figure 4.14 gives data pertaining to a tropical cyclone. The maximum values of Q_δ are close to 2.4 kg/m^2, for wind velocities near the cyclone center in excess of 60 m/s.

CHAPTER 5

THE HYDRODYNAMIC PREDICTION OF CLOUDINESS AND PRECIPITATION

5.1. The first models of precipitation prediction

The first techniques for the prediction of cloudiness and precipitation were worked out by Smagorinsky and Collins [421, 422] and then by Lewis [382]. These methods were based on the use of empirical relations between the vertical velocity $\tau = dp/dt$ and the dew-point deficit Δ on the one hand, and the precipitation and cloudiness on the other.

These same ideas were utilized by Dushkin, Lomonosov, and Lunin [99], Kuznetsov [148], and others to construct graphs relating τ and Δ to weather situations: precipitation, overcast (7-10), partly cloudy (4-6), and clear skies (0-3).

Predictions of τ and Δ make use of equations whose form was established by Shvets and Meleshko [209, 326]. If the turbulent influxes of heat and moisture and the effects of water-vapor condensation are neglected, then the equation for Δ becomes

$$\frac{\partial \Delta}{\partial t} + u\frac{\partial \Delta}{\partial x} + v\frac{\partial \Delta}{\partial y} + bw = 0,$$

where $b = \gamma_a\ [1 - (\varkappa / \varkappa - 1)R_v T_d^2/LT]$, T_d being the dew-point temperature. Shvets and Meleshko used the vorticity-transfer equation

$$\frac{\partial \Omega_a}{\partial t} + u\frac{\partial \Omega_a}{\partial x} + v\frac{\partial \Omega_a}{\partial y} = \Omega_a\frac{\partial w}{\partial z},$$

where $\Omega_a = \Omega_z + 2\omega_z$ is the vertical component of the absolute vorticity ($\Omega_z = \partial v/\partial x - \partial u/\partial y$ being the vertical component of the relative vorticity, and $2\omega_z$ being the Coriolis parameter), and they represented the vertical velocity as a product

$$w = w_s(x, y, t)\,\varphi(z),$$

in order to establish that along the horizontal projection of the trajectory of an air mass, the following product is practically constant (invariant):

$$f\,(\Omega_z + 2\omega_z)^{c_1} = \text{const},$$

where exponent

$$c_1 = \gamma_a \left[\frac{\varkappa}{(\varkappa - 1) T} - \frac{L}{R_v T^2} \right] \frac{\varphi(z)}{\varphi'(z)}$$

can, at a given level (isobaric surface), be assumed to be constant, and f is the relative humidity of the air.

On the basis of an equation describing the transfer of specific humidity, Kuznetsov [148] obtained another invariant (also along the horizontal projection of the air-mass trajectory):

$$E(T_d)(\Omega_z + 2\omega_z)^{c_2} = \text{const},$$

where

$$c_2 = \left(\frac{\varkappa}{\varkappa - 1} \frac{\gamma_a}{T} - \frac{L\gamma_d}{R_v T_d^2} \right) \frac{\varphi(z)}{\varphi'(z)}, \quad \gamma_d = -\frac{\partial T_d}{\partial z}, \quad \varphi' = \frac{d\varphi}{dz},$$

and $E(T_d)$ is the saturation vapor pressure at the dew point T_d.

The predictability of precipitation and cloudiness (i.e., three graduations of cloudiness) using this method is, according to Kuznetsov, 70-75%; the reliability of a precipitation prediction ($\tau \leqslant -10$ mb/12 h and $\Delta \leqslant 2°C$) is, according to Antonov, 63%. If the air temperature at the 850 mb level is lower than -10°C, the precipitation predictability rises to 68%.

Musaelyan [226] is credited with developing a means of analyzing and predicting cloudiness; using Blinova's method [28, 29], he established the distribution of cloudiness over the hemisphere as a function of the vertical-velocity field. The idea of relating the amount and boundary heights of cloudiness to the vertical--velocity field had been presented in an earlier work [177], where it was applied to vertical velocities in the boundary layer.

Marchuk [168] presented a number of proposals for reducing the equations of atmospheric thermodynamics to a closed analytical form for predicting fields of humidity, cloudiness, and precipitation. He took into account both the liquid and solid phases of a cloud, using semiempirical relations for the rates of phase transitions of water.

5.2. A hydrodynamic five-level model for predicting humidity and cloudiness

The method of cloudiness prediction proposed by the authors of [160, 161] was based on an earlier theory [179, 181], whereby cloud elements (water droplets and ice crystals) are assumed to be entrained almost completely by air parcels participating in the turbulent mixing. The initial system of equations describing the transport of heat and moisture thus takes the form

$$\frac{\partial \Pi}{\partial t} + u\frac{\partial \Pi}{\partial x} + v\frac{\partial \Pi}{\partial y} + w\frac{\partial \Pi}{\partial z} = \frac{1}{\rho}\frac{\partial}{\partial z}\rho k\frac{\partial \Pi}{\partial z}, \quad (5.2.1)$$

$$\frac{\partial s}{\partial t} + u\frac{\partial s}{\partial x} + v\frac{\partial s}{\partial y} + w\frac{\partial s}{\partial z} = \frac{1}{\rho}\frac{\partial}{\partial z}\rho k\frac{\partial s}{\partial z}, \quad (5.2.2)$$

$$s = q + \delta, \quad (5.2.3)$$

$$\Pi = T + \gamma_a z + \frac{L}{c_p} q. \quad (5.2.4)$$

For a cloud, in which the water vapor is saturated, the system (5.2.1)-(5.2.4) must be supplemented by the relation

$$q = q_m = 0{,}622\,\frac{E(T)}{p}. \quad (5.2.5)$$

Equations (5.2.1) and (5.2.2) have a significant advantage over the equations of heat and water-vapor influx in their generally accepted form: they are invariant with respect to the condensation of water vapor.

In eqns. (5.2.1) and (5.2.2), the terms allowing for horizontal turbulent mixing have been dropped, on the assumption that its role in the creation of stratiform cloudiness (with typical horizontal scales of hundreds or thousands of kilometers) is small compared to factors such as the advection of heat and moisture, phase transitions of water, and transport of heat and moisture by ordered vertical currents. On the other hand, the role of vertical turbulent mixing in the formation of clouds is always considerable.

Since three velocity components enter into system (5.2.1) and (5.2.2), the equation of motion, the hydrostatic equation, and the continuity equation must be added to it:

$$\begin{aligned}
\frac{du}{dt} &= -\frac{1}{\rho}\frac{\partial p}{\partial x} + 2\omega_z v + \frac{\partial}{\partial z}k\frac{\partial u}{\partial z},\\
\frac{dv}{dt} &= -\frac{1}{\rho}\frac{\partial p}{\partial y} - 2\omega_z u + \frac{\partial}{\partial z}k\frac{\partial v}{\partial z},\\
\frac{\partial \rho}{\partial t} &+ \frac{\partial \rho u}{\partial x} + \frac{\partial \rho v}{\partial y} + \frac{\partial \rho w}{\partial z} = 0,\\
-\frac{1}{\rho}\frac{\partial p}{\partial z} &= g.
\end{aligned} \quad (5.2.6)$$

Here, d/dt indicates the individual derivative.

Equations describing the radiation transfer in the atmosphere and the kinetics of cloud processes, which by right should also be added to the system (5.2.1)-(5.2.6), are not given here, since this chapter will consider a method that ignores the radiative influxes of heat.

In general, the equations of the system (5.2.1)-(5.2.6) must be solved simultaneously. However, serious difficulties arise when implementing the method of cloudiness prediction with the aid of computers, because of the limited size of the machine memory and the limited computing time. Thus it will be advisable to consider the case in which the system (5.2.6) is solved independently of the system (5.2.1)-(5.2.5) in the quasi-geostrophic and quasi-adiabatic approximation. Neglecting phase influxes of heat will, in this case, lead to errors of the order of 10% in forecasting the fields of pressure and wind velocity. The numerical technique used by the author to solve problem (5.2.6) for five levels (1000, 850, 700, 500, and 300 mb) on the computer was worked out by Marchuk and others [172].

In the coordinate system $x_p = x,\ y_p = y,\ \eta = p/p_0,\ t_p = t$ and in the quasi-static approximation, equations (5.2.1) and (5.2.2) are written as

$$\frac{\partial \Pi}{\partial t} = a\frac{\partial^2 \Pi}{\partial \eta^2} + \left(b - \frac{\tau}{p_0}\right)\frac{\partial \Pi}{\partial \eta} - \left(u\frac{\partial \Pi}{\partial x} + v\frac{\partial \Pi}{\partial y}\right), \tag{5.2.7}$$

$$\frac{\partial s}{\partial t} = a\frac{\partial^2 s}{\partial \eta^2} + \left(b - \frac{\tau}{p_0}\right)\frac{\partial s}{\partial \eta} - \left(u\frac{\partial s}{\partial x} + v\frac{\partial s}{\partial y}\right). \tag{5.2.8}$$

Here $a = k(g\eta/RT)^2$; $b = 2k(g\eta/RT)^2[1/\eta) - (1/T)(\partial t/\partial \eta)$; $\tau = dp/dt$ is the analog of the vertical velocity; and R is the specific gas constant of air. As a first approximation, coefficients a and b can be considered functions of η alone. Their values for $k = 10\ m^2/s$ are given in Table 5.1.

Table 5.1. Values of coefficients a and b for $k = 10\ m^2/s$.

η	1	0.85	0.7	0.5	0.3	0.2
$10^4\ a,\ h^{-1}$. . . .	6.12	4.7	3.4	1.9	0.76	0.36
$10^4\ b,\ h^{-1}$. . . .	13.3	13.0	11.5	8.6	5.4	3.6

Initial and boundary conditions. Since it appears at present to be impossible to arrive at a means of predicting such complicated phenomena as cloudiness and precipitation on the basis of theoretical premises alone, empirically established relations have been used to formulate the boundary conditions.

(1) At the initial moment ($t = 0$), the values of functions Π and s are everywhere known:

$$\Pi(x, y, \eta, 0) = \Pi^0(x, y, \eta),\quad s(x, y, \eta, 0) = s^0(x, y, \eta). \tag{5.2.9}$$

Outside the clouds, Π and s are found from measurement data. Inside the clouds, q is assumed to be equal to the maximum specific humidity q_m, and the cloud water content, in the absence of a large quantity of measurements, is found using the formula

$$\delta_0 = 0.201 \cdot 10^{-3} \frac{T}{\eta} \exp\left[17.86\left(1 - \frac{258}{T}\right)\right], \tag{5.2.10}$$

which gives the water content of stratiform clouds as a function of temperature, as indicated by numerous experimental data.

(2) At the lateral boundaries of the region, functions Π and s are taken to remain constant during a 24 h period and equal to their initial values, that is,

$$\Pi(\Omega, t) = \Pi(\Omega, 0), \quad s(\Omega, t) = s(\Omega, 0), \tag{5.2.11}$$

where Ω is the lateral surface of the region within which the formation and development of cloudiness are being considered.

The case where the horizontal fluxes of heat and moisture at surface Ω are assumed to be known is

$$\frac{\partial \Pi}{\partial l} = f_1(\Omega), \quad \frac{\partial s}{\partial l} = f_2(\Omega). \tag{5.2.12}$$

Here l is the horizontal coordinate, and f_1 and f_2 are known functions.

(3) The upper boundary of the region is taken to be the tropopause. Experimental data indicate that beyond the tropopause cloudiness is, as a rule, absent ($\delta = 0$), while the temperature and specific humidity (as well as the specific moisture content) vary only slightly with height. Thus zero fluxes of heat and moisture at the level of the tropopause may be assumed, that is,

$$\frac{\partial T}{\partial \eta} = \frac{\partial s}{\partial \eta} = 0 \quad \text{for } \eta = \eta_{tr}, \tag{5.2.13}$$

where $\eta_{tr} = p_{tr}/p_0$ is the value of η at the tropopause.

For practical calculations, the tropopause was identified with the isobaric surface at 200 mb, that is, $\eta_{tr} = 0.2$.

(4) The lower boundary of the region is taken to be the 1000 mb isobaric surface ($\eta = 1$), identified with the Earth's surface. Since the variation in air temperature close to the ground is determined by advection as well as by nonadvective (transformation) factors, the temperature at $\eta = 1$ can be represented as a known function of time:

$$T(x, y, 1, t^{n+1}) = T(x, y, 1, t^n) + \Delta T^{n+1} \tag{5.2.14}$$

for $\eta = 1$, $n = 0, 1, 2 \ldots$,

$$\Delta T^{n+1} = [\Delta T_a^{n+1} - (0.36 - 0.004 c_g^n)\, \Delta T_a^{n+1}]\, \Delta t,$$

$$\Delta T_a^{n+1} = -\left(u_g \frac{\partial T}{\partial x} + v_g \frac{\partial T}{\partial y}\right)^n \Delta t.$$

Here ΔT^{n+1} is the temperature change during a time interval $\Delta t = t^{n+1} - t^n$, caused by advection (first term on the right-hand side of (5.2.14)) and transformation of the air mass (second term); ΔT_a^{n+1} is the geostrophic advection of heat during a time Δt; and c_g^n is the modulus (absolute magnitude) of the geostrophic wind at the initial time t^n.

The boundary condition for humidity can be written as

$$\Delta T_d^{n+1} = 1.26\,\Delta T^{n+1} \quad \text{for } \eta = 1,\ n = 0,\ 1,\ 2\ldots \tag{5.2.15}$$

Relation (5.2.15) was obtained by analyzing experimental material pertaining to nonperiodic (advective) processes. The correlation coefficient between ΔT_d and ΔT for a 12 h period ranges from 0.80 to 0.86.

For the numerical solution of eqns. (5.2.7) and (5.2.8), with the conditions (5.2.9)-(5.2.15), the splitting method (or method of fractional intervals) was used, with the subsequent aid of a computer.

After the functions Π and s are calculated using the described numerical method for time t^n, their values at t^{n+1}, etc., may be found. The desired prognostic values of the temperature and humidity at each grid point, as well as the cloudiness field at the final moment t^N, are then found from the calculated functions Π_{ijk}^N and s_{ijk}^N with the aid of the relations (5.2.3)-(5.2.5). Next, if it is assumed that the air at a given point in space is saturated (that is, cloudiness is present), the system of equations (5.2.4)-(5.2.5) may be solved using successive approximations to find q_m. A comparison of q_m with the known moisture content s establishes the presence or absence of cloudiness at a given point: if $s \geqslant f q_m$, then the assumption that clouds are present is valid and the air temperature has been found correctly; if $s < f q_m$, there is no cloudiness at that point ($\delta = 0$), and the actual specific humidity is, according to (5.2.3), equal to s, while the air temperature is found directly from (5.2.4). Here, f is a coefficient that is, physically equivalent to the relative humidity of the cloud air (f is calculated with respect to water). The coefficient f is equal to unity for positive temperatures ($T \geqslant 273$ K) and a little less than unity for negative temperatures ($T < 273$ K) because of the presence of ice crystals in the cloud (according to (5.2.5), q_m is calculated with respect to water).

An analysis of experimental data gives f as a function of $T < 273$ K:

$$f = 0.008T - 1.184. \tag{5.2.16}$$

The foregoing numerical method for predicting cloudiness was implemented on a computer. A rectangular grid was used, with a total of 18 x 14 points; the y axis was directed along the Moscow meridian, and the intervals were $\Delta x = \Delta y = 300$ km. The time interval Δt was taken to be 1 h. Calculations were carried out for five levels (1000, 850, 700, 500, and 300 mb). To obtain one 24-h prediction (together with a prediction of the geopotential and the velocity components), 35 min of machine time are needed on an M-220 computer.

Figures 5.1 through 5.4 show typical 24-h cloudiness forecasts for certain isobaric surfaces. An analysis of the figures shows that the forecast cloudiness fields are in satisfactory agreement with actually observed fields. Table 5.2 gives some quantitative data on the predictability of the cloudiness for eight 24-h forecasts at the 850, 700, and 500 mb levels. The notation not explained in the legend for this table is defined in Table 5.3.

To evaluate the predictability, the actual presence of cloudiness at the grid points was determined for the corresponding isobaric surfaces according to the dew-point deficit $\Delta = T - T_d$, and for the 850 mb surface it was also found from cloud data on charts of the weather near the ground. Here, the above-mentioned temperature dependence of the humidity in clouds for $T < 273$ K was taken into account. A forecast was assumed to be valid if for $\Delta \leqslant 24.57 - 0.09T$ or for $\Delta > 24.57 - 0.09T$, forecasts of "cloudy" or "clear", respectively, were obtained. The total predictability of the cloudiness is about 80% according to Table 5.2. The high values (more than 0.6, on the average) of criteria H and Q indicate that the given method is suitable for practical applications.

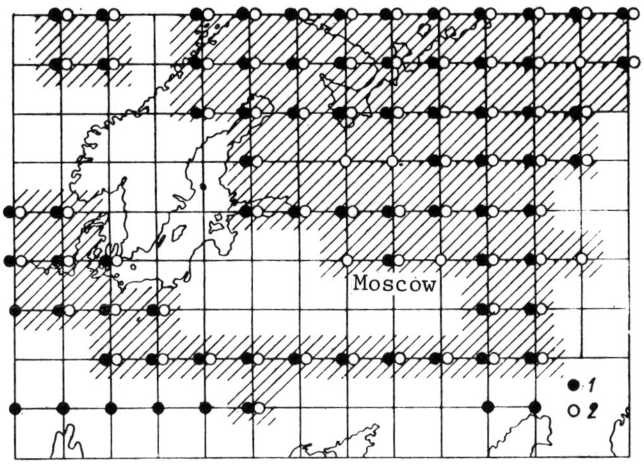

Fig. 5.1. Cloudiness field according to forecast (1) and observational data (2) for 850 mb level at 0300 h on 11 February 1960. Reference information is for 0300 h on 10 February 1960

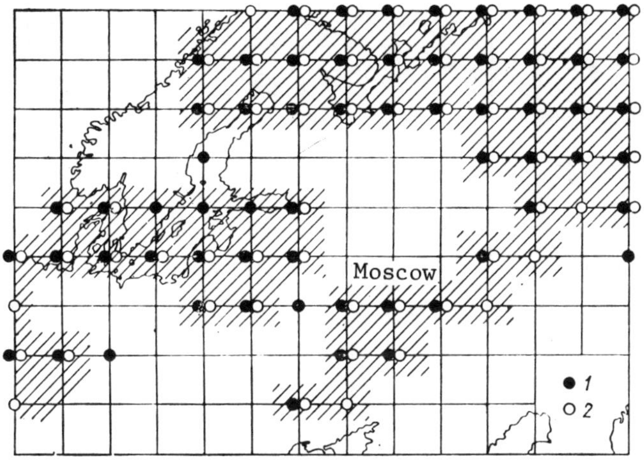

Fig. 5.2. Cloudiness field according to forecast (1) and observational data (2) for 850 mb level at 0300 h on 12 February 1960. Reference information is for 0300 h on 11 February 1960.

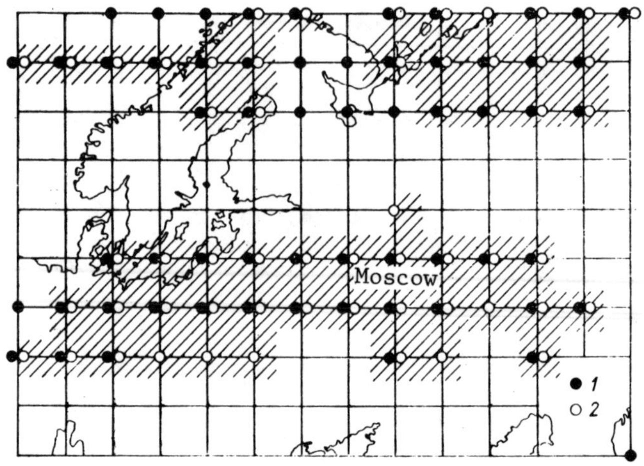

Fig. 5.3. Cloudiness field according to forecast (1) and observational data (2) for 850 mb level at 0300 h on 14 February 1960. Reference information is for 0300 h on 13 February 1960.

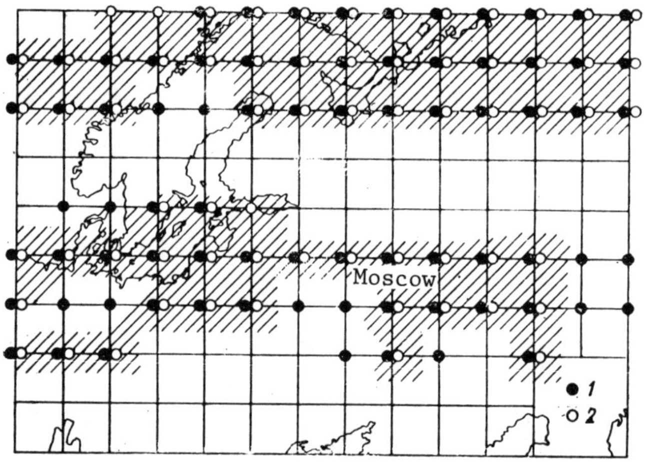

Fig. 5.4. Cloudiness field according to forecast (1) and observational data (2) for 700 mb level at 0300 h on 14 February 1960. Reference information is for 0300 h on 13 February 1960.

Table 5.2. Predictability of cloudiness

Date	Cloudy			Clear		
	N_{02}	N_{22}	U_{22}	N_{01}	N_{11}	U_{22}
10 I 1959	186	148	0.795	234	176	0.753
11 I 1959	165	151	0.916	255	189	0.742
14 I 1960	140	119	0.850	280	187	0.668
15 I 1960	186	153	0.823	234	178	0.761
10 II 1960	176	155	0.880	244	189	0.775
11 II 1960	209	197	0.942	211	162	0.768
12 II 1960	199	176	0.885	221	166	0.752
13 II 1960	199	184	0.925	221	179	0.810
Average	1460	1283	0.880	1900	1426	0.751

Date	U	U_0	ρ	H	Q
10 I 1959	0.772	0.497	0.549	0.547	0.548
11 I 1959	0.810	0.495	0.620	0.623	0.656
14 I 1960	0.730	0.498	0.454	0.462	0.498
15 I 1960	0.788	0.500	0.572	0.572	0.584
10 II 1960	0.820	0.500	0.640	0.640	0.656
11 II 1960	0.855	0.502	0.712	0.710	0.722
12 II 1960	0.814	0.497	0.630	0.630	0.635
13 II 1960	0.864	0.498	0.728	0.728	0.735
Average	0.806	0.500	0.613	0.612	0.630

Note: Here, $U = (N_{11} + N_{22})/N$ is the total predictability, $U_{11} = N_{11}/N_{01}$ is the predictability of "clear" weather, $U_{22} = N_{22}/N_{02}$ is the predictability of "cloudy" weather, $U_0 = (N_{01}N_{10} + N_{02}N_{20})/N$ is the predictability of random forecasts, $\rho = [N_{11} + N_{22}) - (N_{12} + N_{21})]/N$ is the so-called qualitative correlation coefficient, $H = (U - U_0)/(1 - U_0)$ is Bagrov's criterion for reliability of the method, and $Q = 1 - (N_{12}/N_{02}) - (N_{21}/N_{01})$ is Obukhov's criterion for the reliability of alternative predictions.

Table 5.3. Meaning of symbols

Predicted	Actually observed		Total
	Clear	Cloudy	
Clear	N_{11}	N_{12}	N_{10}
Cloudy	N_{21}	N_{22}	N_{20}
Bcero	N_{01}	N_{02}	N

5.3. A ten-level model for predicting frontal precipitation

One of the most comprehensive models for predicting precipitation was worked out in Great Britain [335, 337, 341]. The unknown functions in this case are the geopotential height Φ_p of the isobaric surface p, the horizontal components u, v of the velocity, and the mixing ratio r. An (x, y, p) coordinate system and a stereographic projection are used.

The initial equations are the following:

(a) the equations of motion

$$\frac{\partial u}{\partial t} + m\left(u\frac{\partial u}{\partial x} + v\frac{\partial u}{\partial y} + g\frac{\partial \Phi_p}{\partial x}\right) + \tau\frac{\partial u}{\partial p} - 2\omega_z v = \tag{5.3.1}$$

$$= -g\frac{\partial \tau_x}{\partial p} + k_s\left(\frac{\partial^2 u}{\partial x^2} + \frac{\partial^2 u}{\partial y^2}\right),$$

$$\frac{\partial v}{\partial t} + m\left(u\frac{\partial v}{\partial x} + v\frac{\partial v}{\partial y} + g\frac{\partial \Phi_p}{\partial y}\right) + \tau\frac{\partial v}{\partial p} + 2\omega_z u =$$

$$= -g\frac{\partial \tau_y}{\partial p} + k_s\left(\frac{\partial^2 v}{\partial x^2} + \frac{\partial^2 v}{\partial y^2}\right); \tag{5.3.2}$$

(b) the continuity equation

$$m\left(\frac{\partial u}{\partial x} + \frac{\partial v}{\partial y}\right) + \frac{\partial \tau}{\partial p} = 0; \tag{5.3.3}$$

(c) the heat-influx equation

$$\frac{\partial \Phi^k_{k+1}}{\partial t} + m\left(u\frac{\partial \Phi^k_{k+1}}{\partial x} + v\frac{\partial \Phi^k_{k+1}}{\partial y}\right) + \tau\frac{\partial \Phi^k_{k+1}}{\partial p} - \frac{R}{c_p}\tau\Phi^k_{k+1} =$$

$$= \frac{R}{c_p}\frac{\ln\frac{p_{k+1}}{p_k}}{g}\varepsilon + k_s\left(\frac{\partial^2 \Phi^k_{k+1}}{\partial x^2} + \frac{\partial^2 \Phi^k_{k+1}}{\partial y^2}\right); \tag{5.3.4}$$

(d) the moisture-influx equation

$$\frac{\partial r}{\partial t} + m\left(u\frac{\partial r}{\partial x} + v\frac{\partial r}{\partial y}\right) + \tau\frac{\partial r}{\partial p} = \mu + k_s\left(\frac{\partial^2 r}{\partial x^2} + \frac{\partial^2 r}{\partial y^2}\right). \tag{5.3.5}$$

In these equations, $\tau = dp/dt$ is the analog of the vertical velocity in the (x, y, p) coordinate system; $\tau_x = k\rho(\partial u/\partial z)$ and $\tau_y = k\rho(\partial v/\partial z)$ are the components of the eddy stress created by the vertical mixing; k is the turbulence coefficient (along the vertical); k_s is the coefficient of horizontal turbulent mixing; Φ^k_{k+1} is the geopotential thickness of the atmospheric layer between the isobaric surface p_{k+1} and p_k; ε and μ are the specific influxes of heat and water vapor, respectively, due to condensation or evaporation, the arrival of heat and vapor from the underlying surface, and convection on a subgrid scale; and $m = \sec^2(\pi/H - \varphi/2)$ is the image scale on the stereographic polar projection.

The reader is referred to general works on numerical methods of weather forecasting (Kibel' [119] and Belov [18]), which discuss the transformation of the system of equations upon transition from Cartesian coordinates to isobaric coordinates (as well as cartographic projections and the image scale). The derivation of eqn. (5.3.4) will now be illustrated. The familiar relation

$$\Phi_{k+1}^{k} = \frac{R}{g} \overline{T} \ln \frac{p_{k+1}}{p_k}$$

is used to derive (5.3.4); it relates the mean temperature $\overline{T}$ and the thickness Φ_{k+1}^{k} of the atmospheric layer between surfaces p_{k+1} and p_k. Next, if the heat-influx equation for $\overline{T}$ is written with $\overline{T}$ replaced by Φ_{k+1}^{k} according to the latter relation for $\overline{T}$, eqn. (5.3.4) is obtained.

The fundamental system of equations (5.3.1)-(5.3.5) is supplemented by the relation

$$\frac{\partial \Phi_{1000}}{\partial t} + m\left(u \frac{\partial \Phi_{1000}}{\partial x} + v \frac{\partial \Phi_{1000}}{\partial y}\right) + \tau \frac{\partial \Phi_{1000}}{\partial p} = \\ = m\left(u \frac{\partial \zeta}{\partial x} + v \frac{\partial \zeta}{\partial y}\right) + k_s\left(\frac{\partial^2 \Phi_{1000}}{\partial x^2} + \frac{\partial^2 \Phi_{1000}}{\partial y^2}\right), \tag{5.3.6}$$

reflecting the fact that at the 1000 mb level the vertical velocity equals the vertical component of the velocity produced by the topography of the locality (here $\zeta = \zeta(x, y)$ is the elevation of the place above sea level).

For the numerical implementation of the model, the horizontal interval of the grid was taken to be about 50 km, the total number of grid points being 95 x 63 (the center of the grid was close to 50° N, 10° W). A ten-level model was used, covering the layer of the atmosphere from 1000 to 100 mb. Fucntions Φ_p, u, and v were calculated at the isobaric surfaces 1000, 900, ..., 200, 100 mb, while functions Φ_{k+1}^{k} and r were calculated at levels between these surfaces (950, 850, ..., 150 mb). The air humidity was determined up to the 300-mb level, above which the atmosphere was assumed to be dry.

Beginning with Smagorinsky and his collaborators [389, 421], the condition of so-called adaptation was employed to work out models for predicting precipitation, temperature, and humidity. The adaptation principle reduces to the following. Because of turbulent mixing, and especially convection, the lapse rates cannot be higher than some critical value γ_{cr}. In a number of works, γ_{cr} was assumed to be equal to the dry-adiabatic lapse rate γ_a in unsaturated air and the moist-adiabatic lapse rate γ_a' in saturated air (clouds). The authors of [337] propose some improvements: they assume that $\gamma_{cr} = \gamma_a$ for a relative humidity $f < 0.5$, while for $f > 0.5$ they compute γ_{cr} using the formula

$$\gamma_{cr} = \gamma_a\left(\frac{1-f}{1-f_{cr}}\right) + \gamma_a'\left(\frac{f-f_{cr}}{1-f_{cr}}\right), \tag{5.3.7}$$

where $f_{cr} = 0.5$, f being the relative humidity in fractional units. Therefore, for $f > 0.5$, γ_{cr} varies between γ_a and γ_a', gradually approaching the latter value as $f \to 1$.

If in some layer the calculation gives $\gamma > \gamma_{cr}$, then in this layer γ is always assumed to equal γ_{cr}, since the unstable state existing in the layer gives rise to convection (intense turbulence), bringing ("adapting") γ close to γ_{cr}.

The following relations are used to take the heat of condensation into account and to determine the mass of the condensed water vapor:

$$r + \delta r = r_m(T + \delta T,\ p) \tag{5.3.8}$$

$$c_p\,\delta T + L\,\delta r = 0, \tag{5.3.9}$$

where δr and δT are the increments of the mixing ratio and temperature during time δt.

The first of these relations indicates that the relative humidity in saturated air cannot exceed 100%, while all the excess water vapor (above saturation) goes into the liquid state. The second relation signifies that all the heat released during condensation goes to raise the temperature.

If the saturation mixing ratio $r_m(T + \delta T,\ p)$ is represented as the series

$$r_m(T + \delta T,\ p) = r_m(T,\ p) + \left(\frac{\partial r_m}{\partial T}\right)_p \delta T + \ldots,$$

in which just the linear term in δT is retained, (5.3.8) can be rewritten as

$$\delta r = r_m(1 - f) + \left(\frac{\partial r_m}{\partial T}\right)_p \delta T, \tag{5.3.10}$$

where $f = r/r_m$ is the relative humidity.

Equations (5.3.9) and (5.3.10) yield

$$\delta r = \frac{r_m(1-f)}{1 + \frac{L}{c_p}\left(\frac{\partial r_m}{\partial T}\right)_p}. \tag{5.3.11}$$

The derivative $(\partial r_m/\partial T)_p$ is determined with the aid of the Clausius-Clapeyron equation

$$\left(\frac{\partial r_m}{\partial T}\right)_p = \frac{L r_m (T, p)}{R_v T^2}, \quad (5.3.12)$$

where T and p are the temperature and pressure at time t, and R_v is the specific gas constant of the water vapor.

Equations (5.3.10)-(5.3.12) for a known f at a time t are used to calculate increments δT and δr, as well as $T + \delta T$ and $r + \delta r$ at time $t + \delta T$. Then the relative humidity

$$f = \frac{r + \delta r}{r_m (T + \delta T, p)}$$

at time $t + \delta t$ is found and the new value of f substituted into relations (5.3.10) and (5.3.11), etc. The iteration process is repeated until f is close to 100%. Sample calculations indicate that only one or two iterations are necessary to arrive at the value of f with an accuracy of about 1%.

Precipitation is defined by the following conditions: in the cloud, water vapor condenses until saturation is reached (f = 100%); the liquid water forming via condensation falls out as precipitation.

At ground level the heat-balance equation is

$$Q_T + LQ_v = R - Q_{soil},$$

where $Q_T = \rho_0 k_T v_0 c_p \Delta T$ and $Q_v = \rho_0 k_T v_0 \Delta r$ are, respectivelt, the turbulent fluxes of heat and water vapor (here $\Delta T = T_0 - T_{950} - g\Phi_{950}/2c_p$, $\Delta r = r_0 - r_{950}$ T_0, ρ_0, v_0, and r_0 are the temperature, air density, wind velocity, and mixing ratio at the 100-mb level, T_{950} and r_{950} are the temperature and mixing ratio at the 950-mb level, and k_T is the coefficient of heat and moisture exchange), R is the radiation balance of the Earth's surface, and Q_{soil} is the heat flux into the soil.

The heat-exchange (moisture-exchange) coefficient is found with the aid of the formula

$$k_T = \begin{cases} (1.25 + 0.1\,\Delta T) \cdot 10^{-3} & \text{for } \Delta T > 0, \\ 0.1\,(1.25 + 0.1\,\Delta T) \cdot 10^{-3} & \text{for } \Delta T < 0. \end{cases}$$

The radiation balance R is found from the expression

$$R = \alpha \frac{1 - A}{1 + B} I_0 (0.6 + 0.2 \sin h_s) \sin h_s - 0.05\beta I_0,$$

where A = 0.25 is the albedo, B = 0.1 is a parameter allowing for the heat flux into the soil, I_0 is the solar constant, h_s is the height of the sun, and α and β are factors that take into account the effect of the cloudiness on the total solar radiation and the effective radiation of the ground:

$$\alpha = (1 - 0.7n_L)(1 - 0.7n_m)(1 - 0.4n_u),$$
$$\beta = 1 - 0.6n_L - 0.3n_m - 0.1n_u.$$

Here, n_l, n_m, and n_u are the amounts of cloudiness in the lower, middle, and upper levels, respectively.

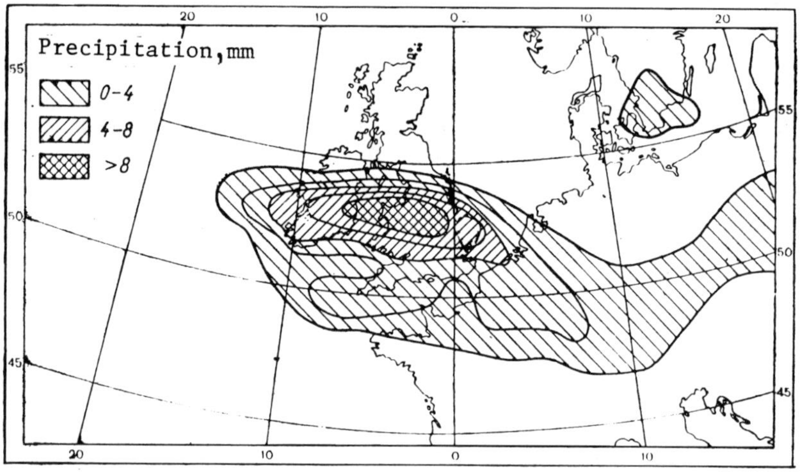

Fig. 5.5. Amounts of precipitation predicted according to 10-level numerical model for a period from 0600 to 1800 h (GMT) on 1 December 1961 (isohyets drawn every 2 mm).

Other relations used to implement the method are omitted. Figure 5.5 presents the calculated precipitation amounts from frontal clouds for the period 0600-1800 h on 1 December 1961.

A comparison of Figs. 5.5 and 5.6 indicates that the region of heaviest precipitation (> 8 mm) is predicted satisfactorily by the calculation method. However, the regions of observed moderate (4-8 mm) and, especially, slight (0-4 mm) precipitation are considerably larger than the calculated regions. Taking the orography of the British Isles into account does not essentially change the calculated results: the region of moderate precipitation becomes somewhat larger, but the region of slight precipitation in the east becomes smaller.

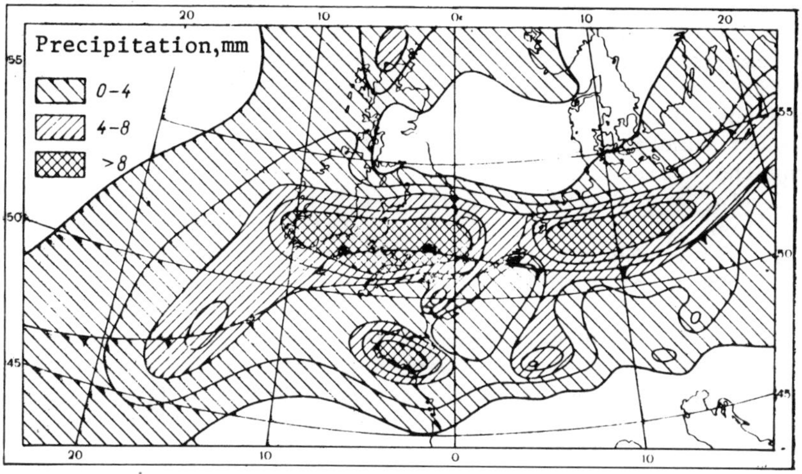

Fig. 5.6. Actual amounts of precipitation falling from frontal clouds during period 0600-1800 h (GMT) on 1 December 1961 (isohyets drawn every 2 mm).

5.4. A numerical model for predicting the formation of stratus in the atmospheric boundary layer

Gerrity [360] worked out a model for predicting stratus cloudiness. He introduced equations of the atmospheric boundary layer, to complement the equations describing the transport of heat and moisture. The boundary layer is divided into the ground layer, in which the fluxes of heat, water vapor, and momentum are constant with height, and the Ekman layer, where the geostrophic wind intensifies linearly with height; the turbulence coefficient is not a function of height. The motion in the boundary layer is steady, but the local derivatives are also retained in the heat--transport and moisture-transport equations. In the ground layer, a number of dependences (on the Rossby and Richardson numbers) established in similarity theory must be taken into account. In particular, the following formulas are used for the angle of deviation α_0 of the wind near the Earth from the isobar and for the geostrophic resistance [drag] coefficient u_*/ug (where u_* is the friction velocity):

$$\alpha_0 = 0.625\,(\lg \mathrm{Ro})^2 - 12.750 \lg \mathrm{Ro} + 80.625,$$
$$u_*/u_g = (7.625 - 0.625 \lg \mathrm{Ro}) \cdot 10^{-2},$$

where $R_0 = ug/2\omega_z z_0$ is the Rossby number.

The model of the atmospheric boundary layer, in conjunction

with a scheme for the short-term prediction of the pressure field in the free atmosphere, forms a basis for the method of predicting the temperature, humidity, and cloudiness (as well as the wind velocity) in the boundary layer that has been applied to weather surveying in aviation since 1969.

For numerical application of the model, the grid interval along the horizontal is taken to be 160 km, and the time interval is 15 min, and the interval along the vertical is variable, from 50 m close to the ground to 450 m in the upper part of the boundary layer (giving a total of 12 levels in the boundary layer). The topography is allowed for in the model.

Although the numerous other relations involved will not be reproduced here, some results of the cloudiness predictions will be presented. These results were obtained for the eastern USA at 1200 h. The reference information pertained to 1200 h on 6 February 1964 (example A), 0000 h on 23 February 1965 (example B), and 1200 h on 24 January 1965 (example C).

A comparison of predicted and actually observed sky conditions in example A showed the following: at 16 out of 22 points (grid nodes) where precipitation or cloudiness of 8–10 had been forecast, the predicted weather was indeed observed. At 3 points cloudiness of 3–7 was observed, and at 3 point cloudiness of 0–2. At 3 out of 11 points where cloudiness of 3–7 had been forecast, cloudiness of 8–10 (or precipitation) was observed, while at 4 points the cloudiness ranged from 3 to 7, and at 3 points from 0 to 2. At one point where cloudiness of 0–2 was forecast, cloudiness of 3–7 was observed.

Thus, the total predictability of cloudiness in example A is $(16 + 4 + 0)/34 = 59\%$. In examples B and C, the predictabilities (for the same three graduations of cloudiness) are, respectively, $(14 + 2 + 6)/33 = 67\%$ and $(17 + 2 + 0)/36 = 53\%$. Here, the numerators give the number of points at which cloudiness of 8–10, 3–7, or 0–2 was both predicted and observed, while the denominators give the total number of points for which cloudiness forecasts were made.

The amount of cloudiness was determined using a relation between the cloudiness and the relative humidity f established empirically by Smagorinsky [421]: for clouds of the lower level, f values from 62 to 77% correspond to a cloudiness of less than 3; f values from 77 to 87% correspond to cloudiness of 3–7; $f > 87\%$ corresponds to cloudiness of 8–10 (with precipitation if $f > 90\%$.

5.5. A numerical model for predicting cloudiness and precipitation

Dymnikov and Guseva [101] formulated some general requirements to be met by models for predicting fields of humidity, cloudiness, and precipitation:

(1) The model should describe all the indicated fields simultaneously.

(2) The system of equations should, as far as possible, be

put into such a form as to eliminate small differences of large quantities (this requirement applies even more to the initial and boundary conditions of the problem).

(3) Phase influxes should be taken into account in the model (especially at the beginning of condensation).

(4) The techniques for numerical application of the model should not be too complicated (in particular, it is desirable not to impose the condition of monotony, that is, the condition that the sign of the unknown quantity remain unchanged during the entire prediction period).

The authors of [101] also mention that the requirements to be met by objective analyses of the initial data, by the accuracy of calculating different kinds of sources, by the methods used to solve the transport equations, etc., are in general different for models used to predict fields of humidity, cloudiness, and precipitation.

The foregoing requirements having been met, Dymnikov and Gusev so transformed the initial system of equations

$$\frac{dq}{dt}=\frac{1}{\rho}\frac{\partial}{\partial z}\rho k\frac{\partial q}{\partial z}-\nu, \tag{5.5.1}$$

$$\frac{d\delta}{dt}=\frac{1}{\rho}\frac{\partial}{\partial z}\rho k\frac{\partial \delta}{\partial z}+\nu, \tag{5.5.2}$$

$$\frac{dT}{dt}=\frac{1}{\rho}\frac{\partial}{\partial z}\rho k\left(\frac{\partial T}{\partial z}+\gamma_a\right)-\gamma_a w+\frac{L}{c_p}\nu \tag{5.5.3}$$

that the fundamental prediction equation contained the quantity

$$\Phi=q-q_m+\left(1+\frac{L}{c_p}\frac{\partial q_m}{\partial T}\right)\delta. \tag{5.5.4}$$

Since in clouds, q is close to q_m ($q \approx q_m$) and $\delta > 0$, it follows that $\Phi > 0$ there; outside of clouds, $\delta = 0$ and $q < q_m$. so that $\Phi < 0$, that is, the function Φ changes sign, when solving the equation containing Φ, the need to use monotonic numerical methods is no longer stipulated.

The equation for Φ will now be derived. If the specific-humidity deficit $\Delta = q - q_m$ is introduced,

$$\frac{d\Delta}{dt}-\frac{1}{\rho}\frac{\partial}{\partial z}\rho k\frac{\partial \Delta}{\partial z}=\frac{dq}{dt}-\frac{dq_m}{dt}-$$
$$-\frac{1}{\rho}\frac{\partial}{\partial z}\rho k\frac{\partial q}{\partial z}+\frac{1}{\rho}\frac{\partial}{\partial z}\rho k\frac{\partial q_m}{\partial z} \tag{5.5.5}$$

can be written. Since q_m is a function just of T and p, it follows that

$$\frac{dq_m}{dt}=\frac{\partial q_m}{\partial T}\frac{dT}{dt}+\frac{\partial q_m}{\partial p}\frac{dp}{dt}.$$

If this relation and equations (5.5.1) and (5.5.3) are taken into account, expression (5.5.5) can be rewritten as

$$\frac{d\Delta}{dt}-\frac{1}{\rho}\frac{\partial}{\partial z}\rho k\frac{\partial\Delta}{\partial z}=-\nu-\frac{\partial q_m}{\partial T}\frac{dT}{dt}-$$

$$\frac{\partial q_m}{\partial p}\frac{dp}{dt}+\frac{1}{\rho}\frac{\partial}{\partial z}\rho k\frac{\partial q_m}{\partial z}=-\nu-\beta\left(\frac{1}{\rho}\frac{\partial}{\partial z}\rho k\frac{\partial T}{\partial z}+\frac{L}{c_p}\nu-\right.$$

$$\left.-\gamma_a w+\frac{\gamma_a}{\rho}\frac{\partial\rho k}{\partial z}\right)-\frac{\partial q_m}{\partial p}\frac{dp}{dt}+\frac{1}{\rho}\frac{\partial}{\partial z}\rho k\frac{\partial q_m}{\partial z},$$

where $\beta = \partial q_m/\partial T$.

After some simple transformations are made and some negligible terms dropped, the latter equation becomes

$$\frac{d\Delta}{dt}=\frac{1}{\rho}\frac{\partial}{\partial z}\rho k\frac{\partial\Delta}{\partial z}-\left(1+\beta\frac{L}{c_p}\right)\nu+$$

$$+\left(1+\beta\frac{L}{c_p}\right)\left[\frac{c_p}{L}\left(\gamma_a-\gamma_a'\right)w-\frac{c_p}{L\rho}\frac{\partial}{\partial z}\rho k\left(\gamma_a-\gamma_a'\right)\right]. \quad (5.5.6)$$

In a cloud, the specific-humidity deficit $\Delta = 0$. By setting terms in (5.5.6) containing Δ equal to zero, an expression for the rate of water-vapor condensation in a cloud is obtained:

$$\nu=\frac{c_p}{L}\left[\left(\gamma_a-\gamma_a'\right)w-\frac{1}{\rho}\frac{\partial}{\partial z}\rho k\left(\gamma_a-\gamma_a'\right)\right]. \quad (5.5.7)$$

Multiplying eqn. (5.5.2) by $(1 + \beta L/c_p)$ and ignoring the dependence of β on the coordinates and time yields

$$\frac{d}{dt}\left(1+\beta\frac{L}{c_p}\right)\delta=\frac{1}{\rho}\frac{\partial}{\partial z}\rho k\frac{\partial}{\partial z}\left(1+\beta\frac{L}{c_p}\right)\delta+\left(1+\beta\frac{L}{c_p}\right)\nu. \quad (5.5.8)$$

Adding the left- and right-hand sides of eqns. (5.5.6) and (5.5.8) yields the desired (approximate) equation for function Φ:

$$\frac{d\Phi}{dt}=\frac{1}{\rho}\frac{\partial}{\partial z}\rho k\frac{\partial\Phi}{\partial z}+\left(1+\beta\frac{L}{c_p}\right)\left[\frac{c_p}{L}\left(\gamma_a-\gamma_a'\right)\left(w-\frac{1}{\rho}\frac{\partial\rho k}{\partial z}\right)\right], \quad (5.5.9)$$

where the dependence of γ_a on height is neglected, which is equivalent to the previous assumption of constancy of parameter β.

In coordinates (x, y, p), eqn. (5.5.9) becomes

$$\frac{\partial\Phi}{\partial t}+u\frac{\partial\Phi}{\partial x}+v\frac{\partial\Phi}{\partial y}+\tau\frac{\partial\Phi}{\partial p}=\frac{\partial}{\partial p}\frac{g^2p^2}{R^2\overline{T}^2}k\frac{\partial\Phi}{\partial p}-$$
$$-\frac{c_p}{L}\left(1+\beta\frac{L}{c_p}\right)\left[(\gamma_a-\gamma_a')\frac{R\overline{T}}{gp}\tau+\frac{\partial}{\partial p}\frac{gp}{R\overline{T}}k(\gamma_a-\gamma_a')\right]-d. \tag{5.5.10}$$

Here, $\tau = dp/dt$, $\overline{T}$ is some mean value of the temperature, and d is a term describing the fallout of precipitation (the cited paper does not give more details about this term).

In (x, y, p) coordinates, the equation describing the heat influx becomes

$$\frac{\partial T'}{\partial t}+u\frac{\partial T'}{\partial x}+v\frac{\partial T'}{\partial y}+\tau\frac{\partial T'}{\partial p}=$$
$$=\frac{\gamma_a-\gamma}{gp}R\overline{T}\tau+\frac{\partial}{\partial p}\frac{g^2p^2}{R^2\overline{T}^2}k\frac{\partial\theta}{\partial p}+\frac{L\nu}{c_p}, \tag{5.5.11}$$

where $T' = T - \overline{T}$ and $\gamma = -\partial\overline{T}/\partial z$.

The system (5.5.10) and (5.5.11) is supplemented by the continuity equation

$$\frac{\partial u}{\partial x}+\frac{\partial v}{\partial y}+\frac{\partial\tau}{\partial p}=0 \tag{5.5.12}$$

and the equations of motion.

Taking (5.5.12) into account, eqn. (5.5.10) can be rewritten in "semidivergent" form:

$$\frac{\partial\Phi}{\partial t}+\frac{1}{2}u\frac{\partial\Phi}{\partial x}+\frac{1}{2}\frac{\partial\Phi u}{\partial x}+\frac{1}{2}v\frac{\partial\Phi}{\partial y}+\frac{1}{2}\frac{\partial\Phi v}{\partial y}+\frac{1}{2}\tau\frac{\partial\Phi}{\partial p}+$$
$$+\frac{1}{2}\frac{\partial\Phi\tau}{\partial p}=\frac{\partial}{\partial p}\frac{g^2p^2}{R^2\overline{T}^2}k\frac{\partial\Phi}{\partial p}-\frac{c_p}{L}\left(1+\beta\frac{L}{c_p}\right)\times$$
$$\times\left[(\gamma_a-\gamma_a')\frac{R\overline{T}}{gp}\tau+\frac{\partial}{\partial p}\frac{gp}{R\overline{T}}k(\gamma_a-\gamma_a')\right]-d. \tag{5.5.13}$$

At the lateral boundaries of the territory for which the prediction is being made, the condition that functions Φ and T are constant in time is imposed:

$$\frac{\partial\Phi}{\partial t}=\frac{\partial T}{\partial t}=0.$$

At the upper limit of the atmosphere, the derivatives of Φ and T with respect to height go to zero:

$$\frac{\partial\Phi}{\partial p}=\frac{\partial T}{\partial p}=0.$$

At the lower boundary of the atmosphere, the turbulent fluxes of heat and moisture are found with the aid of relations established in boundary-layer physics.

Equation (5.5.13) is solved using the method of fractional intervals (splitting). To predict the fields of u, v, τ, and T, a prediction model using the general equations of hydrothermodynamics for a limited territory centered on Novosibirsk is employed . The grid includes 22 x 36 points with intervals $\Delta x = \Delta y$ = 300 km and five standard levels (1000, 850, 700, 500, and 300 mb), the values of Φ, T, and τ being determined at intermediate levels (925, 775, 600, and 400 mb).

Now some data on the 24-h predictability will be presented (Tables 5.4 and 5.5). Inspection of the tables shows that 60-80% of the time, the error in predicting the dew-point deficit does not exceed 3°C. An evaluation of the precipitation fields according to data from 32 points showed that 63% of the time, the predicted zones of precipitation agreed with the actual ones (in some instances, the predictability is as high as 80%).

Table 5.4. Frequency (%) of errors in predicting dew-point deficit according to seven points

Level, mb	Prediction error, °C			No. of cases
	0–3	4–6	>7	
850	82	14	4	66
700	61	22	17	67
500	60	12	28	67

Dymnikov and Ishimova [102] noted that the accuracy of the calculating methods is very important when predicting humidity, cloudiness, and precipitation. Since the fields of air humidity and cloud moisture content are highly variable in time and space, the use of methods with second-order accuracy relative to the spatial coordinates to solve the moisture-transport equations introduces sizable errors. In [102] a calculating method of fourth-order accuracy was employed, with a smoothing operator also of fourth order. The variation with latitude of the scale of the stereographic projection was taken into account, as well as the interrelated effect of the boundary layer and the orography on vertical motions of the atmosphere (this interesting problem was formulated in [76]).

Two approaches are used to describe the moisture transport

and the evolution of the humidity field. In model A, the previously introduced function Φ (see (5.5.4)) is used. It satisfies the equation

$$\frac{\partial \Phi}{\partial t} + m^2 u \frac{\partial \Phi}{\partial x} + m^2 v \frac{\partial \Phi}{\partial y} + \tau \frac{\partial \Phi}{\partial p} = \qquad (5.5.14)$$

$$= -\left(1 + \frac{L}{c_p}\frac{\partial q_m}{\partial T}\right)\left[\frac{c_p}{L}\left(\gamma_a - \gamma_a'\right)\frac{RT}{gp}\tau - \mu\right].$$

Table 5.5. Frequency (%) of errors in predicting dew-point deficit at Novosibirsk

Prediction method	Level, mb	Prediction error, °C			No. of cases
		0–3	4–6	>7	
Numerical	850	80	20	0	10
	700	50	30	20	10
Synoptic	850	75	0	25	8
	700	44	33	23	9

In model B, the transport equation is expressed in terms of the specific humidity q:

$$\frac{\partial q}{\partial t} + m^2 u \frac{\partial q}{\partial x} + m^2 v \frac{\partial q}{\partial y} + \tau \frac{\partial q}{\partial p} = -\nu. \qquad (5.5.15)$$

Here, $\tau = dp/dt$ is the analog of the vertical velocity in the isobaric coordinate system (x, y, p), μ is a function describing the rate of fall of precipitation, ν is the rate of water-vapor condensation, $m = (1 + \sin 60°)/(1 + \sin \varphi)$ is the scale factor of the stereographic projection, and φ is the geographic latitude.

The moisture-transport equations are supplemented by the heat-influx equation, in which the phase influx of heat is taken into account; the equations of motion and continuity; and the hydrostatic equation.

The influence of turbulent mixing is allowed for only in the boundary condition at the lower limit of the atmosphere: here, the vertical velocity is not set equal to zero, but rather is assumed to be $w = w_{fr} + w_{or}$.

The following expression is used to determine the component w_{fr} due to friction:

$$w_{fr} = \frac{1}{2\omega_z \rho_0}\left(\frac{\partial \tau_y}{\partial x} - \frac{\partial \tau_x}{\partial y}\right),$$

where $\tau_x = \rho_0 u_*^2 \cos(\alpha_g + \alpha_0)$, $\tau_y = \rho_0 u_*^2 \sin(\alpha_g + \alpha_0)$ are the x and y components of the eddy frictional stress; α_0 is the angle of deviation from geostrophic of the wind close to the ground; α_g is the angle the geostrophic wind makes with the x axis; and u_* is the dynamic velocity (u_* and α_0 are calculated using relations established in boundary-layer physics with the aid of dimensional analysis and similarity theory).

Component w_{or} due to the orography is in turn represented as a sum:

$$w_{\mathrm{or}} = w'_{\mathrm{or}} + w''_{\mathrm{or}},$$

where

$$w'_{\mathrm{or}} = m^2 u \frac{\partial \zeta}{\partial x} + m^2 v \frac{\partial \zeta}{\partial y}.$$

This is the generally accepted expression for the vertical velocity, arising because air cannot penetrate the Earth's surface. The second component (w''_{or}) exists because of the non-linear interaction between the boundary layer and the orographically irregular surface. For this component, only the first (rough) approximation [76] is used:

$$w''_{\mathrm{or}} = a c_g^2 \Delta \zeta,$$

where $a \approx 10^4$ s, c_g is the magnitude of the geostrophic wind, and $\partial^2\zeta/\partial x^2 + \partial^2\zeta/\partial y^2$ is the Laplacian of the elevation ζ of the Earth's surface. At the lateral boundaries of the region, all the quantities involved are assumed to be invariable in time (that is, $\partial q/\partial t = \partial T/\partial t = \partial \Phi/\partial t = \partial z/\partial t = 0$, where z is the height of the isobaric surface). At the upper boundary (that is, at the 200-mb level), the condition $\partial T/\partial p = 0$ (isothermy) is assumed.

The condensation rate ν in model B is determined from the condition that saturation is reached in the cloud ($f = q/q_m = 1$), the increments of temperature and specific humidity being related by the expression

$$T^{n+1} - T^{n+1/2} = \frac{L}{c_p}\left(q^{n+1} - q^{n+1/2}\right),$$

where $q^{n+1} = q_m(T^{n+1}, p)$.

In model A, the following situation is assumed for the prediction of precipitation: when a certain threshold value of the water content δ_{thr} is reached, the moisture that has accumulated in the layer falls into the underlying layer.

Numerical experiments were carried out using the five-level model, on a rectangular 33 x 33 grid with intervals of 375 km; the grid center was at Novosibirsk. The reference data in this case were values of the geopotential and dew-point deficit obtained by objective analysis using the method of the Soviet Gidromettsentr. The values of u, v, and z were determined at the 1000-, 850-, 700-, 500-, and 300-mb levels, while the values of T, τ, Φ, and q pertained to intermediate levels at 925, 775, 600, and 400 mb. The cloud water content at the initial moment was assumed to be zero (because of the lack of systematic measurements of the water content).

Table 5.6 illustrates the quality of a prediction of the dew-point deficit on the basis of three synoptic situations. The frequency of the prediction errors is seen to be practically the same for both models. At the lowest level (925 mb), model A provides a somewhat better predictability of the deficit (prediction errors of less than 3°C are more frequent than for Model B). At the other levels, however, the predictability of deficits to 3°C or less does not differ for the two models. Moreover, using model A, errors greater than 7°C are more frequent at all levels.

Table 5.6. Frequency (%) of errors in predicting the dew-point deficit 24 h ahead, according to model A (numerator) and model B (denominator).

Level, mb	Prediction error, °C		
	0—3	4—6	>7
925	82/73	14/25	4/2
775	79/76	20/23	1/1
600	67/67	29/32	4/1
400	65/68	29/29	6/3

A comparison of the regions of heavy precipitation predicted 24 h ahead, according to reference data for 0300 h on 22 January 1978, with actually observed regions indicates that the two models give approximately the same results (the predicted and observed precipitation regions coincided 55-60% of the time).

Data in [102] show that the use of a finite-difference method of fourth-order accuracy improves the quality of the prediction: the frequency of low errors in predicting the dew-point deficit (less than 3°C) at the upper levels (775 mb and above) is 8-12% higher than with the method having second-order accuracy. Taking the orography into account improves the predictability of the geopotential at lower levels by 5-10%: the relative error in predicting the geopotential at the 100 and 850 mb surfaces 24 h ahead is 0.80 allowing for orography, and 0.88-0.91 neglecting it; the corresponding correlation coefficients are, respectively, 0.64--0.69 and 0.55-0.61.

The function Φ introduced by Dymnikov is intended, in his opinion, to eliminate prediction errors caused by "a difference between large quantities" (a reference to the method of calculating the cloud water content based on the concept of the moisture content) [101]. However, it is quite evident that the introduction of function Φ does not remove this difficulty. In fact, function Φ includes a difference between two large quantities (q and q_m). Consequently, the initial field of Φ may well contain sizable errors, which will naturally affect the determination of the second term of this function (the cloud water content). The predictions made using models A and B, discussed briefly above, verify this conclusion: model A, which is based on the equation for function Φ, does not appear to offer any advantages over model B, whose initial equation is the equation for the specific humidity.

Several sources of prediction error can be noted. The vertical currents (the main cloud-producing factor) are, in the paper discussed above, determined from the equation of heat influx. A comparison of the vertical velocities calculated with the aid of the heat-influx equation and the continuity equation indicated that the regions of ascending and descending motions agree quite well with one another; however, there are significant differences between the absolute values of the vertical velocities. In [102] the effect of turbulent influxes of heat and moisture is not taken into account, even though, as earlier studies had indicated, these influxes can play an important part not only in the boundary layer but throughout the entire troposphere. A certain amount of error is also introduced by not taking into account the cloud water content at the initial moment.

On the whole, however, Dymnikov and his co-workers have contributed substantially to the development of methods for predicting fields of temperature, cloudiness, humidity, and precipitation. Further such development, along with an increase in the accuracy of air temperature and humidity measurements and greater systematization and precision when measuring the cloud water content and precipitation intensity, will make predictions of the fields of the major meteorological quantities more reliable.

5.6. A synoptic-hydrodynamic method for predicting precipitation

Scientists at the Soviet Gidromettsentr (B. D. Uspenskii, A. N. Mertsalov, I. A. Petrichnko, and others) worked out a five-level synoptic-hydrodynamic method of predicting steady and showery precipitation 12 and 24 h in advance [246, 247, 305-307]. This method has the following features:

(a) The pressure field is forecast simultaneously according to a three-level quasi-geostrophic method and a synoptic-hydrodynamic method (the relative error of a prediction of the geopotential does not exceed 0.5-0.7).

(b) A comparatively small horizontal grid interval (150 km) and a small vertical interval (10 mb) are employed, so as to discern more effectively the features of the reference fields of the air temperature and humidity, and also to portray more accurately the vertical profiles of these quantities and of the convection parameters.

(c) The velocity of large-scale vertical motions is predicted, taking into account the form of the baric field (cyclone, anticyclone), and the actual wind is used to construct the prognostic trajectories of air parcels (at the 925-, 850-, 700-, 500-, and 300-mb levels).

The air temperature and humidity are forecast at the ground and at 850, 700, 500, and 300 mb. An objective analysis of the temperature and humidity fields (as well as of the fields of pressure and geopotential height) is carried out with the aid of optimal interpolation.

The final result of the application of this method using a computer is a prediction of the amount and duration of steady and showery precipitation 12 and 24 h in advance.

This method for forecasting steady precipitation is based on Dyubyuk's formula [103] for the amount of precipitation:

$$Q_{\text{pr}} = \frac{1}{g} \int_0^t \int_{p_{\text{d}}}^0 \frac{dq_m}{dt} \, dp \, dt, \tag{5.6.1}$$

where q_m is the saturation specific humidity, and p_c is the pressure at the condensation level.

The working formula for the three-level method has the form

$$Q_{\text{pr}} = 1.5\,\Delta q_{850} + 1.8\,\Delta q_{700} + \Delta q_{500}. \tag{5.6.2}$$

where Δq is the individual variation in the specific humidity at the corresponding isobaric surface. The latter is determined as the difference between the values of q_m at the beginning and end points of condensation, with the help of an aerological diagram on which the stratification curves are plotted, taking into account the main factors involved (advection, vertical motions, and transformation).

The duration of the steady precipitation t_{hp} is found with the aid of the empirical relation

$$t_{\text{hp}} = t\left[1 - \frac{12\,(T - T_d)_{850}}{|\tau_{850}|}\right], \tag{5.6.3}$$

where $(T - T_{\text{d}})_{850}$ is the dew-point deficit at the 850-mb level, predicted allowing for advective and transformation factors; $\tau_{850} = dp_{850}/dt$ is an analog of the vertical velocity (at the 850-mb level); and t is the forecasting period (12 or 24 h). From formula (5.6.3) it follows that the precipitation lasts longer if the dew-point deficit is lower or if the absolute value of

τ_{850} is higher. An analogous formula can be written for the 700-mb level (except that the coefficient is 10 instead of 12).

During the above series of studies, a method for predicting the intensity and duration of showery precipitation was also worked out. This method is considered in detail in the monograph by Orlova [236].

Some data on the predictability of steady precipitation, when the synoptic-hydrodynamic method is used, will now be presented. In December 1975 and January 1976, ten series of predictions were made and evaluated, the prediction periods being 12, 24, and 36 h (in each series, the number of points for which predictions were made fluctuated from 35 to 63). The precipitation forecasts were evaluated on a circular test area 100 km in radius. The mean (over all ten series) predictabilities of the amount of precipitation were 81, 76, and 72%, respectively, for prediction periods of 12, 24, and 36 h (for fluctuations on individual days of 65–91%).

Dyubyuk's formula. In 1947, Dyubyuk [103] proposed using formula (5.6.1) to calculate the amount of precipitation. The derivation of this formula will now be examined.

To do this, first the equations of water-vapor transport and cloud water content (see Section 2.1) are written:

$$\rho \frac{dq}{dt} = \frac{\partial}{\partial z} \rho k \frac{\partial q}{\partial z} - \nu, \tag{5.6.4}$$

$$\rho \frac{d\delta}{dt} = \frac{\partial}{\partial z} \rho k \frac{\partial \delta}{\partial z} + \nu - \frac{\partial Q_d}{\partial z}, \tag{5.6.5}$$

where Q_d is the flow of droplets and ice crystals due to gravity. It is directed vertically downward and hence is negative:

$$Q_d = -\frac{4}{3} \pi \rho_d n \int_0^\infty v_d(r) f(r) r^3 \, dr. \tag{5.6.6}$$

Here, $v_d(r)$ is the fall velocity of a droplet of radius r, $f(r)$ is the density of the droplet-radius distribution, and n is the total number of droplets per unit volume.

By adding the left- and right-hand sides of eqns. (5.6.4) and (5.6.5),

$$\rho \frac{dq}{dt} + \frac{\partial \rho \delta}{\partial t} + \frac{\partial \rho u \delta}{\partial x} + \frac{\partial \rho v \delta}{\partial y} + \frac{\partial \rho w \delta}{\partial z} = \tag{5.6.7}$$

$$= \frac{\partial}{\partial z} \rho k \frac{\partial q}{\partial z} + \frac{\partial}{\partial z} \rho k \frac{\partial \delta}{\partial z} - \frac{\partial Q_d}{\partial z}.$$

is obtained. Here the continuity equation has been used in order to recast the left-hand side of eqn. (5.6.5) into the form it has on the left-hand side of (5.6.7).

Next, eqn. (5.6.7) is integrated over the height from the lower cloud boundary (z_l) to the upper (z_u):

$$\left(-Q_d + \rho k \frac{\partial q}{\partial z} + \rho k \frac{\partial \delta}{\partial z} - \rho w \delta\right)_u - \left(-Q_d + \rho k \frac{\partial \delta}{\partial z} - \rho w \delta\right)_l -$$

$$-\left(\rho k \frac{\partial q}{\partial z}\right)_l = \int_{z_L}^{z_B} \rho \frac{dq}{dt} dz + \int_{z_L}^{z_u} \left(\frac{\partial \delta^*}{\partial t} + \frac{\partial u \delta^*}{\partial x} + \frac{\partial v \delta^*}{\partial y}\right) dz, \qquad (5.6.8)$$

where $\delta^* = \rho\delta$ is the absolute cloud water content.

The precipitation intensity I_{pr} at an arbitrary level in the cloud is represented as the sum

$$I_{pr} = -Q_d + \rho k \frac{\partial \delta}{\partial z} - \rho w \delta. \qquad (5.6.9)$$

Here, the first term is the flow of droplets due to gravity (it is always positive), the second term is the turbulent droplet flow (positive if the water content rises with increasing height), and the third term is the droplet flow due to vertical motions of the air (positive for downdrafts).

At the lower boundary of the cloud, as implied by (5.6.8), the precipitation intensity can be written as

$$I_{pr} = -\int_{z_L}^{z_u} \rho \frac{dq}{dt} dz - \int_{z_L}^{z_u} \frac{\partial \delta^*}{\partial t} dz - \int_{z_L}^{z_u} \delta^* \left(\frac{\partial u}{\partial x} + \frac{\partial v}{\partial y}\right) dz -$$

$$-\int_{z_L}^{z_u} \left(u \frac{\partial \delta^*}{\partial x} + v \frac{\partial \delta^*}{\partial y}\right) dz - \left(\rho k \frac{\partial q}{\partial z}\right)_L, \qquad (5.6.10)$$

where the first term on the left-hand side of (5.6.8), which will evidently be small in comparison with the other terms, has been neglected.

Relation (5.6.10) shows that the precipitation intensity depends on several factors. The values of the first term taken into account by Dyubyuk fluctuate, according to calculations, from hundredths of a mm/h to 1 or 2 mm/h. It is easy to see that several of the other terms on the right-hand side of (5.6.10) have the same order of magnitude (this was first pointed out by Petrichenko [246]). For instance, for a cloud with a thickness $z_u - z_l = 5000$ m having a mean (over the cloud) water content $\delta^* = 1.0$ g/m^3 and a wind velocity divergence of 10^{-5} s^{-1}, the third, divergent term contributes 0.18 mm/h to the total precipitation intensity, a contribution comparable to that of the first term. The last, turbulent term, even for the mean (over the entire troposphere) gradient of the specific humidity ($-\partial q/\partial z \approx 10$ g/kg/10 km = 10^{-6} m^{-1}), is not so small (for $\rho = 1$ kg/m^3 and $k = 10$ m^2/s, this term will be equal to 0.036 mm/h) that it can be neglected.

The foregoing considerations make it clear that Dyubyuk's formula can indeed be used as a first approximation.

CHAPTER 6

MESOSCALE ATMOSPHERIC MOTIONS AND CLOUDINESS

6.1. Some experimental data on mesoscale clouds

Satellite observations of clouds have shown that cloud systems with horizontal dimensions ranging from several kilometers to 100 or 200 kilometers often form the atmosphere. Such systems are known as *mesoscale systems*. The cloud elements making up a mesoscale system generally have the shape of irregular hexagonal cells; cloud banks are also observed, together with transitional forms between cells and banks. The data show that mesoscale cloud systems arise under the influence of convective motions developing in layers with unstable stratification or over a nonuniform underlying surface. For highly developed convection the cloud systems consist of cumulonimbus, which assume the shape of individual large masses, huge banks, or spirals.

Cells. The first investigations of the conditions under which convective motions (circulation) appear in the form of cells were carried out by Bénard (1900) and Rayleigh (1916). Bénard studied this phenomenon in the laboratory; Rayleigh studied it theoretically. The linear theory of convection devised by Rayleigh made it possible to formulate two fundamental conclusions:

(1) The regime of motion is a function of a dimensionless quantity that later came to be known as the Rayleigh number:

$$\mathrm{Ra} = \frac{g}{\bar{T}} \frac{\gamma - \gamma_a}{\nu k} h^4.$$

Here g is the free-fall acceleration, $\bar{T}$ is the mean temperature of the layer, γ is the lapse rate, γ_a is the dry-adiabatic lapse rate (replaced by the moist-adiabatic lapse rate γ_a' inside the cloud), and k are, respectively, the molecular viscosity and the turbulent temperature conductivity, and h is the thickness of the convective layer.

(2) For Rayleigh numbers higher than some critical value Ra_{cr}, periodic wave motions of the cell type appear in a layer of liquid or gas, the amplitude of the wave motions increasing with time. If $\mathrm{Ra} < \mathrm{Ra}_{cr}$, the amplitude of the wave motions does not vary with time.

The theory based on the linearized equations of motion could not explain why, for $\mathrm{Ra} > \mathrm{Ra}_{cr}$, periodic motions are observed with amplitudes that do not increase with time even when the motions stop being stable. The answers to these questions were found only within the framework of the nonlinear theory. It was ascertained that as a result of nonlinear factors, Ra_{cr} increases to $10^5 \div 5\cdot10^5$. On the other hand, in order to evaluate Ra under atmospheric conditions, ν and k should be replaced by turbulence coefficients of the order of 10-100 m^2/s. Since the multiplier $\gamma - \gamma_a'$ is of the order of $10^{-4} - 10^{-3}$°C/m, Ra must be $10^4 - 10^6$ for values of h from 1 to 3 km.

Therefore, for a convective layer in the atmosphere 1-3 km thick, since $Ra \leqslant Ra_{cr}$, there may well exist stable periodic motions of the convective-cell type.

An analysis of the solution, together with experimental data, point to two different kinds of convective cells: *open* and *closed*. In an open cell, ascending motions and clouds are located at the periphery, while in the middle of the cell air descends and cloudiness is absent; in a closed cell the converse is true. Above an unstable convective layer (where $\gamma > \gamma_a$), there is a blocking layer ($\gamma < 0$).

According to experimental data analyzed by Vel'tishchev, the mean thickness of the convective layer, starting from the underlying surface, proved to be 1.74 km for an open cell and 1.53 km for a closed cell (for, respectively, standard deviations of 0.62 and 0.44 km and 98 and 37 cases studied). The mean value of γ in the convective layer was 0.82°C/100 m in open cells and 0.81°C/100 m in closed cells (for σ_γ values of 0.15 and 0.12°C/100 m). The thickness of the blocking layer was the same for open and closed cells (0.62 km), but $\bar{\gamma}$ was different: -0.16°C/100 m for open cells and -0.86°C/100 m for closed cells.

The horizontal diameter D of an open cell lies between 11 and 100 km (according to data for 635 cases) with a maximum frequency (47%) in the interval from 31 to 40 km. The diameters of closed cells varied from 11 to 80 km (data for 386 cases), with a maximum frequency (47%) in this same interval (31-40 km).

The ratio h/D, according to experimental data, ranged from 1/35 to 1/7, with an average value of 1/16. According to Rayleigh's theory, this ratio should be about 1/3. It was suggested that the large difference between the experimental and theoretical values of h/D may be due to a difference between the eddy viscosities in the horizontal (ν_{hor}) and vertical (ν_{vert}) directions.

The formation of open and closed cells was explained by assuming different curvatures of the lapse rates. Open cells originate in layers where the lapse rate decreases with height, that is, $\partial\gamma/\partial z < 0$, while closed cells appear in layers where $\partial\gamma/\partial z > 0$. For instance, the mean value of $\partial\gamma/\partial z$ obtained from the data was $-2.2\cdot10^{-6}$°C/m^2 for open cells and $0.2\cdot10^{-6}$°C/m^2 for closed cells. This conclusion is borne out by observations of the geographical distribution of cloud cells. Open cells appear most frequently over warm ocean currents during the cold half of the year, when the difference ΔT between the air and water temperatures is positive and equal, on the average, to 3-4°C (under these conditions, γ is high close to the water surface and it decreases with height, that is, $\partial\gamma/\partial z < 0$). Closed cells, on the other hand, are generally observed over cold ocean currents, where $\Delta T < 0$ and is equal, on the average, to about -1°C. This feature of the cells can be utilized in analyzing the synoptic situation according to satellite data: regions where open cells are observed are occupied by cold air masses moving over a warmer underlying surface, whereas closed-cell regions correspond to warm air masses moving over a cooler surface.

Regular convective cells, as the theory implies, are produced when there is a slight breeze (strictly speaking, in a calm). Observations indicate that the wind speed close to the ground when cells form does not exceed 5-7 m/s. According to both theory and observations, the most favorable conditions for the appearance of stable cells are created in anticyclones, where the absolute vorticity is low.

Banks of convective clouds. In cases where, in a convective layer, the wind speed varies sharply with height (that is, the vertical gradient of the wind velocity is high), individual cells become united into a *cloud bank*. The distance between cloud elements along the bank will be several times smaller than the distance between adjacent cloud banks. The wavelength (the cloud-bank width plus the cloudless space) fluctuates from 5 to 25 km over land (with a mean value of 9.1 km) and from 6 to 50 km over the ocean (with a mean value of 15.8 km). On the average, a convective layer in the upper part of which cloud banks form will be about 2.3 km thick, and rarely will it be thicker than 3.5 km; the lapse rate in this layer is, on the average, 0.89°C/100 m, and it decreases with height ($\partial\gamma/\partial z = -1.3 \cdot 10^{-6}\,°C/m^2$). Above the convective layer is the blocking layer (mean thickness 0.76 km, $\overline{\gamma} = -0.36°C/100$ m).

The formation of cloud banks is accompanied by an intensification of the wind close to the ground and in the lower troposphere. The wind speed increases with height throughout the convective layer (mean wind-velocity gradient $\partial u/\partial z = 2.6 \cdot 10^{-3}\ s^{-1}$), but the wind direction changes only slightly (by 6.9°/km, on the average). Cloud banks are oriented along the wind direction: the difference between the wind direction at the cloud level and the direction of the cloud bank does not, on the average, exceed 2°.

A more detailed analysis of wave motions of the atmosphere will not be given here, since, according to present-day ideas, the main factors in the genesis of wave clouds are turbulent exchange, large-scale vertical motions, and advection. It is only after these factors have brought the air close to saturation or clouds have formed that the wave motions are able to cause the above-mentioned features of cloud structure (billows, shelves, etc.) to appear.

6.2. A numerical model of convection in an unstable layer

An analysis of observations with satellites has shown that the temperature field is nonuniform not only on the land, but also on the ocean surface, where the regions of temperature deviation from the mean values have the form of bands and alternating patches. This patchiness remains to a depth of several tens of meters and for periods of time as long as tens of days. The nonuniformity of the temperature field in the surface layer of the ocean causes temperature deviations in the atmosphere, which in turn are a source of convective motions.

Vel'tishchev and Zhelnin [57, 58] formulated a theory of the two-dimensional convective motions arising in a layer with uniformly distributed internal heat sources, and they constructed numerical models of such motions. A layer of infinite extent along the horizontal and of thickness h is considered (where $-h/2 \leqslant z \leqslant h/2$). The distribution of the mean temperature $\overline{T}$ with height and its variation with time are assumed to be given. The equation for $\overline{T}(z,t)$, for a constant turbulent temperature conductivity k and a constant rate of variation of $\overline{T}$ with time ($\partial\overline{T}/\partial t = \eta = \text{const}$), takes the form

$$\frac{\partial \overline{T}}{\partial t} = k\frac{\partial^2 \overline{T}}{\partial z^2} \quad \text{or} \quad k\frac{d^2\overline{T}}{dz^2} - \eta = 0. \tag{6.2.1}$$

By integrating (6.2.1) twice, we obtain

$$\overline{T}(z,t) = \overline{T}_0 - \gamma z + \frac{\eta}{2k}\left(z^2 - \frac{h^2}{4}\right), \tag{6.2.2}$$

where $\gamma = -\partial\overline{T}/\partial z = -\Delta\overline{T}/h$ is the lapse rate of the mean temperature, $\Delta\overline{T} = \overline{T}_{h/2} - \overline{T}_{-h/2}$ is the difference between the temperatures at tha upper and lower layer boundaries, $\overline{T}_0$ is the temperature in the middle of the layer, and $\overline{T}_0 = \overline{T}_0(0) + \eta t$.

Let $\overline{u}$, $\overline{v}$, $\overline{T}$, $\overline{\rho}$, and $\overline{p}$ be meteorological quantities pertaining to an unperturbed, large-scale flow, which vary only with the height ($\overline{u} = \overline{u}(z)$, $\overline{v} = \overline{v}(z)$, $\overline{T} = \overline{T}(z)$, $\overline{\rho} = \overline{\rho}(z)$, $\overline{p} = \overline{p}(z)$), and let u', v', w', ρ', p', and Θ' be the deviations of the corresponding quantities produced by the convective motion (Θ' is the deviation of the potential temperature).

Next, the simplifications assumed in free-convection theory (Boussinesq approximation) are applied to the initial system of equations of hydrothermodynamics. If, in this simplified system of equations written for the deviations, a change is made to dimensionless quantities by using scales of length h, velocity νh^{-1}, temperature $(\gamma - \gamma_a)h$, time $h^2\nu^{-1}$, and pressure $\overline{\rho}h^{-2}\nu^2$, then this system can be rewritten in the following form:

$$\frac{du}{dt} = -\frac{\partial p}{\partial x} + \nabla^2 u - R_1 \frac{\partial u}{\partial x} - R_2 \frac{\partial u}{\partial y} - R_3 w,$$

$$\frac{dv}{dt} = -\frac{\partial p}{\partial y} + \nabla^2 v - R_1 \frac{\partial v}{\partial x} - R_2 \frac{\partial v}{\partial y} - R_4 w,$$

$$\frac{dw}{dt} = -\frac{\partial p}{\partial z} + \mathrm{Pr}^{-1}\,\mathrm{Ra}\,\Theta + \nabla^2 w - R_1 \frac{\partial w}{\partial x} - R_2 \frac{\partial w}{\partial y},$$

$$\frac{d\Theta}{dt} = (1 - Nz)\,w + \mathrm{Pr}^{-1}\,\nabla^2 \Theta - R_1 \frac{\partial \Theta}{\partial x} - R_2 \frac{\partial \Theta}{\partial y},$$

$$\frac{\partial u}{\partial z} + \frac{\partial v}{\partial y} + \frac{\partial w}{\partial z} = 0,$$

where all the independent and dependent quantities are dimensionless (for instance, $x = X/h$, $z = Z/h$, $\Theta = \Theta'/(\gamma - \gamma_a)h$, $p = p'/\bar{\rho}\nu^2)h^2$, $u = (u'/\nu)h)$; ν is the kinematic eddy viscosity, $\mathrm{Pr} = \nu/k$ is the Prandtl number (assumed to be 0.71); Ra is the Rayleigh number (see Section 6.1); $N = \eta h k^{-1}(\gamma - \gamma_a)^{-1}$ is a dimensionless parameter characterizing the unsteady nature of the field of the mean temperature; $\gamma = \partial\bar{T}/\partial z$; $R_1 = h\bar{u}\nu^{-1}$, $R_2 = h\bar{v}^{-2}$, $R_3 = h^2\nu^{-1}(\partial\bar{u}/\partial z)$, $R_4 = h^2\nu^{-1}(\partial\bar{v}/\partial z)$ are analogs of the Reynolds number,

$$\nabla^2 = \frac{\partial^2}{\partial x^2} + \frac{\partial^2}{\partial y^2} + \frac{\partial^2}{\partial z^2}; \quad \frac{d}{dt} = \frac{\partial}{\partial t} + u\frac{\partial}{\partial x} + v\frac{\partial}{\partial y} + w\frac{\partial}{\partial z};$$

u, v, w are the dimensionless components of the velocity deviation with respect to the dimensionless x, y, z; and Θ and p are the dimensionless deviations of the potential temperature and the pressure.

The unknown functions satisfy the boundary conditions

$$u = v = w = \Theta = 0 \quad \text{for} \quad z = -\frac{1}{2} \quad \text{and} \quad z = \frac{1}{2},$$

that is, at the upper and lower boundaries of the layer the convective motions are damped.

At the lateral boundaries, all the functions are assumed to be periodic. The extent of the region in the x direction is 2.34 times the thickness of the convective layer, and in the y direction it is 4.032 times this thickness. Numerical calculations were carried out on a computer for a grid of 25 x 25 x 9 points in the x, y, and z directions, respectively.

At the initial moment, convective (disturbed) motion is absent; to excite it, an arbitrary perturbation of the potential temperature is specified:

$$\Theta(x, y, z, 0) = \Theta_0(x, y, z).$$

Without describing in detail the method of calculation used, some of the results obtained will be presented.

An analysis of these results showed that the unsteady nature of the mean-temperature field, that is, the parameter N, affects considerably the form of the convective motions. For values of N from 3.3 to 11.8 and Rayleigh numbers slightly higher than critical, the motions assume the form of two-dimensional billows, regardless of the Prandtl number and the initial conditions. In contrast to the classical two-dimensional billows observed for $N = 0$ (steady field of the mean temperature), the horizontal distribution of the vertical velocity inside each two-dimensional cell is asymmetrical. For $N > 0$, a region with an updraft will be only half the size of a region with a downdraft, while the maximum values of the vertical velocity in the former region will be about twice those in the latter region. For $N < 0$, the ratios between the sizes of these regions and the maximum vertical velocities in them will be the reciprocals of those for $N > 0$.

As the Rayleigh number increases, the two-dimensional convective currents become three-dimensional, and this will occur sooner, the higher the values of the Prandtl number and parameter N. For instance, for Pr = 0.71 this transition occurs at Ra = 16,000 and Ra = 10,000, when $\eta = 1.5°C/day$ and $\eta = 3°C/day$, respectively. For $N \neq 0$, the structure of the three-dimensional motions differs considerably from the structure for $N = 0$. When $N > 0$, there are downdrafts in the middle of the cells and updrafts at the periphery, the downdraft area being much greater than the updraft area; the maximum positive vertical velocities have absolute values about three times as great as the maximum negative velocities. For $N < 0$, the nature of the motion and the velocity distribution will be the opposite of those for $N > 0$. The motion for $N > 0$ corresponds to open cells; for $N < 0$, to closed cells.

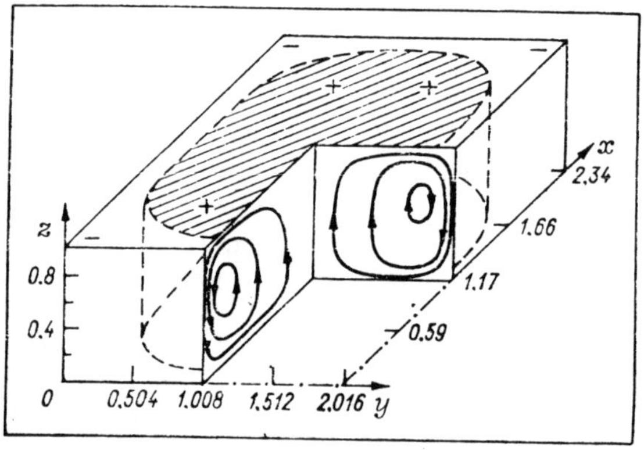

Fig. 6.1. Three-dimensional motions in a closed cell

Figure 6.1 shows a cutaway diagram of three-dimensional motions in a closed cell. The solid curves indicate streamlines in the xz and yz planes ($x = X/h$, $y = Y/h$, and $z = Z/h$ are dimensionless coordinates). Dashed lines indicate isolines of zero vertical velocity; the region of ascending motion is shaded; plus and minus signs signify positions of velocity maxima for ascending and descending motions, respectively. Seen from above, a cell only remotely resembles a hexagon. It should be noted that the maximum values of w are observed not in the middle of the cell, but closer to the downdraft region.

Since a temperature rise ($\eta > 0$, $N > 0$) with time is observed in cold air masses moving over a warm underlying surface, it follows from the foregoing that in such air masses convective motions have the nature of open cells, in a narrow peripheral zone of which there will be intense updrafts, accompanied by the formation of a ring of cumulus and cumulonimbus. As a warm air mass moves over a cold underlying surface, the air mass cools ($\eta < 0$, $N < 0$) and closed cells form. A sizable (central) part of a closed cell has in it comparatively weak upward motions, which produce vast fields of stratocumulus, the elements of which are separated by narrow clear gaps because of the downward motions at the periphery of the closed cells.

All the foregoing results were obtained for a velocity of undisturbed flow that does not vary with height ($\bar{u}$ = const, $\bar{v}$ = = const). Observations in the atmosphere, as well as laboratory experiments with fluids, have shown that motions with a banded structure (as seen from above) arise in an unstably stratified flow with shear. It has been established by Asai [333], Lipps [384], and others that in a one-dimensional large-scale shear flow, convective motions have predominantly the form of staggered eddies (vertexes) with a horizontal axis. The rotational senses of adjacent eddies are opposite, so that the resultant motion is essentially three-dimensional: particles move along helixes inside an eddy.

Models employing a one-dimensional large-scale flow satisfactorily describe the conditions of formation of the mesoscale cloud banks forming in a thin convective layer, within which the wind direction varies slightly [376]. If, on the other hand, the convective layer is quite thick, a substantial variation of the wind direction inside it is more likely, and a model of convection in a two-dimensional large-scale flow will be necessary. The first such models were worked out by Hill [365] and Asai [332]. Hill studied two-dimensional perturbations in an ideal fluid, and Asai studied three-dimensional perturbations in a viscous fluid. Hill's main finding was that the convective motions take the form of billows, oriented along the wind-shear vector. Asai found that the billows travel with a phase velocity equal to the wind-vector component normal to the shear vector.

In [57, 58] a more complete formulation of the problem of convection in a two-dimensional large-scale flow was solved, by numerical integration of the hydrothermodynamical equations of a viscous incompressible fluid. Assuming that the Coriolis force can be neglected and that the eddy viscosity ν is constant, the equations of large-scale motion can be integrated to obtain:

$$\bar{u} = \frac{1}{2\nu\rho}\frac{\partial \bar{p}}{\partial x} z^2 + Az + B, \qquad \bar{v} = \frac{1}{2\nu\rho}\frac{\partial \bar{p}}{\partial y} z^2 + Cz + D,$$

where $-\partial\bar{p}/\partial x$ = const and $-\partial\bar{p}/\partial y$ = const are the x and y components of the baric gradient of the unperturbed pressure, and A, B, C, D

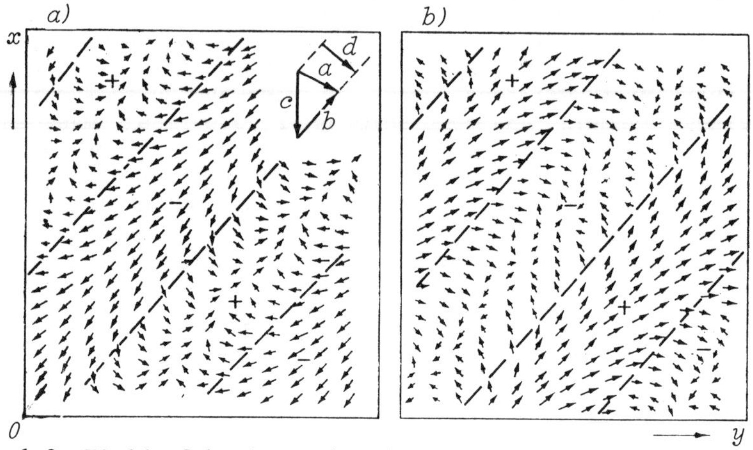

Fig. 6.2. Field of horizontal velocity of convective motion in Couette flow close to lower *(a)* and upper *(b)* boundaries of unstable layer.

are constants of integration. These formulas describe the velocity distribution in a Poiseuille flow, which has been well investigated in hydrodynamics. In the particular case of $\partial\bar{p}/\partial x = \partial\bar{p}/\partial y = 0$, the velocity profile is linear, and the flow is known as a Couette flow.

The profile of the unperturbed temperature inside the unstable layer was assumed to be linear, and the buoyancy parameter $\alpha = g/\bar{T}$ was taken to be constant. Figure 6.2 shows the results of the numerical solution of the problem. Here, arrows portray the horizontal velocities of convective motion to a certain scale. The small diagram at the top of Fig. 6.2a shows (to the same scale as the velocities of convective motion) the wind-velocity vectors in the unperturbed flow at the lower *(a)* and upper *(c)* levels of the layer, as well as the vector increment (shear) of the wind velocity *(b)* in the large-scale,unperturbed flow. The dashed lines in Fig. 6.2 are the isolines of zero vertical velocity, a plus sign corresponding to updraft regions ($w > 0$) and a minus sign to downdraft regions ($w < 0$). Convective motions, viewed from above, have the appearance of billows, the billow axes being in the direction of the wind-shear vector. In a one-dimensional large--scale flow, the billows are stationary. When the wind direction changes, creating a two-dimensional unperturbed flow, the billows travel with a phase velocity equal to the normal to-the-billow-axis component *(d)* of the wind velocity of the unperturbed flow.

An analysis of the theoretical results indicated that the distribution of temperature deviations (from $\bar{T}$) has the same phase as the vertical-velocity distribution: in a region of upward motions the temperature deviations are positive, and in a downdraft region they are negative. Close to the lower boundary of the convective layer, the pressure is higher in a downdraft region ($w < 0$) and lower in a region of upward motion ($w > 0$). Close to the upper boundary of the layer, the pressure is the opposite.

A calculation was made for a Poiseuille flow (selecting the coefficients in the formulas so that the direction of the wind shear remained constant with height), assuming a supercritical value of the Rayleigh number (Ra = 4500) and a Prandtl number Pr = 1. The results of this calculation were qualitatively similar to the results for a Couette flow summarized above.

According to the estimates of Asai, Nakasui, and others, the Boussinesq approximation (used to approximate the Navier-Stokes equations in studies of convective motions) for typical values of the meteorological quantities introduces almost no errors for layer thicknesses not exceeding 3 km ($h \leqslant 3$ km). For $h \approx 10$ km, on the other hand, the Boussinesq approximation leads to errors of about 30%.

Since the Coriolis force was neglected when constructing the theory, the results obtained are applicable to perturbations having characteristic lifetimes not longer than $2\pi/2\omega_z$, that is, not longer than about 24 h.

Taking these two considerations into account, the authors of [57] compared the theoretical findings with experimental data for cloud systems consisting of convective clouds arranged to form quasi-rectilinear bands; the depth of these clouds did not exceed 4-5 km, and they had a typical lifetime of about a day. Such cloud systems generally form at fronts during the warm season.

In practice, the rate of displacement of the cloud bands is determined as follows. The rectilinear part of a band is found on a cloud chart, constructed from television or infrared satellite photos, and the band axis is marked. Then, radiosonde or satellite data are used to estimate the height of the upper cloud boundary and the mean (for the cloud band) wind velocity at the standard level closest to the cloud summit. Next, if the component of the mean wind velocity along the normal to the cloud-band axis is determined, then according to theory this component will also be equal to the phase velocity of displacement of the cloud band. According to data obtained with radiosondes and with the *Meteor 10* satellite on 26 August 1972, the cloud band was expected to shift 130 km in 12 h; the actual shift was 150 km. An analysis of 27 instances of frontal cloud bands situated over the European USSR showed the standard deviation of the predicted shift of a cloud band from the actual shift during a 12-h period to be 65 km (relative error of 20%). The direction of motion of the cloud band was in all cases correctly determined.

Heat fluxes. In the theory of heat transfer, the Nusselt number is used as a dimensionless characteristic of the heat flux:

$$\mathrm{Nu} = \frac{-k\dfrac{\partial \overline{T}}{\partial z} + \overline{wT}}{-k\dfrac{\Delta \overline{T}}{h}},$$

namely, the ratio of the total heat flux (turbulent plus convective) to the turbulent flux in the undisturbed state. Here k is the eddy temperature conductivity, and $\Delta\overline{T}$ is the temperature difference in the undisturbed flow at the upper and lower boundaries of a layer of thickness h. Figure 6.3 indicates how satisfactorily the calculated values of Nu fit the experimental data. At several points, the curve is seen to change its slope. Close to these points, the heat flux varies sharply. Most of these discrete variations in the heat flux are related to a change in the nature of the convective motions.

Table 6.1 presents some values of $\beta = \partial(\mathrm{Nu\ Ra})/\partial\,\mathrm{Ra}$, which serves as a measure of the dependence of the heat flux on the Rayleigh number, in various intervals of variation of Ra, for Pr = 0.71 and for different values of $N = \eta h^2/k\Delta T$. The sharp variation of β for $N = 0$ and Ra ≈ 13,000 is connected with the appearance of a slight bimodality in the vertical-velocity distribution, and for Ra ≈ 25,000 it corresponds to a transition to

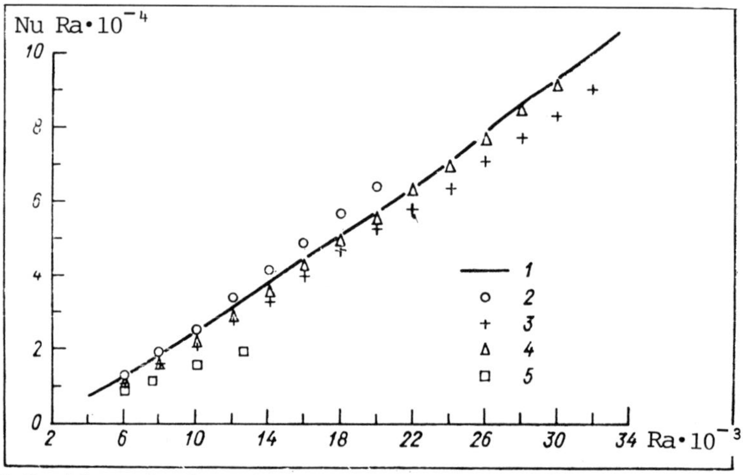

Fig. 6.3. Dimensionless heat flux H = Nu Ra as function of Rayleigh number Ra for Pr = 0.71 and $N = 0$.

Calculations of: *(1)* Vel'tishchev and Zhelnin, *(2)* Willis (1972), *(3)* Willis and Deardorff, *(4)* Brown, *(5)* Krishnamurti.

completely three-dimensional convective motion.
For $N > 0$ or $N < 0$, the variation in β is related to a transition from two-dimensional billows to open or closed cells, and from the

latter to oscillatory motions. Since the product βk represents a kind of effective temperature conductivity k_{eff} (turbulent plus convective), the value of k_{eff} in a convective layer can be estimated on the basis of cloud observations.

Table 6.1.

<table>
<tr><th rowspan="2">N</th><th colspan="15">Ra.10⁻³</th></tr>
<tr><th>4.5</th><th>6</th><th>8</th><th>10</th><th>12</th><th>14</th><th>16</th><th>18</th><th>20</th><th>22</th><th>24</th><th>26</th><th>28</th><th>30</th><th>32</th></tr>
<tr><td>0</td><td colspan="2">2.68</td><td colspan="3">2.86</td><td colspan="6">3.25</td><td colspan="4">3.50</td></tr>
<tr><td>3.3—5.7</td><td colspan="7">2.84</td><td colspan="8"></td></tr>
<tr><td>7.3—11</td><td colspan="3">2.70</td><td colspan="6">3.47</td><td colspan="6">4.35</td></tr>
</table>

6.3. Ensembles of cumulus clouds

Convective motions, arising as a result of vertical instability, are an important vehicle of energy exchange in the atmosphere. The contribution of the energy of instability to the overall energy balance in the tropics is comparable to the contribution of the energy of large-scale motions. Consequently, it is of interest to find some way of determining the vertical fluxes of heat, momentum, water vapor, and liquid water on the basis of data characterizing the large-scale motions; such data can be obtained from synoptic charts or from soundings of the atmosphere. The solution of the problem, stated in this form, is called "parametrization of phenomena having a subgrid scale" (in the given case of moist convection). This problem has been tackled by many investigators, especially during recent years: Ooyama [408], Ogura and Cho [403], Arakawa and Schubert [331], Yanai, Esbensen, and Chu [437], Lopez [386], Gray [361], Lewis, Ogura, and Gidel [381], Shvets and Shneerov [327], Fal'kovich [309], and others.

Several methods of parametrization of the moist convection have been proposed. These may be arbitrarily divided into three groups.

In the first group, the concept of convective adaptation is used: as soon as, during the calculations, the lapse rate and the relative humidity at some grid point exceed the critical values, they are replaced by the latter. The critical values of γ and f given by different authors fluctuate between, respectively, the dry-adiabatic and moist-adiabatic lapse rates and humidities of 70 and 100%. Here, it is assumed that the kinetic energy of the small-scale eddies originating from convection is dissipated instantaneously, and that all the liquid water falls as precipitation.

The second group of parametrization methods involves the use of schemes of penetrating convection, based on the concept of conditional instability of the second kind. In these schemes the appearance of convection, propagating throughout the troposphere, is associated not only with the temperature-humidity stratification but also with the influence of dynamic factors. The amount of heat liberated during the development of moist convection is assumed to be proportional to the vertical velocity at the upper boundary of the atmospheric boundary layer. These schemes differ in the form of the height dependence of the proportionality factor.

The third group includes methods based on the construction of models of convective-cloud ensembles. Although such models are more difficult to implement (both with regard to constructing the algorithm and with regard to saving machine time), still they provide a fuller and more profound description of the physical mechanism of moist convection and of the formation of an ensemble of convective clouds. Therefore, the concepts on which the third group of parametrization methods is based will now be considered in more detail.

In the model of a cumulus ensemble constructed in [437], the vertical flows of air and liquid water, the amounts of condensed and evaporated moisture, and the other quantities are defined in terms of the characteristics of the large-scale motion. The authors generalized the experience gained prior to 1973 on the modeling of individual clouds and cloud groups, taking entrainment into account. To calculate the influxes and fluxes of heat and moisture, as well as the vertical profiles of the indicated quantities, observations of the characteristics of cloud clusters in the vicinity of the Marshall Islands were used.

In [331] the parametrization method was based on a spectral representation of the cloud ensemble. The model proposed by these authors makes it possible to study not only the overall characteristics of cloud clusters, but also the features of different cloud forms, which differ primarily in their vertical depth, this being in turn dependent on the thickness of the layer of penetration of moist convection. However, in [331, 437] several not wholly justified simplifications were made. For example, the cloud water content was assumed to be zero close to the cloud base, which rules out precipitation. In addition, the derivative of the cloud water content along the vertical was taken to be proportional to the water content itself, the proportionality factor being a function of the height.

In [327] the initial premises and the system of equations are close to those in [437], but the boundary conditions and some of the other relations are formulated more correctly than in [437].

The initial equations of the problem are the averaged equations describing the heat influx

$$c_p\left[\frac{\partial\overline{\Theta}}{\partial t}+\left(\overline{u\frac{\partial\Theta}{\partial x}}+\overline{v\frac{\partial\Theta}{\partial y}}\right)+\frac{\partial\overline{\Theta}\overline{\tau}}{\partial p}\right]=\overline{\varepsilon}_r+L\overline{\nu}-c_p\frac{\partial}{\partial p}(\overline{\Theta'\tau'}), \qquad (6.3.1)$$

and the mositure influx

$$\frac{\partial \bar{q}}{\partial t} + \left(\overline{u \frac{\partial q}{\partial x}} + \overline{v \frac{\partial q}{\partial y}}\right) + \frac{\partial \bar{q}\bar{\tau}}{\partial p} = -\bar{\nu} - \frac{\partial}{\partial p}\left(\overline{q'\tau'}\right), \tag{6.3.2}$$

where $\Theta = T + \gamma_a z$ is the potential temperature; q is the specific humidity; $\tau = dp/dt \approx -g\rho w$ is the vertical velocity of the air with respect to an isobaric surface; u, v are the velocity components in the horizontal x and y directions; ν is the specific rate of water-vapor condensation; ε_r is the specific radiative heat influx; L is the latent heat of condensation; c_p is the specific heat of air for p = const; a bar indicates averaging over a horizontal section through the system of cumulus clouds; and a prime indicates a deviation from the mean value to mesoscale convection.

The left-hand sides of eqns. (6.3.1) and (6.3.2) can be calculated using the data of the large-scale (average) motion, assuming that mesoscale convection motions do not affect the horizontal transport (that is, $u' = v' = 0$, $\overline{u(\partial\Theta/\partial x)} = \bar{u}(\partial\bar{\Theta}/\partial x)$, etc.). If the left-hand side of eqn. (6.3.1) is designated as $\bar{\varepsilon}_\Theta$ and the left-hand side of (6.3.2) as $\bar{\varepsilon}_q/L$ and these are added together, then

$$\bar{\varepsilon}_\Theta + \bar{\varepsilon}_q - \bar{\varepsilon}_r = -\frac{\partial}{\partial p}\left(\overline{h'\tau'}\right), \tag{6.3.3}$$

is obtained. Here, $h' = c_p\Theta' + Lq'$ is the deviation of the total specific heat (including sensible and latent heat) from the mean value. By integration of (6.3.3) with respect to the pressure p from the cloud base pressure p_1, where $\tau' = 0$, an expression is obtained for the convective flux of the total specific heat:

$$-\frac{1}{g}\overline{h'\tau'} = \frac{1}{g}\int_{p_1}^{p}(\bar{\varepsilon}_\Theta + \bar{\varepsilon}_q - \bar{\varepsilon}_r)\,dp. \tag{6.3.4}$$

Now the convective fluxes can be expressed in terms of the cloud parameters, if it is assumed that deviations h', Θ', q', and τ' are due exclusively to mesoscale convection.

Let σ_i be the cross-sectional area of the ith cloud of isobaric surface p, and let h_i and τ_i be the mean values of h and τ over this area. The vertical flow of the air mass through section σ_i will then be

$$m_i = -\frac{1}{g}\int_{(\sigma_i)} \tau_i\, d\sigma = -\frac{1}{g}\tau_i\sigma_i. \tag{6.3.5}$$

If the quantities characterizing the medium surrounding the cloud are designated as $\tilde{\tau}$, $\tilde{\Theta}$, $\tilde{q}$, and $\tilde{h}$, then the mean values can be written as

$$\overline{\Theta}=\sum_i \Theta_i\sigma_i+\widetilde{\Theta}(1-\sigma)=\widetilde{\Theta}+\sum_i\left(\Theta_i-\widetilde{\Theta}\right)\sigma_i, \tag{6.3.6}$$

where $\sigma=\sum_i \sigma_i$ is the total relative area of the cloud sections, the horizontal area through which the flow is calculated being assigned a value of unity.

Similarly,

$$\bar{\tau}=\tilde{\tau}+\sum_i(\tau_i-\tilde{\tau})\,\sigma_i \tag{6.3.7}$$

can be written. Then, since

$$\overline{\Theta'\tau'}=\overline{\Theta\tau}-\overline{\Theta}\,\overline{\tau}\text{ and }\overline{\Theta\tau}=\widetilde{\Theta}\tilde{\tau}(1-\sigma)+\sum_i\Theta_i\tau_i\sigma_i, \tag{6.3.8}$$

it follows that if relations (6.3.5)-(6.3.7) are taken into account and it is assumed on the basis of observational data that $\sigma_i^2\ll 1$ and $|\tilde{\tau}|\ll\tau_i$, , the formula obtained is

$$-\frac{1}{g}\overline{\Theta'\tau'}=\sum_i m_i\left(\Theta_i-\widetilde{\Theta}\right), \tag{6.3.9}$$

while, similarly,

$$-\frac{1}{g}\overline{q'\tau'}=\sum_i m_i\left(q_i-\tilde{q}\right), \tag{6.3.10}$$

$$-\frac{1}{g}\overline{h'\tau'}=\sum_i m_i\left(h_i-\tilde{h}\right). \tag{6.3.11}$$

In view of relations (6.3.9) and (6.3.10), the fundamental equations (6.3.1) and (6.3.2) take the form

$$L\sum_i\nu_i+c_pg\frac{\partial}{\partial p}\sum_i m_i\left(\Theta_i-\widetilde{\Theta}\right)=\bar{\varepsilon}_\Theta-\bar{\varepsilon}_r, \tag{6.3.12}$$

$$-L\sum_i\nu_i+Lg\frac{\partial}{\partial p}\sum_i m_i\left(q_i-\tilde{q}\right)=\bar{\varepsilon}_q. \tag{6.3.13}$$

Now the equations describing the balance of mass, heat, moisture, and water content for the ith cloud will be obtained. To do this, the familiar theorem

$$\int_{\sigma_i}\operatorname{div}(f\mathbf{c})\,d\sigma=\oint_l fc_n\,dl, \tag{6.3.14}$$

linking the integral of div $(f\mathbf{c})$ over area σ_i to the integral over the contour l, bounding area σ_i is employed. Here c_n is the component of velocity c normal to contour l, and f is an arbitrary function. Next, contour l is divided into l_1 and l_2, so that on l_1 the normal component $c_n > 0$, and thus $\oint\limits_{l_1} c_n dl = \mu_i^{(1)} > 0$, while on l_2 component $c_n < 0$, so that $\oint\limits_{l_2} c_n dl = -\mu_i^{(2)} < 0$. The first of these quantities ($\mu_i^{(1)}$) characterizes the rate of air inflow to the ith cloud, while the second ($\mu_i^{(2)}$) gives the rate of air outflow from the ith cloud.

If the continuity equation div $\mathbf{c} + \partial\tau/\partial p = 0$ (where $\mathbf{c}$ is the horizontal velocity) is integrated over σ_i and theorem (6.3.14) for $f \equiv 1$ is invoked, the expression

$$\mu_i^{(1)} - \mu_i^{(2)} + g\frac{\partial m_i}{\partial p} = 0. \tag{6.3.15}$$

is obtained. Again applying the theorem of the mean $\left(\oint\limits_{l_1} fc_n\,dl = \mu_i^{(1)}\bar{f}\right.$ and $\left.\oint\limits_{l_2} fc_n\,dl = -\mu_i^{(2)}f_i\right)$, the following system of equations describing the balance of heat, water vapor, and liquid water in the ith cloud is obtained:

$$\mu_i^{(1)}\widetilde{\Theta} - \mu_i^{(2)}\Theta_i + g\frac{\partial(m_i\Theta_i)}{\partial p} + \frac{L}{c_p}\nu_i = 0, \tag{6.3.16}$$

$$\mu_i^{(1)}\tilde{q} - \mu_i^{(2)}q_i + g\frac{\partial(m_iq_i)}{\partial p} - \nu_i = 0, \tag{6.3.17}$$

$$-\mu_i^{(2)}\delta_i + g\frac{\partial(m_i\delta_i)}{\partial p} + \nu_i + \varphi_i = 0, \tag{6.3.18}$$

where δ_i is the specific water content of the cloud and precipitation (if the latter is present), and φ_i is the droplet runoff due to gravity. The latter is related as follows to the mass - weighted mean fall velocity $\tilde{v}$ of droplets and the specific water content δ (see Section 2.1.1):

$$\varphi = \frac{1}{\rho}\frac{\partial}{\partial z}(\rho\tilde{v}\delta) = -g\frac{\partial}{\partial p}(\rho\tilde{v}\delta), \tag{6.3.19}$$

where ρ is the air density.

The system of equations (6.3.4), (6.3.12), (6.3.13), (6.3.15)-(6.3.18) contains eight variables: $h_i, m_i, \Theta_i, q_i, \nu_i, \mu_i^{(1)}, \mu_i^{(2)}$ and δ_i. The quantity φ_i entering into (6.3.18) is found with the aid of relation (6.3.19), in which $\tilde{v}$ is calculated using one of the empirical formulas, for instance,

$$\tilde{v} = 2.86 \cdot 10^3 (g\rho_d\delta)^{0.125} \sqrt{\frac{\rho_0}{\rho}}. \tag{6.3.20}$$

where $\tilde{v}$ is in cm/s, and δ is in g/g.

The missing eighth equation is obtained on the basis of the completely justified assumption that inside the cloud the water vapor is saturated:

$$q_i = q_m(\Theta_i, p). \tag{6.3.21}$$

Hence

$$q_i = q_m(\widetilde{\Theta}, p) + \left(\frac{\partial q_m}{\partial T}\right)_{p=\text{const}} (\Theta_i - \widetilde{\Theta}). \tag{6.3.22}$$

Since $h = c_p\Theta + Lq$, one obtains, in view of (6.3.22),

$$\Theta_i - \widetilde{\Theta} = \frac{1}{c_p(1+\Gamma)}(h_i - \tilde{h}_m), \tag{6.3.23}$$

$$q_i - \tilde{q}_m = \frac{\Gamma}{L(1+\Gamma)}(h_i - \tilde{h}_m), \tag{6.3.24}$$

where

$$h_m = c_p\widetilde{\Theta} + Lq_m(\widetilde{\Theta}, p), \quad \Gamma = \frac{L}{c_p}\left(\frac{\partial q_m}{\partial T}\right)_{p=\text{const}},$$

$$\tilde{q}_m = q_m(\widetilde{\Theta}, p).$$

The authors of [327] also introduce the mean values of the unknown functions for the entire cloud system $\left(m = \sum_i m_i,\ \nu = \sum_i \nu_i,\ \mu_1 = \sum_i \mu_i^{(1)},\ \mu_2 = \sum_i \mu_i^{(2)},\ \Theta_c = \left(\sum_i m_i\Theta_i\right)\Big/ m,\ q_c,\ h_c \text{ and } \delta_c\right)$ and for these they obtain equations similar in structure to eqns. (6.3.4), (6.3.12), (6.3.13), and (6.3.15)-(6.3.18). During the numerical solution of the system, the following assumptions are made:

(1) A thin outflow layer exists, within which $h_i = \tilde{h}_m$; from (6.3.23) and (6.3.24) it then follows that there

$$\Theta_i = \widetilde{\Theta} \text{ and } q_i = \tilde{q}_m.$$

(2) In the outflow layers, the cloud water content is equal to the mean value for the entire ensemble: $\delta_i = \delta_c$

As a boundary condition, δ_c is set equal to zero at the level where $\nu = 0$.

The system of equations is solved using the iteration method, where, in the first iteration it is assumed that $\delta_c = 0$. As initial data, the vertical profiles of $\tilde{\Theta}$, $\tilde{h}$. $\bar{\varepsilon}_p$, $\bar{\varepsilon}_\Theta$, $\bar{\varepsilon}_q$ and $-(\overline{h'\tau'})/g$ are used; these were determined in [437] for some initial moment.

Figures 6.4-6.7 give some calculated characteristics of a cloud ensemble. The ordinate in these figures is the ratio $\xi = p/p_0$ (for p_0 = 1000 mb). Figure 6.4 illustrates the vertical

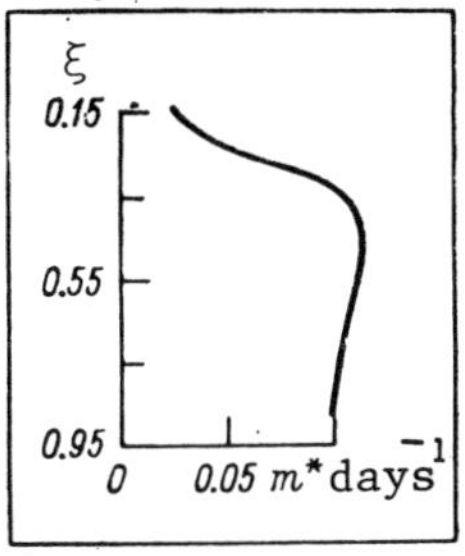

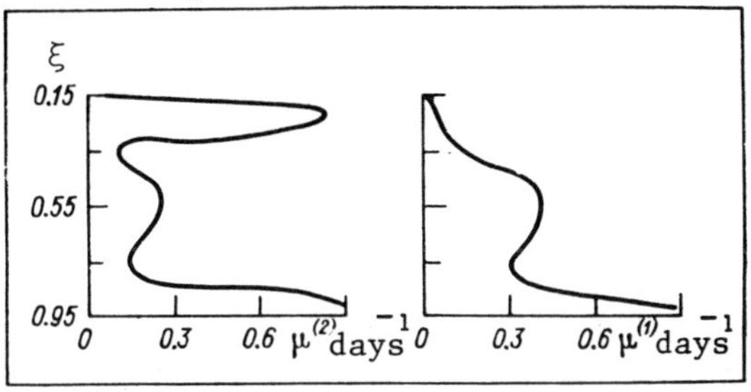

Fig. 6.4. Average (over cloud ensemble) vertical flow of air mass $m^* = mg/p_0$.

Fig. 6.5. Average ratios of inflow $\mu^{(1)}$ and outflow $\mu^{(2)}$ of air.

mass flow, averaged over the cloud ensemble, as a function of height. Note that in formula (6.3.5), as in all subsequent formulas, σ_i is a dimensionless quantity equal to the ratio of the section of the ith cloud to the total area intersecting the cloud ensemble. Therefore, m_i has units of kg/m^2 s), while $m^* = mg/p_0$

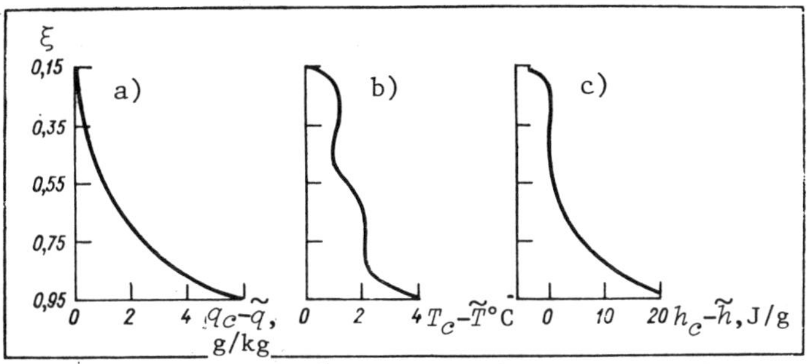

Fig. 6.6. Average (over cloud ensemble) increases (relative to surroundings) in specific humidity (a), temperature (b), and total specific heat (c).

has units of s^{-1} (see Fig. 6.4). According to Fig. 6.4, the air--mass flow is practically constant from the cloud base, situated at the 950-mb level, up to the 350-mb level (7-8 km), above which the flow decreases.

In the lower and middle troposphere of the tropical latitudes, because of convection an air mass equal to 10-15% of the mass of the vertical column M_0 will be transported in 24 hours: $m = m^* (p_0/g) = m^* M_0 = 0.10M_0 - 0.15M_0$. Thus, it can be stated that the mesoscale convection completely renews the air in the vertical column every 7-10 days. The rate of inflow of air to the cloud (see Fig. 6.5) is highest near the cloud base and in the middle troposphere, while the outflow rate is highest close to the cloud base and in the upper troposphere. The resultant transport ($\mu^{(1)} - \mu^{(2)}$) is close to zero in the lower part of the cloud, positive in the middle, and negative (outflow exceeds inflow) in the upper part of the cloud. At all levels of the cloud, according to Fig. 6.6, the surroundings are warmer (by about 4°C close to the base and by 1.5-2°C throughout the rest of the cloud); the specific humidity close to the cloud base is 4-6 g/kg higher than the saturation specific humidity in the surroundings. The specific water content of a cloud (see Fig. 6.7) is about 0.2 g/kg in the lower and middle parts of the cloud, and it rises to 0.3 g/kg in the upper part; in the lower part, large raindrops predominate.

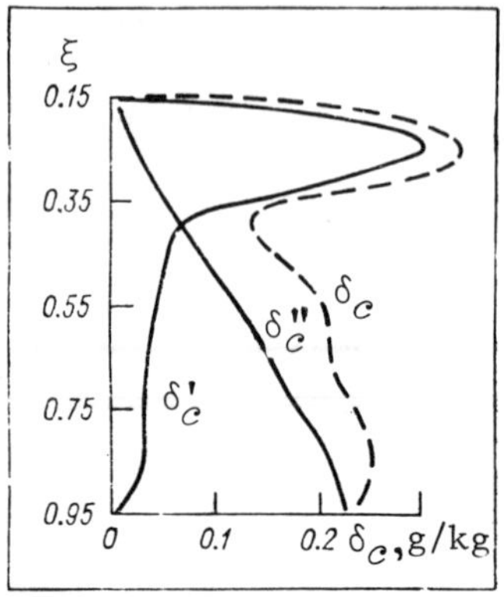

Fig. 6.7. Average (over the cloud ensemble) specific water content of cloud δ_c and of its component parts: δ_c' is the contribution of cloud droplets; δ_c'' is the contribution of raindrops.

In [403] the statistical characteristics of an ensemble of cumulus in the tropics were studied. The spectral function of the cloud-height distribution is written as

$$s_v(p) = -f(p)\tau_v(p_1),$$

where $\tau_v(p_1)$ is the isobaric vertical velocity at the level of the cloud base p_1, and $f(p)$ is the spectral function; the product $f(p)\Delta p$ represents the likelihood that the cloud is situated in the interval $p - \Delta p/2 \leqslant p \leqslant p + \Delta p/2$.

With the aid of observational results for the summertime over the Marshall Islands, the authors of [403] determined functions $s_v(p)$ and $f(p)$ for the heights of the summits of convective clouds. The spectral function $s_v(p)$ was constructed for the tropical zone, where waves moving from east to west are observed.

In [309] the rates of inflow and outflow of the air mass were determined by numerically integrating the above system of equations, and in addition the vertical mass flux m^* in an ensemble of cumulus clouds was found from observational data collected in the tropical Atlantic (1972). The inflow rate reaches a maximum in the lower troposphere, after which it decreases monotonically with height. The rate of outflow has two maxima: one at 950-850 mb in the lower troposphere and one at 300-200 mb in the upper troposphere. In the middle troposphere (from 750 to 450 mb), on the other hand, practically no outflow of mass from the cloud ensemble is observed, which may indicate that clouds of average depth are absent there, and that there is no entrainment (inflow and outflow) in the interior of very thick cumulonimbus.

6.4. The albedo of a cumulus field

The energetics of the atmosphere is greatly affected by radiative influxes of heat, especially by the radiative heat influx to the Earth's surface. The latter depends significantly on the amount and thickness (depth) of the cloudiness.

The cloudiness characteristics (size and number of droplets, heights of cloud boundaries, thickness and quantity of clouds) are highly variable in time and space. Because of this, and due to the lack of information concerning a number of parameters, it is almost impossible to determine theoretically the heat influxes at points on a regular grid in the presence of cloudiness (although attempts have been made [225, 284]). Consequently, experimental data on the optical properties of clouds are of great interest,

The influx of short-wave radiation ε_s to some layer in which clouds are located is related as follows to the total radiation flux Q_{ub} and the albedos at the upper (r_{ub}) and lower (r_{lb}) boundaries of this layer:

$$\varepsilon_s = Q_{ub}[(1 - r_{ub}) - P(1 - r_{ub})], \qquad (6.4.1)$$

where P is the coefficient of transmission of the total radiation by the cloud layer. In [78] the quantities entering into (6.4.1) were determined with the aid of observational material (obtained with aircraft in 1971-1974) on the fluxes of solar radiation over the Ukraine during the warm season, primarily for Cu hum. and Cu med. (90% of measurements) and in part for Cu cong. (10% of measurements). Each measurement represents an average over a path from 70 to 140 km long.

The albedo of the layer r_{ub} is related as follows to the albedo r_{cl} of the clouds themselves:

$$r_{ub}(n) = r_{ub}(0)(1 - n) + r_{lb}n, \qquad (6.4.2)$$

where n is the amount of clouds (in fractions of a unit), and $r_{ub}(0)$ is the albedo of the cloudless atmosphere at the level of the upper cloud boundary.

In the absence of clouds the albedo of the free atmosphere varies slightly (for a 15-20% albedo of the Earth's surface, data indicate that the albedo in the 1-3 km layer decreases, on the average, by 0.3% for every km of height), so that $_{ub}(0)$ can be set equal to $_{lb}$. From relation (6.4.2) it follows that

$$r_{CL}(n) = \frac{r_{ub}(n) - r_{lb}(1 - n)}{n}. \qquad (6.4.3)$$

The height of the sun influences the albedo and the other radiation characteristics of a cloud layer, as do the value of n and the cloud thickness. However, this influence is significant only for sun heights less than 30°. For the observations being discussed here, the height of the sun was greater than 30° in 94% of the cases. The mean value of r_{1b} was 14.5%, and albedos $r_{1b} < 13\%$ and $r_{1b} > 20\%$ were recorded only in 20% of the cases (minimum values $r_{1b} \approx 5$-6% were observed in spring, shortly after the disappearance of the snow cover).

Table 6.2 gives some values of the albedo $r_{ub}(n)$ for different amounts of cloudiness. It is seen that with an increase in the cloud cover n, the mean, maximum, and minimum values of $r_{ub}(n)$ all increase.

Table 6.2. Statistical data on albedo of upper boundary of cloud layer r_{ub} (%).

Parameter	n									
	0	1–2	2–3	3–4	4–5	5–6	6–7	7–8	8–9	9–10
Mean	14.5	18.8	19.1	20.7	22.9	25.0	27.2	31.0	43.6	54.5
Max.	22.0	25.3	25.7	35.2	43.2	43.3	39.5	41.1	66.8	78.8
Min.	5.0	11.2	10.9	11.0	13.7	14.5	19.3	21.4	21.5	39.0
σ	2.1	3.1	2.9	3.7	4.9	5.3	5.4	6.3	12.2	10.4
v	8.7	16.0	15	18	21	21	20	20	28	19
δ	1.4	2.3	2.3	2.8	3.6	4.0	4.5	5.6	9.2	8.2
N	242	38	63	60	68	57	41	29	19	28

Note. Here σ is the standard deviation, v is the variation coefficient, δ is the arithmetic mean of the absolute deviations from the mean value, and N is the number of observations.

An analysis of the experimental data indicates that the relationship between $r_{ub}(n)$ and n can be approximated satisfactorily by the expression

$$r_{ub}(n) - r_{ub}(0) = Q_{ub}(n) - Q_{ub}(0) = \frac{n}{0.826 - 0.060n}, \qquad (6.4.4)$$

where the albedo is in percent, and n is in points on the scale.

The coefficient of correlation between the actual values of $r_{ub}(n)$ and those calculated using expression (6.4.4) is quite high (0.91 ± 0.01). The standard deviation of the values of $r_{ub}(n)$ calculated using (6.4.4) from the actual values is 4.5%, the arithmetic mean of the absolute deviations of these values being 3.3%.

The albedo $r_{c1}(n)$ of the clouds themselves, determined using (6.4.3), remains almost constant (31-34%) for an increase in n to 6-7, after which it rises rapidly, approaching $r_{ub}(n)$ at $n = 10$. This variation of $r_{c1}(n)$ can be explained as follows: When the cloudiness is slight ($n \leqslant 6$), the distances between clouds are much greater than the cloud depth, and individual clouds behave independently of one another with regard to radiation. As n increases, the geometrical dimensions of individual clouds, including the cloud depth, increase (for the observations in question, the mean cloud depth varied from 0.40 km for $n = 1$ to 1.94 km for $n = 10$). The radiation scattered by the sides of individual clouds reaches the surfaces of other clouds and is rescattered. Consequently, the reflectivity of the cloud layer as a whole rises, as does the absorptivity, because the length of the path of solar rays in the cloud layer increases. In such a case, the albedo $r_{ub}(0)$ of the cloudless region at the upper cloud boundary cannot be assumed to be equal to the albedo r_{1b} of the atmosphere at the lower boundary (thus formula (6.4.3) is not accurate for high n).

An analysis of the experimental data shows that the albedo $r_{ub}(n)$ of a cloud layer depends slightly on the cloud depth (for instance, for $n = 4$-6 the coefficients of correlation between $r_{ub}(n) - r_{ub}(0)$ and the cloud depth are only 0.19 ± 0.10). This is because the albedo of the clouds themselves varies considerably with a rise in cloud depth only for shallow depths (radiation is scattered mainly in the upper part of the cloud). For instance, the cloud albedo increases from 40 to 72% as the cloud depth rises from 100 to 400 m; a further deepening of the cloud to 700 m raises the albedo by only 10%. Since in the observations being discussed the cloud depths did not, as a rule, exceed 400 m, a variation in this depth did not significantly affect the albedo r_{c1} of the clouds themselves, and this did not significantly affect the albedo of the cloud layer as a whole either.

6.5. Cloud formation in a mountainous region

Irregularities in the relief of the land disturb the air flow. The disturbed motions constitute a system of waves, on the ascending branch of which clouds may form.

A great number of studies have been made of wave motions caused by irregularities in the Earth's surface, the first of which, by Kochin, Dorodnitsyn, and Kibel', have become classic studies.

Kozhevnikov [132, 133] considered the movement of a stream of air around a long ridge of cylindrical shape. Far ahead of the ridge, the oncoming flow is assumed to be undisturbed, and the airstream properties do not depend on coordinate y, directed along the ridge (plane problem).

If the influence of turbulence and the Coriolis force are neglected and steady-state motion is assumed, the system of nonlinear equations incorporating the simplifications of free-convection theory can be written as follows:

$$u\frac{\partial u}{\partial x}+w\frac{\partial u}{\partial z}=-RT_1\frac{\partial}{\partial x}\left(\frac{p'}{\tilde{p}}\right), \tag{6.5.1}$$

$$u\frac{\partial w}{\partial x}+w\frac{\partial w}{\partial z}=-RT_1\frac{\partial}{\partial z}\left(\frac{p'}{\tilde{p}}\right)+\alpha T', \tag{6.5.2}$$

$$\frac{\partial u}{\partial x}+\frac{\partial w}{\partial z}=0, \tag{6.5.3}$$

$$u\frac{\partial T'}{\partial x}+w\frac{\partial T'}{\partial z}=(\gamma-\gamma_a)\,w. \tag{6.5.4}$$

Here, u and w are, respectively, the velocity components in the x (normal to the ridge) and z (vertically upward) directions; p' and T' are the deviations of the pressure and temperature from the values $\tilde{p}$ and $\tilde{T}$ in an undisturbed flow; T_1 is the mean characteristic temperature; γ is the lapse rate in an undisturbed flow; γ_a is the dry-adiabatic lapse rate; $\alpha = g/T_1$ is the buoyancy parameter; g is the free-fall acceleration; and R is the specific gas content of air.

The system (6.5.1)-(6.5.4) was obtained from the usual system of equations of thermohydrodynamics on the basis of the following simplifications:

$$\frac{1}{\rho}\frac{\partial p}{\partial x}=\frac{RT}{p}\frac{\partial p}{\partial x}=RT\frac{\partial \ln p}{\partial x}=$$
$$=RT\frac{\partial}{\partial x}\left[\ln\tilde{p}+\ln\left(1+\frac{p'}{\tilde{p}}\right)\right]\approx RT\frac{\partial}{\partial x}\left(\frac{p'}{\tilde{p}}\right),$$
$$\frac{1}{\rho}\frac{\partial p}{\partial z}=RT\left[\frac{\partial\ln\tilde{p}}{\partial z}+\frac{\partial}{\partial z}\left(\frac{p'}{\tilde{p}}\right)\right],$$

since $\tilde{p} = \tilde{p}(z)$ is a function of z only.

The pressure $\tilde{p}$ in the undisturbed flow satisfies the hydrostatic equation

$$\frac{\partial p}{\partial z}=-g\tilde{\rho}=-\frac{g\tilde{p}}{R\tilde{T}},$$

so that

$$\frac{1}{\rho}\frac{\partial p}{\partial z}=RT\frac{\partial}{\partial z}\left(\frac{p'}{\tilde{p}}\right)-g\frac{T}{\tilde{T}}=RT\frac{\partial}{\partial z}\left(\frac{p'}{\tilde{p}}\right)-g-\alpha T'.$$

In eqn. (6.5.4) the derivative $\partial p'/\partial z$ was also neglected in comparison with $\partial\tilde{p}/\partial z$, and in the coefficients of (6.5.1) and (6.5.2) the temperature T was replaced by its mean constant value T_1.

The continuity equation (6.5.3) permits the introduction of the stream function ψ:

$$u=\frac{\partial\psi}{\partial z},\qquad w=-\frac{\partial\psi}{\partial x}. \tag{6.5.5}$$

The reader will recall that the ψ = const curves represent streamlines, with the velocity vector tangent to them at all points. The vorticity component perpendicular to the xz plane is

$$\Omega_y = \frac{\partial u}{\partial z} - \frac{\partial w}{\partial x} = \frac{\partial^2 \psi}{\partial x^2} + \frac{\partial^2 \psi}{\partial z^2} = \Delta\psi. \tag{6.5.6}$$

The boundary conditions for the unknown functions have the following form:

(a) At a sufficiently great distance in front of the mountain, the velocity and temperature are undisturbed, that is,

$$\left.\frac{\partial \psi}{\partial x}\right|_{x=-\infty} = 0, \quad T'\big|_{x=-\infty} = 0, \quad \left.\frac{\partial \psi}{\partial z}\right|_{x=-\infty} = U_0, \tag{6.5.7}$$

where U_0 is the velocity of the undisturbed flow

(b) The condition of "smoothing" is imposed on the surface of the mountain:

$$\psi\big|_{z=h(x)} = 0, \tag{6.5.8}$$

meaning that one of the streamlines coincides with the mountain profile; here, $h(x)$ is the mountain profile in the xz plane.

Now, if $p'/\tilde{p}$ is eliminated from eqns. (6.5.1) and (6.5.2), and relation (6.5.5) is used, the result is the system of equations

$$\alpha \frac{\partial T'}{\partial x} = \frac{\partial(\psi, \Delta\psi)}{\partial(x, z)},$$

$$(\gamma - \gamma_a)\frac{\partial \psi}{\partial x} = \frac{\partial(\psi, T')}{\partial(x, z)}, \tag{6.5.9}$$

where $\Delta = \partial^2/\partial x^2 + \partial^2/\partial z^2$ is the Laplacian, and $\partial(A, B)/\partial(x, z) = (\partial A/\partial x)(\partial B/\partial z) - (\partial A/\partial B)(\partial z/\partial x)$ is the Jacobian.

The first integrals of the system (6.5.9) have the form

$$T' = f(\psi) + (\gamma - \gamma_a) z, \tag{6.5.10}$$

$$\Delta\psi = f_1(\psi) + \alpha z \frac{df}{d\psi},$$

where f and f_1 are arbitrary functions of the variable ψ, whose exact form is determined unambiguously by conditions (6.5.7) and by the dependence of the wind velocity U_0 on height in an undisturbed flow.

In the general case, when the velocity U_0 of the undisturbed flow is a function of height, the equation for ψ is nonlinear. However, in the particular case where U_0 = const, the equation for the stream function assumes the form of the linear inhomogeneous Helmholtz equation

$$\Delta\psi + k^2\psi = -k^2 U_0 z, \quad (6.5.11)$$

where

$$k^2 = \frac{g}{T_1}\frac{\gamma_a - \gamma}{U_0^2}, \quad (6.5.12)$$

and the equation for T' has the form

$$T' = k^2 (U_0/\alpha)(\psi + U_0 z). \quad (6.5.13)$$

The solution of the Helmholtz equation is found by separation of variables in the form of a series expansion of cylindrical functions. In addition to conditions (6.5.7) and (6.5.8), the condition of an especially rapid damping of the flow perturbations counter to the mainstream is imposed:

$$\lim_{\substack{r\to\infty \\ \varphi=\pi}} r^{m+\frac{1}{2}}\psi' = 0, \quad m = 0, 1, 2, 3, \ldots, \quad (6.5.14)$$

where r and φ are polar coordinates (r is the distance from the mountain axis, and φ is the polar angle).

The form of the solution depends on the sign of the stratification parameter k^2. If the stratification is unstable ($\gamma > \gamma_a$), then $k^2 < 0$, and the motion has the nature of ordinary damping. For stable stratification ($\gamma < \gamma_a$), on the other hand, $k^2 > 0$ and the motion will be oscillatory.

The parameter $\sqrt{g(\gamma_a - \gamma)/T_1} = kU_0$ was introduced in 1925 by the Finnish scientist Väisälä. The quantity

$$\frac{1}{2\pi}\sqrt{g(\gamma_a - \gamma)/T_1} = \frac{kU_0}{2\pi}$$

is the frequency f_n of natural oscillations of an air volume, originally at rest in a stably stratified layer of the atmosphere:

$$f_n = kU_0/2\pi.$$

Parameter k is inversely proportional to the wavelength λ_n of the trajectory of motion of an air particle:

$$\lambda_n = U_0/f_c = \frac{2\pi}{k}. \quad (6.5.15)$$

The quantity f_n is called the Väisälä-Brunt frequency.

The Helmholtz equation was solved by Long (1953), assuming the presence of a solid horizontal surface bounding the flow at the top. It was also solved by Lira (1943) for the case of a linearized equation, and by Mörbt (1959) for a linearized boundary condition. In recent decades a series of studies were carried out by Dorodnitsyn, Kibel', and their students (Zeitunyan, Pekelis, and Musaelyan). The nature of the motion was found to depend considerably on the shape of the object around which the air flows.

Kozhevnikov [132] investigated the case of a mountain profile in the vertical xz plane that has the shape of a semicircle of radius r_0. The solution of eqns. (6.5.11) and (6.5.13) for the boundary conditions (6.5.7), (6.5.8), and (6.5.14) will, in this case, be

$$\frac{\psi}{U_0 r_0} = \left[\frac{I_1(\eta_0 R)}{I_1(\eta_0)} - R\right] \sin\varphi + \sum_{n=1}^{N} b_n \left[Y_n(\eta_0 R) - a_n I_n(\eta_0 R)\right] \sin n\varphi, \quad (6.5.16)$$

$$\frac{T'}{(\gamma_a - \gamma) r_0} = -\left(\frac{\psi}{U_0 r_0} + R \sin\varphi\right), \quad (6.5.17)$$

where $R = r/r_0$; $a_n = Y_n(\eta_0)/I_n(\eta_0)$; I_n and Y_n are, respectively, a Bessel function of the first order and a Weber function; and

$$\eta_0 = k r_0 = 2\pi r_0/\lambda_n = 2\pi \frac{f_n r_0}{U_0}. \quad (6.5.18)$$

While solving (6.5.16), it was assumed that only a finite number N of coefficients b_n differ from zero, while all the other $b_n = 0$.

In the particular case where $\gamma \to \gamma_a$ (the stratification is almost negligible), the stream function becomes

$$\lim_{\gamma \to \gamma_a} \psi = \lim_{\eta_0 \to 0} \psi = -U_0 \left(\frac{r_0^2}{r} - r\right) \sin\varphi. \quad (6.5.19)$$

This result is identical to the classical solution of the problem of potential flow around a circular cylinder. The streamlines in the given case are curved symmetrically over the mountain, and with increasing distance from it they rapidly become horizontal. The limiting case $r_0 \to 0$ corresponds to both η_0 and r_0 approaching zero. Consequently, from (6.5.19) it follows that

$$\lim_{r_0 \to 0} \psi = U_0 r \sin\varphi = U_0 z. \quad (6.5.20)$$

The foregoing results pertain to $N = 4$.

Figures 6.8 and 6.9 show the dimensionless characteristics of a field of motion disturbed by a mountain: streamlines $\psi/U_0 r_0$, the temperature deviations $T' = T'/(\gamma_a - \gamma) r_0$, and vertical velocity w/U_0.

R

Orographic disturbances of a flow have the following properties: (1) They have a wave nature, with considerable damping with increasing distance from the mountain, especially in the $\varphi = \pi$ direction.(2) The intensity of the disturbances increases with an increase in parameter η_0, reaching a maximum at intermediate levels, while close to the ground it is small, in accordance with the boundary conditions.(3) The flow disturbances are most developed in the sector between 50 and 70°, as a result of condition (6.5.14). (4) The streamlines experience reversal, and the regions of ascent and descent have almost everywhere a large vertical extent comparable to the height of the mountain, the transitions between them being quite sudden. The frequency of line reversal increases with an increase in η_0.

The deviations of the vertical velocity will now be considered in greater detail. Right next to the lee slope of the mountain, there is a region of downdrafts. Here $|w|$ reaches a maximum; further away from the mountain $|w|$ decreases, and the sign of w alternates. At the centers of regions located between 40 and 50°, the ratio w/U_0 varies as follows with increasing distance from the mountain: -0.9, +0.4, -0.3, +0.3, etc., for $\eta_0 = 1$; -3.5, +2.0, -1.6, +1.4, etc., for $\eta_0 = 2$; -8.0, +7.2, -5.4, +4.6, etc., for $\eta_0 = 3$. For specified values $r_0 = 1$ km and $\gamma = 6.08$°C/km, the velocity U_0 will be equal to 11.8, 5.9, and 3.9 m/s, for values of

$$\eta_0 = \frac{r_0}{U_0}\sqrt{\frac{g}{T_1}(\gamma_a - \gamma)} = 2\pi\frac{r_0}{\lambda_c}, \tag{6.5.21}$$

equal to 1, 2, and 3, respectively. Therefore, at the centers of regions close to the mountain, the highest absolute values of the downdraft velocity will correspond to -10.6, -20.6, and -30.8 m/s, while at the centers of the updraft regions closest to these the velocities will be 4.7, 11.8, and 27.7 m/s, respectively, for η_0 = 1, 2, and 3.

Kozhevnikov [133] not only calculated the velocity components, he also estimated the temperature variation caused by disturbances of the velocity field. For these same values (r_0 = 1 km and γ = = 6.08+C/km), the highest temperature deviations T' at the centers of regions with closed isotherms (see Figs. 6.8 and 6.9) and the dimensionless deviations τ' are

	$\eta_0 = 1$			$\eta_0 = 2$		
T',°C	—3.40	+2.65	—1.89	—3.40	+6.43	—4.16
τ'	—0.9	+0.7	—0.5 etc.	—0.9	+1,7	—1,1 etc.

The temperature perturbations depend greatly on the stability of the oncoming flow and on the height of the mountain. With an increase in $(\gamma_a - \gamma)$, the absolute values of τ', and especially of T', increase as well. It should be noted that as η_0 increases, the deviations τ' and T' go through a maximum not in the region near the ground, but rather in the next region of the disturbances of T.

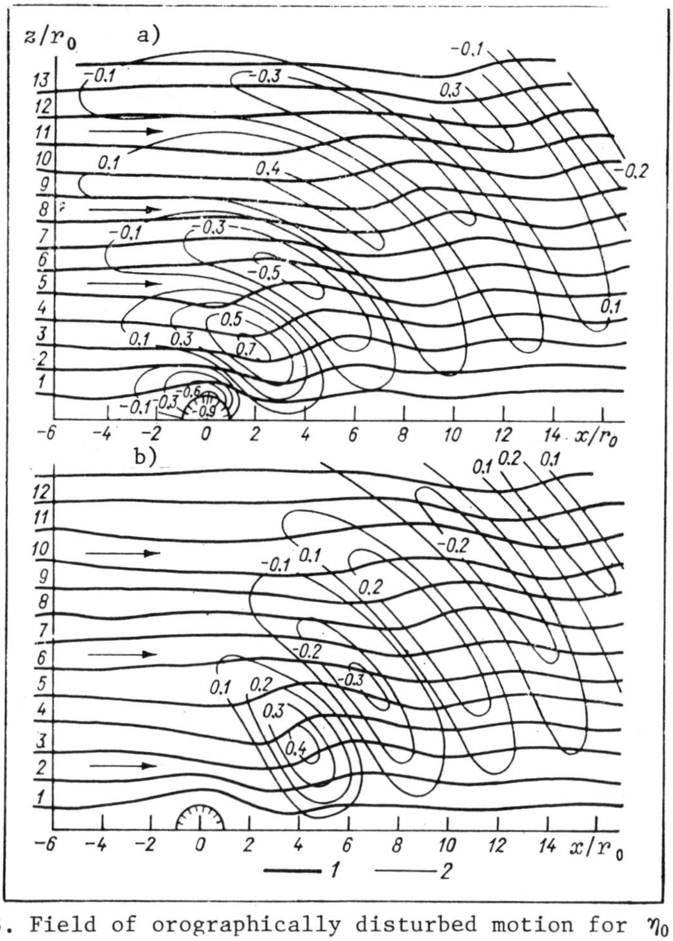

Fig. 6.8. Field of orographically disturbed motion for $\eta_0 = 1$: in Fig. 6.8a, (1) streamlines $\psi/u_0 r_0$ = const, (2) isotherms τ' = = const;
in Fig. 6.8b, (1) streamlines $\psi/u_0 r_0$ = const, (2) isolines w/u_0 = = const.

An important conclusion following from the foregoing estimates of w and T' and from Figs. 6.8 and 6.9 is noteworthy. It is generally assumed that in a region of upward motions the air becomes cooled and clouds form, while in a region of downdrafts the air warms up and clouds dissipate. The data being examined, however, not only do not verify this widely held opinion, but in fact they indicate that just the opposite is true: in downdraft regions ($w < 0$) negative temperature deviations ($T' < 0$) prevail, whereas in regions of upward motions ($w > 0$) positive deviations ($T' > 0$) are predominant. This is because over an irregular terrain, the temperature varies because of advective and convective influxes of heat (terms on the left-hand side of eqn. (6.5.4)), which play

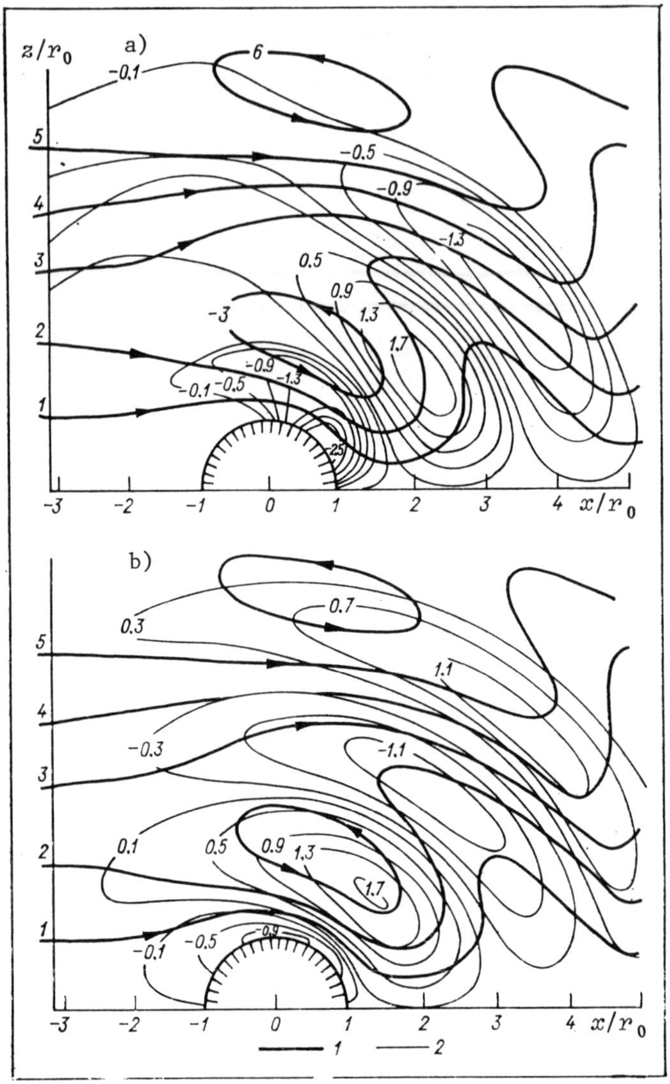

Fig. 6.9. Field of orographically disturbed motion for $\eta_0 = 2$: in Fig. 6.9a, (1) streamlines $\psi/u_0 r_0$ = const; (2) isolines w/u_0 = = const;
in Fig. 6.9b, (1) streamlines $\psi/u_0 r_0$ = const; (2) isotherms τ' = const.

a decisive role in the process, and because of vertical motions (term on the right-hand side of eqn. (6.5.4)).

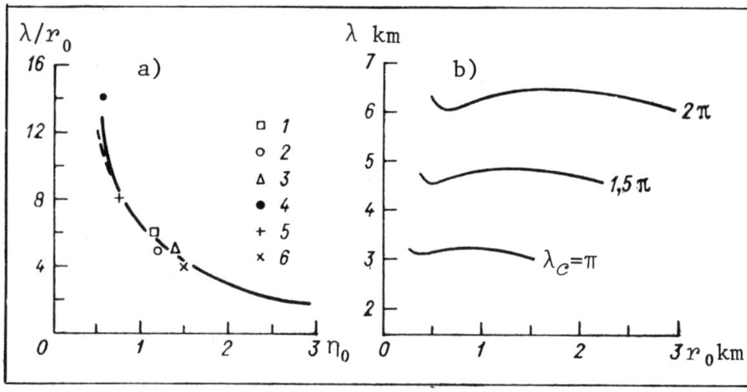

Fig. 6.10. Length of lee waves: λ/r_0 as function of η_0 (a) and λ as function of r_0, for three values of λ_n (b) Dates of observations: (1) 16 February 1952, (2) 28 June 1958, (3) 31 July 1958, (4) 18 April 1960, (5) 3 May 1962, (6) 2 August 1962.

In a disturbed region of the atmosphere, its static stability varies. With the aid of the relation

$$-\frac{\partial T}{\partial z} = \gamma - \frac{\partial T'}{\partial z} = \gamma - (\gamma_a - \gamma)\, r_0 \frac{\partial \tau'}{\partial z}, \qquad (6.5.22)$$

we see that for $\gamma = 6.08°C/km$, inside the region of heightened stability the lapse rates ($-\partial T/\partial z$) decrease to no less than 4.19°C/km for $\eta_0 = 1$ and to 2.3°C/km for $\eta_0 = 2$, whereas inside the region of heightened instability the lapse rates go up to no less than 7.97 and 13.64°C/km, respectively.

One of the most remarkable features of the velocity field is the presence of closed eddies, which arise at a value of η_0 between 1 and 2. For $\eta_0 = 1$ there are, according to Fig. 6.8, no closed eddies, while for $\eta_0 = 2$ over a mountain there will be at least two closed eddies, and even higher up there will be several open eddies. For $\eta_0 = 3$ over a mountain there will be more than six closed eddies, and in addition at some distance downstream from the mountain, extending right down to the ground, there is still another eddy. The motion of the air in the first six eddies is counterclockwise, but in the eddy near the ground it is clockwise. The appearance of these eddies is accompanied by considerable enhancement of all the disturbances. For $\eta_0 = 3$, velocity components u and w are more than 7 times as great at U_0, and in the eddies over the mountain u and w reach their highest values in the rear and upper parts; here, however, the velocity and temperature change the most from point to point.

In the eddies with counterclockwise circulation, the temperature deviations $T' > 0$; consequently, conditions in them are unfavorable for cloud formation. In the eddy with clockwise circulation, $T' < 0$, so that conditions there are favorable for cloud formation.

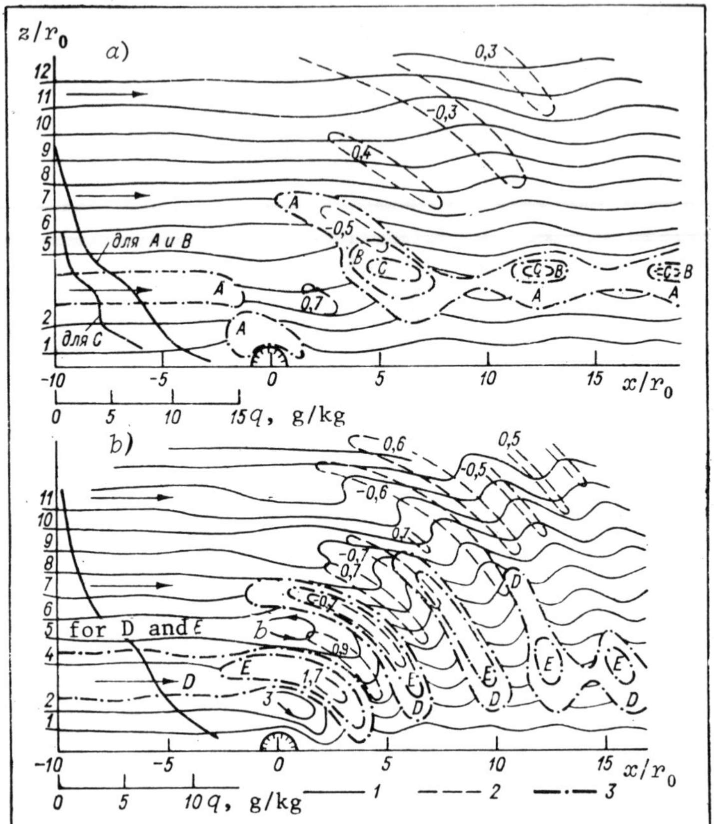

Fig. 6.11. Streamlines $\psi/u_0 r_0$ (1), temperature deviations τ' (2), and cloud boundaries (3) for $\eta_0 = 1$ (a) and $\eta_0 = 2$ (b).

Inspection of Figs. 6.8 and 6.9 indicates that the wavelength λ (the distance between two successive crests) may vary greatly, especially in the vicinity of a closed eddy. As one moves along a streamline, ratio λ/r_0, as a rule, decreases to half and approaches some minimum value in the lee part of the flow, this minimum apparently being the same (if η_0 = const) for the entire stream of air. For η_0 values of 0.5, 1, 2, and 3, the minimum ratios λ/r_0 will be, respectively, 12.7, 6.3, 3.0, and 2.0. Since as $\eta_0 \to 0$ the ratio $\lambda/\gamma_0 \to 0$ (because in this case only one crest is observed), the dependence of the λ/r_0 minima on η_0 can be approximated by the function

$$\lambda/r_0 = a\eta_0^{-b} \exp(-c\eta_0), \qquad (6.5.23)$$

where a = 7.527, b = 0.7204, and c = 0.1780. In view of (6.5.21), which states that $\eta_0 = 2\pi r_0/\lambda_n$, relation (6.5.23) can also be used to determine λ as a function of r_0 and λ_n. The results of the calculations of λ/r_0 and λ are shown in Fig. 6.10. According to Fig. 6.10 b, the length of the lee waves depends only slightly on r_0, the main factor determining λ being λ_n, the characteristic of the undisturbed (oncoming) flow: from relations (6.5.12) and (6.5.15) it follows that with an increase in U_0 and γ (the stratification approaches dry-adiabatic), λ_n increases as well, and along with it the length λ of the lee waves.

Kozhevnikov [133] was apparently the first to try to calculate the field of orographic cloudiness. He did not, to be sure, use the equation of moisture transport, but instead made the elementary assumption that the specific humidity remains constant along a streamline. The cloud boundaries were found from the points on the streamline where, for a given q, the air became saturated ($q = q_m$) because of T'. The variation of q from one streamline to another was assumed to be given, in the undisturbed flow ahead of the mountain.

Figure 6.11 shows the results of calculating the cloudiness fields. Each example differs from the others by the particular idiosyncrasies of the oncoming flow. In examples A, B, D, and E, the vertical profiles $q(z)$, shown at the left of the figure, and the values γ = 6°C/km and r_0 = 1 km, are the same; whereas in examples A and B, $\eta_0 = 1$; in examples D and E, $\eta_0 = 2$. Examples A and D differ from examples B and E in that for A and D the air temperature near the ground is 20°C, while in B and E it is 2°C higher, that is, the relative humidity is lower. Example C was calculated for $\eta_0 = 1$ and for values of γ, r_0, T_1, and $q(z)$ recorded during observations in the Crimean upland on 13 May 1962.

A comparison of the cloudiness fields with the fields of the velocity and the temperature deviations shows that the relation between these is far from single-valued. The cloudiness is influenced markedly by the distribution with height of the specific humidity and temperature in the oncoming flow. In example A the sky is overcast ahead of the mountain, there is a billow over the mountain, and there is a continuous cloud cover of fluctuating thickness to leeward of the mountain. In example B, which differs from A only in that the relative humidity near the ground is lower, clouds appear only in the lee waves, and, beginning with the second wave, they are typically Ac lent. The thickness of the billows is reduced even more in example C, which differs from A and B in that q is lower and varies more sharply with height.

The cloudiness field is changed significantly if η_0 varies. In examples D and E (Fig. 6.11 b), there is no billow close to the mountain; further downstream, there is no continuous cloud cover (as in example A), while the billows in the lee waves are much more developed along the vertical, the bend of their vertexes toward the main flow being quite pronounced. In example E, the first and second billows, as seen by an observer located some distance ahead of the mountain, can resemble multilayer Ac lent., which are sometimes seen in the mountains.

In order to ascertain the quantitative characteristics of lee waves and clouds, Dyubyuk organized expeditions in Southern Crimea. In addition to observations in the layer near the ground and radiosonde measurements in the atmosphere, these expeditions included cloud photography designed to determine λ_{exp}, the distance between the crests of two successive billows. In the theory being considered, parameters U_0 and γ are assumed to be independent of height; therefore, for the theory to be applied, the mean values of U_0 and γ in the troposphere had to be found.

Table 6.3. provides a comparison of λ_{exp} and λ_{cal}. The last two examples in the table were borrowed from Döös [352], who analyzed the results of lee-wave studies carried out with aircraft (California) and the *Tiros 1* satellite (Patagonian Cordillera). Considering the approximate nature of the theory, as well as the errors in determing γ, U_0, and λ_{exp}, the fit between λ_{cal} and λ_{exp} is satisfactory. It was also established that in some cases there is agreement between the oscillation amplitudes. For instance, in the Sierra Nevada, observations gave the amplitude of the first lee wave as 0.76 km, while calculation gave 0.92 km; for the Crimea on 2 August 1962, the corresponding values were 0.30 and 0.24 km. However, there were also some cases of sharp divergence between the theoretical predictions and the experimental data. For example, for 13 May 1962, theory predicted no more than three cloud bands, and in reality no less than eight bands appeared; for 2 August 1962, theory did not predict cloud formation, but it was indeed observed. Such discrepancies are attributable to the roughness of the theory and to the errors in measuring the temperature, and especially the humidity, of the air, to which cloudiness is very sensitive.

Table 6.3. Parameters of lee waves.

Date	T_1,K	γ,K/km	U_0,m/s	r_0,km	η_0,km	λ_{cal},km	λ_{exp},km
			Crimea				
28 VI 1958	281	6.8	7.8	0.8	1.06	4.8	4.4
31 VII 1958	281	5.3	10.0	1.1	1.37	6.2	5.5
13 V 1962	273	7.0	11.0	0.8	0.73	6.6	6.5
2 VIII 1962	268	6.3	11.7	1.2	1.17	6.5	6.0
			Sierra Nevada				
16 II 1952	250	6.0	31.0	2.8	1.11	16.0	16.8
			Argentina				
18 VI 1960	273	5.0	20.0	0.8	0.53	9.2	11.3

Borovikov and Mazin [36] used data from aircraft observations over the western part of the European USSR in September-October 1964 to compare the direction of banks of St-Sc (α_{cl}) with the wind direction (α_w) at the cloud level. The frequency P of the difference α_{cl} - α_w (total number of cases 20) is as follows:

$\alpha_{cl} - \alpha_w$. . .	<—25	—25... —16	—15... —6	—5... 5	6... 15	>15
P, %	15	0	30	40	15	0

According to these data, in 85% of the cases the direction of the cloud banks deviated from the wind direction toward one side or the other by no more than 15° (practically within the limits of measurement accuracy). The authors of [36] pointed out the curious fact that whenever α_{cl} differed from α_w by more than 15°, the cloud banks were shifted to the left of the wind direction.

The ratio of the distance between banks to the cloud depth fluctuates (according to the data of 13 observations) from 2.5 to 23.3, but 76% of the time this ratio was between 5 and 15.

Bibikova [27] detected a rare species of altocumulus, namely "screw clouds", or "helical clouds", on the lee side of the main ridge of the Caucasus, in the vicinity of Elbrus, at a height of 9-11 km under the stable layer (γ = 0.25°C/100 m) with a considerable wind shear (from 6 to 12 m/s in 100 m), in a jet stream almost perpendicular to the ridge (the wind speed at the jet axis reached 45 m/s). The lengths of two screw clouds were 16 and 9 km, the distance between them (normal to the stream) was 7 km, and the screw pitch (first cloud) was 0.6 km.

CHAPTER 7

VERTICAL STRUCTURE OF A CLOUD FIELD

7.1. Correlations between cloudiness characteristics

The presence of clouds has considerable effect on the structure of the temperature and humidity fields. In recent decades it has become possible to use data from aircraft soundings of the atmosphere to ascertain the statistical characteristics of cloud heights, temperature fields, and humidity fields close to cloud boundaries.

The autocorrelation and cross-correlation functions are calculated using the relations

$$r_{ff}(z_k, z_l) = \frac{B_{ff}(z_k, z_l)}{\sigma_f(z_k)\,\sigma_f(z_l)}, \tag{7.1.1}$$

$$r_{f\varphi}(z_k, z_l) = \frac{B_{f\varphi}(z_k, z_l)}{\sigma_f(z_k)\,\sigma_\varphi(z_l)}. \tag{7.1.2}$$

Here $\sigma_f^2(z_k) = \frac{1}{N}\sum_{i=1}^{N}[f_i(z_k) - \bar{f}(z_k)]^2$ is the variance of quantity f at level z_k; $\sigma_f(z_k)$ is the standard deviatior; $B_{ff}(z_k, z_l)$ and $B_{f\varphi}(z_k, z_l)$ are the autocorrelation and cross-correlation functions, defined by the formulas

$$B_{ff}(z_k, z_l) = \frac{1}{N}\sum_{i=1}^{N} f_i'(z_k)\, f_i'(z_l), \tag{7.1.3}$$

$$B_{f\varphi}(z_k, z_l) = \frac{1}{N}\sum_{i=1}^{N} f_i'(z_k)\, \varphi_i'(z_l), \tag{7.1.4}$$

where $f_i'(z_k) = f_i(z_k) - \bar{f}(z_k)$ is the deviation (variation) of f from the mean value $\bar{f}$ at level z_k for the ith application; $\varphi_i'(z_l) = \varphi_i(z_l) - \bar{\varphi}(z_l)$ is the deviation of φ from the mean value $\bar{\varphi}$ at level z_l; N is the number of observations (i = 1, 2, ..., N). The following quantities were taken as f and φ: the temperature T, the specific humidity q, the height of the lower cloud boundary h_1, the height of the upper cloud boundary h_2, and the height h_3 of the top of the inversion layer above the cloud.

Malkevich and Polovina [167] were among the first to try to evaluate the correlations among the fields of temperature, humidity, and cloudiness. They made use of data from network aircraft soundings of stratus and stratocumulus clouds (110 soundings, during the cool and warm seasons).

Devyatova [85-87] used a considerable volume of data to investigate the structure of Sc-St clouds. She analyzed the results of aircraft soundings of the atmosphere in the Moscow region over a period of 13 years (1956-1968), a total of 8755 flights. In 397 cases (about 5% of the total), solid single-layer stratocumulus and stratus were observed, with an upper-boundary height not greater than 2.5 km. In 88% of the cases, the clouds were under the temperature-inversion layer, while in 78% of the cases the beginning of inversion coincided with the upper cloud boundary. Only the latter 290 cases were used to calculate the statistical characteristics of the cloudiness, temperature, and humidity fields to be given below.

During the cold season (October-March), which is the most favorable time for the formation of Sc-St, the frequency of occurrence of the clouds stayed about the same (23 cases in October, 39 in November, 33 in December, 36 in January, 42 in February, and 27 in March - a total of 200 cases). During the warm season, which in view of the temperature conditions included some of the soundings in October and November, Sc-St clouds formed most frequently in early spring and in autumn (90 cases in all). Stratocumulus accounted for 79%, and stratus 21%. Note that since the conditions for formation of these clouds are the same, it is justifiable to consider them a single species. According to the classification, St-Sc are wave clouds, not layer clouds (frontal Ns-As-Cs are known as layer clouds).

According to Table 7.1, the mean heights h_1 of the lower cloud boundary and h_2 of the upper boundary over Moscow during the warm season are 300-400 m greater than during the cold season: $\bar{h}_1$ increases from 0.51 to 0.90 km, and $\bar{h}_2$ from 0.86 to 1.22 km. However, the cloud depth $\bar{h}_2 - \bar{h}_1$ remains about the same for the two seasons: 0.35 km during the cold season, and 0.32 km during the warm.

Table 7.1. Statistical characteristics of clouds and inversions, Moscow ($\Delta\bar{h}$ is absolute error in determining $\bar{h}$).

	h km	σ_h km	$\Delta\tilde{h}$ km	$\bar{T}$°C	σ_T°C	$\tilde{q}$, g/kg	σ_q, g/kg
			Cold season				
h_1	0.51	0.37	0.03	−8.6	4.95	2.0	0.83
h_2	0.86	0.38	0.03	−11.0	4.70	1.8	0.72
h_3	1.13	0.40	0.03	−5.7	5.73	2.0	0.95
			Warm season				
h_1	0.90	0.57	0.06	0.1	5.75	4.2	1.79
h_2	1.22	0.54	0.06	−2.0	5.53	3.9	1.64
h_3	1.46	0.58	0.06	2.3	5.86	3.9	1.90

The standard deviation σ_h is comparable to the mean cloud depth, and during the warm season may even exceed it by almost a factor of two. These data indicate the high variability in time of h_1, h_2, and h_3.

The data in Table 7.1 agree with current physical thinking about the influence of various factors on the cloud height. According to the author of [192], one of the formulas for the height h_1 of the lower boundary has the form

$$h_1 = D(T_0 - \tau_0),$$

where $T_0 - \tau_0$ is the dew-point deficit close to the ground (at a height of 2 m); D is a multiplier depending mainly on the dew point τ_0 (an increase in τ_0 is accompanied by an increase in D) and on the lapse rate γ between the ground and the cloud (an increase in γ leads to a decrease in D). Since the dew-point deficit $T_0 - \tau_0$, and the dew point τ_0 itself, are considerably higher in summer than in winter, while the lapse rates γ remain practically the same (according to Devyatova [85], the mean values of γ are 0.76°C/100 m during the warm season and 0.81°C/100 m during the cold season), it follows from this formula that the height h_1 must be greater during the warm season that during the cold. The data of Table 7.1 are in accordance with this conclusion. A lifting of the boundaries of St-Sc clouds from winter to summer is also observed in other places, for instance, at Riga and Sverdlovsk [318], and at Hamburg, Lindenberg, and Helsinki [35].

7.1.1. The vertical structure of the temperature field

The temperature distribution in a cloud and close to the cloud boundaries has been studied in detail by Zak [107], Borovikov et al. [35], Polovina [253], Khrgian [318], and others. In addition, Devyatova [85, 86] has analyzed the most complete data available on this subject. Aircraft soundings carried out over Moscow from 1934 to 1969 (a total of 17,967 soundings) revealed single-layer St-Sc clouds in 1068 cases, that is, a frequency of about 6% for these clouds in the Moscow region. Devyatova divided the vertical temperature profiles into six types.

Types I and II are characterized by a temperature drop from the ground up to the cloud summit. Above the cloud there is a temperature inversion, which begins either at the upper cloud boundary (type I) or somewhat below it (100 m, on the average 1/M - type II). The frequency of types I and II together is 83%, whereas for the period from 1958 to 1969, when the cloud boundaries were specified precisely with the aid of the electrical--contact stylus of an airborne A-10 meteorograph, type II profiles accounted for only 5% (out of 394 events). Type V (frequency 5%) differs from types I and II only in that the above-cloud temperature inversion begins somewhat (240 m, on the average) above the upper cloud boundary. Thus, in 88% of the cases St-Sc clouds are situated below the inversion. For a profile of type I, the mean depth of the temperature is 280 m during the cold season and 250 m during the warm season (the maximum frequency of the inversion depth is in the interval from 110 to 200 m). In the type III profile (frequency 3%), the temperature drops with height, both in the cloud and above it. For types IV and VI (frequencies of 5 and 4%, respectively), there is typically an inversion throughout the cloud (type IV) or most of the cloud (type VI); above the cloud, the temperature drops with increasing height.

The following table gives the mean values of the lapse rates (°C/100 m) under the cloud (γ_1), in the cloud (γ_2), and in the inversion layer above the cloud (γ_3), together with the standard deviations of these quantities for type I:

Season	γ_1	γ_1	γ_2	σ_{γ_1}	γ_3	σ_{γ_3}
Warm (N = 95)	0.76	0.30	0.71	0.23	—2.15	1.61
Cold (N = 235)	0.81	0.53	0.72	0.32	—2.60	1.94

The maximum frequency of γ values under and in a cloud lies in the interval from 0.0 to 1.0°C/100 m (frequency in a cloud is 91-92%; under a cloud, 70-80%); the frequency of γ values from 1.0 to 2.0°C/100 m is 6-7% in a cloud for all seasons, while under a cloud it is 17% in the warm season and 25% in the cold season. The range of variation of γ in the inversion layer above the cloud is somewhat greater; here the frequencies P of the various gradations of γ are the following:

γ, °C/100 m	0.0... —1.0	—1.0... —2.0	—2.0... —3.0	—3.0... —4.0	—4.0... —5.0	<—5.0
P (%) during warm season	28	30	19	12	4	7
P (%) during cold season	17	34	19	10	8	12

Table 7.2 gives the autocorrelation functions for the temperatures at the boundaries of the cloud and the inversion layer for profiles of type I.

Table 7.2. Autocorrelation matrix of temperature.

	h_1	h_2	h_3
		Warm season	
h_1	1	0.98	0.89
h_2	0.96	1	0.88
h_3	0.81	0.85	1
	Cold season		

7.1.2. *The height of the cloud boundaries*

A close correlation exists between the deviations of all three heights h_1, h_2, h_3 (Table 7.3). It is seen from the table that the correlation coefficients r_{hh}, calculated using the formula

$$r_{h_ih_j} = \frac{1}{N} \frac{\sum_{k=1}^{N} \left(h_i^{(k)} - \bar{h}_i\right)\left(h_j^{(k)} - \bar{h}_j\right)}{\sigma_{h_i}\sigma_{h_j}},$$

are positive and quite high. For example, $r_{h_1h_2} = 0.88$ for the cold

Table 7.3. Coefficients of correlation r_{hh} between heights h_1, h_2, h_3.

	h_1	h_2	h_3
		Warm season	
h_1	1	0.96	0.92
h_2	0.88	1	0.95
h_3	0.74	0.84	1
	Cold season		

season and 0.96 for the warm season. Consequently, if one of the cloud boundaries rises or sinks, then, as a rule, so does the other boundary, as well as the upper boundary of the inversion layer; however, the attendant variation in the cloud depth $\Delta h = h_2 - h_1$ will be insignificant, as is seen from the low values of the coefficients of correlation between Δh and h_1 and between Δh and h_2

(the summer values of these coefficients are only 0.19 and 0.20).

Table 7.4. Coefficients of cross-correlation r_{T_h} between heights h_1, h_2, h_3 and temperature at these heights, and also between water contents δ and temperature at different heights.

	Cold season					Warm season				
	h_1	h_2	h_3	δ_1	δ_2	h_1	h_2	h_3	δ_1	δ_2
T_{h_1}	−0.33	−0.16	−0.15	0.38	0.34	−0.69	−0.67	−0.62	—	0.22
T_{h_2}	−0.34	−0.26	−0.24	0.31	0.24	−0.66	−0.68	−0.64	—	0.20
T_{h_3}	−0.48	−0.41	−0.27	0.29	0.28	−0.71	−0.72	−0.62	—	0.97
T_{δ_1}				0.39	0.28				—	—
T_{δ_2}				0.58	0.31				—	0.29

The coefficients of correlation between the heights h_1, h_2, h_3 and the temperatures at these heights turned out to be negative and low in absolute value, especially for the cold season (Table 7.4). The negative values of r_{T_h} indicate that in most instances a lifting of the cloud boundaries is accompanied by a drop in temperature, and a lowering of h_i by a temperature rise. With regard to the lower boundary, such a relationship between h_1 and T_{h_1} is certain: here γ is, as a rule, greater than zero, so that an increase in height h_1 is accompanied by a drop in temperature T_{h_1}, and a decrease in h_1 by a rise in T_{h_1}. For the upper boundary, the results ($r_{T_h} < 0$) contradict the commonly held opinion that St-Sc clouds form beneath a temperature inversion already present (supposedly due to subsidence). Here, it is assumed that the upper cloud boundary is lifted because of the inflow of water vapor from below, and that it penetrates the inversion layer.

Given this explanation of the development of St-Sc clouds, the correlation between h_2 and T_{h_2} should be positive, that is, an increase in h_2 is accompanied by a temperature rise. However, since $r_{T_h} < 0$ even at the upper cloud boundary, it must be concluded that St-Sc clouds form via the evolution of stratiform cloudiness (Ns-As) in a field of descending vertical currents. For such a mechanism of St-Sc formation, the upper cloud boundary is in most cases lowered since the air descends, while the temperature rises (that is, $r_{T_h} < 0$).

The statistical relationships established, as well as their qualitative physical interpretation, can be corroborated by the formula for the local temperature variation in a cloud:

$$\frac{\partial T}{\partial t} = w\left(\gamma - \gamma_a'\right),$$

where γ_a' is the moist-adiabatic lapse rate. Since in a cloud it is usually true that $\gamma < \gamma_a'$, for a downdraft ($w < 0$), when the cloud boundaries naturally descend ($\Delta h_i < 0$), the temperature in the cloud, including that close to the cloud boundaries, rises with time ($\partial T/\partial t > 0$). This corresponds to a positive deviation from the mean of the temperature at the cloud boundaries ($Th_i > 0$).

Consequently, vertical motions are the main cause of the negative statistical correlation between h_i and Th_i. However, there is no single-valued relationship between the deviations of h_i and Th_i ($r_{Th} \neq -1$), for two reasons: first, a moist-unstable state ($\gamma > \gamma_a'$) is possible in the cloud, in which case for $w < 0$ the derivative $\partial T/\partial t < 0$; second, the temperature and the height of the cloud boundaries are affected by factors other than vertical motions (turbulent mixing, condensation heat, and radiation - although radiation does not determine the sign of r_{Th}, it must still be taken into account).

Table 7.4 also gives the coefficients of correlation between the cloud water contents close to the upper boundary (δ_1) and in the middle (δ_2) and the temperatures at these levels and at heights h_i. These data confirm the previous conclusion of a positive correlation between δ and the temperature in the cloud. Although correlation coefficients are not very high (maximum value 0.58), the cloud water content is affected by such factors as the height of the point above the cloud base, γ, w, cloud depth, in addition to being influenced by the temperature.

7.1.3. The humidity field

The statistical structure of the humidity field was also studied in [86]. The data in Table 7.5 reveal a close correlation between the specific humidities at the lower and upper cloud boundaries.

Table 7.5. Autocorrelation matrix of specific humidity r_{qq}

	2 m	h_1	h_2	h_3
		Warm season		
2 m	1	0.84	0.84	0.53
h_1	0.90	1	0.96	0.77
h_2	0.88	0.96	1	0.76
h_3	0.70	0.81	0.82	1
	Cold season			

The correlation coefficient is seen to be as high as 0.96 during the cold and warm seasons. The table also gives coefficients of the correlation between the specific humidity close to the ground (at a height of 2 m) and at heights h_1. Here r_{qq} is also greater than zero, although it is somewhat lower than the coefficients of correlation between q_{h_1} and q_{h_2}. Such a close correlation between the deviations of q at different levels can be attributed to the influence of vertical motions:

$$\frac{\partial q}{\partial t} = -w\frac{\partial q}{\partial z} - \left(u\frac{\partial q}{\partial x} + v\frac{\partial q}{\partial y}\right),$$

as well as to the advective influx of water vapor (the second term of the right-hand side of the above equation). Since $w\ \partial q/\partial z$, and the advection do not, as a rule, change sign with height (at least, up to heights of 1.5-2 km), these factors cause deviations of q of the same sign, that is, they cause $r_{qq} > 0$.

Table 7.6 reveals a close correlation between the temperatures and specific humidities at both the lower and upper cloud boundaries: coefficients of 0.94 and 0.95 at h_1 and h_2, respectively, during the cold season, and 0.96 and 0.95 during the warm season. Such a correlation is easily explained: since the water vapor in the cloud is close to saturation, an increase in T will be accompanied by a rise in q, and vice versa. A close correlation is also observed between T at level h_1 and q at h_2 (r_{qT} = 0.91 in winter and r_{qT} = 0.93 in summer) or between T at h_2 and q at h_1 (r_{qT} = = 0.92 in winter and r_{qT} = 0.95 in summer), which is also easily explained, since the temperatures at the cloud boundaries correlate closely with one another (see Table 7.2).

Table 7.6. Cross-correlation function r_{qT} between specific humidity and temperature at various levels (T_0 and q_0 are temperature and specific humidity at level 2 m above ground)

	q_0	q_{h_1}	q_{h_2}	q_{h_3}	q_0	q_{h_1}	q_{h_2}	q_{h_3}
	Cold season				Warm season			
T_0	0.92	0.79	0.78	0.58	0.82	0.67	0.66	0.33
T_{h_1}	0.86	0.94	0.91	0.73	0.77	0.96	0.93	0.80
T_{h_2}	0.86	0.92	0.95	0.79	0.76	0.95	0.95	0.79
T_{h_3}	0.65	0.78	0.77	0.75	0.61	0.81	0.79	0.72

Table 7.6 shows the coefficients of correlation between T and q close to the ground (2 m) to be quite high: 0.92 in winter and 0.82 in summer. The first of these figures is close to the winter values obtained by other investigators. For instance, Koprova and Boldyrev [143] used sounding data for the 1000-mb level over Berlin to get r_{qT} values of 0.92 in winter and only 0.47 in summer. Kudryavtsev obtained coefficients of correlation between the variations in temperature and dew point during a 12 h

period (Arkhangelsk, 1963) of 0.99 in January and 0.65 in July. Popov [257] also points out that r_{qT} close to the ground is higher in winter and lower in summer.

In all the above studies, r_{qT} was determined from sounding data for all days regardless of the cloudiness. However, Komarov [140] separated the soundings during clear weather (cloudiness of 0–2) from those during the overcast weather (cloudiness of 8–10) and found the seasonal correlation coefficients r_{qT} to be very different for clear weather (close to the ground, they were 0.94 in January and 0.47 in July) and fairly close for overcast weather (0.97 in January and 0.78 in July close to the ground; 0.80 in January and 0.75 in July at the 850-mb level). Inspection of Table 7.6, which gives the values of r_{qT} only for overcast skies, indicates that in this case the correlation coefficients close to the ground during the cold (0.92) and warm (0.82) seasons are not so different, as compared with the previous data obtained without differentiating between the amounts of cloudiness.

The high values of r_{qT} in winter, and for overcast skies in summer as well, are explained by the fact that during the winter (and for overcast skies, the year round) the relative humidity is, as a rule, quite high (the air is close to saturation). Thus, a temperature increase is accompanied by a rise in the specific humidity. At the same time, in summer the amount of water vapor reaching the surface of the ground (especially dry ground) is insufficient to keep the air close to saturation: the relative humidity is, in general, low. Consequently, a rise in temperature (during the daytime, for instance) will by no means always be attended by an increase in specific humidity. The outflow of water vapor into higher layers due to daytime convection is also known to play a role here. In more southerly latitudes, due to the lack of soil moisture during the warm season, even in overcast weather the evaporation may not always ensure a rise in specific humidity for a rise in temperature. This explains why the correlation coefficient r_{qT} near the ground for summer in Kharkov was only 0.55 for a solid St-Sc cloud cover. For this same reason, the correlation coefficient r_{qT} close to the ground at Ashkhabad for the whole year was only about half the values at Moscow, Berlin, and Arkhangelsk.

7.1.4. Double-layer cloudiness

In [87] Devyatova calculated the statistical characteristics of St-Sc cloudiness consisting of two layers. Data of aircraft soundings in the Moscow region over 15 years (1956-1970) were considered. Only in 95 out of 9352 cases was double-layer St-Sc cloudiness observed, with an upper boundary no higher than 2.5 km and with no clouds above or below these clouds (the cloudiness in the upper level was at most 3). During the cold season (October--March) 65 cases were observed, and during the warm season only 30.

The upper cloud layer was solid in 90 instances, and the lower was solid in only 56 out of 95.

Tables 7.7 and 7.8 give the mean depths Δh of the first and second cloud layers, as well as the lapse rates γ in the layers.

Table 7.7. Mean depths Δh of cloud layers and clear layer between them (km).

Season	First layer		Second layer		Clear layer	
	Δh	$\sigma_{\Delta h}$	Δh	$\sigma_{\Delta h}$	Δh	$\sigma_{\Delta h}$
Warm	0.30	0.20	0.33	0.20	0.54	0.31
Cold	0.28	0.21	0.41	0.32	0.53	0.33

Table 7.8. Mean lapse rate γ and its standard veviation σ_γ (°C/100 m).

Season	First layer		Second layer	
	γ	σ_γ	γ	σ_γ
	Under cloud layer			
Cold	0.67	0.36	0.50	0.40
Warm	0.74	0.30	0.53	0.34
	In cloud layer			
Cold	0.40	0.64	0.54	0.42
Warm	0.56	0.30	0.47	0.46
	Inversions above cloud layer			
Cold	–0.93	0.85	–2.30	1.91
Warm	–0.63	0.53	–1.38	1.07

During the cold season the depths of the upper and lower cloud layers in 75% of the cases do not exceed 400 m, and they are never greater than 1000 m. During the warm season the depth of the lower layer is in 90% of the cases 600 m or less, and it never exceeds 750 m; in 73% of the cases the depth of the upper layer does not exceed 400 m, and it is never more than 1500 m. In 16% of the cases during the cold season the clear intervening layer is less than 200 m deep, in 78% of the cases it is not deeper than 600 m, and in all cases it is less than 1500 m. During the warm season the clear layer is in 37% of the cases less than 200 m deep, in 33% of the cases it is between 400 and 600 m deep, and in 14% of the cases it is between 600 and 1400 m.

According to Table 7.8, the mean data show a temperature inversion ($\bar{\gamma} < 0$) above both the first and second cloud layers. During the cold season and inversion was observed over the lower cloud layer in 25 cases, and over the upper layer in 52 cases, out of 65; during the warm season the corresponding figures are 12 and 25 cases out of 30. The depth of the inversion layer during the cold season ranged from 100 to 780 m above the lower cloud layer and from 50 to 900 m above the upper layer; during the warm season, the corresponding ranges were from 90 to 520 m and from 70 to 730 m.

Before these results a discussed, additional information will be given concerning the autocorrelation functions for the temperature at the boundaries of the cloud layers and at the upper boundaries of the inversions (Table 7.9). As in the case of single-layer cloudiness, the correlation coefficients between the temperature deviations at the boundaries (h_1 and h_2, h'_1 and h'_2) of the upper and lower cloud layers are positive and close to unity (for instance, $r_{T_{h_1}T_{h_1}}$ is 0.95 for the cold season and 0.94 for the warm season). There is also a close positive correlation between the temperatures at the boundaries of the lower (h_1 or h_2) and upper (h'_1 or h'_2) cloud layers: $r_{TT} \approx 0.60$–0.85 (for instance, $r_{T_{h_1}T_{h'_1}}$ is 0.64 in the cold season and 0.83 in the warm season).

The data in Tables 7.8 and 7.9 tend to refute the hypothesis of a decisive role for radiation in forming the temperature field of a double-layer cloudiness as well. Actually, according to the estimates of Feigel'son [311], the temperature deviation from the initial distribution during a 10-h period at various levels, due to radiation is the following:

z,km	0	1	2	3	4	5	6	7	8	9
g/kg	0	0	0	0	–0.1	–0.3	0	0	–0.1	–0.1

Table 7.9. Autocorrelation matrix of temperature (here h_1, h_2, h_3 are, respectively, heights of boundaries of lower cloud layer and height of upper boundary of overlying inversion; h'_1, h'_2, h'_3 are corresponding heights for upper cloud layer).

	h_1	h_2	h_3	h_1	h_2	h_3
			Cold season			
h_1	1	0.83	0.86	0.64	0.60	0.66
h_2	0.97	1	0.74	0.82	0.77	0.78
h_3	0.57	0.74	1	0.67	0.68	0.67
h'_1	0.83	0.87	0.77	1	0.95	0.89
h'_2	0.84	0.85	0.81	0.94	1	0.91
h'_3	0.71	0.77	0.86	0.82	0.85	1
			Warm season			

During the calculations it was assumed that the lower cloud layer lies between 1 and 2 km, while the upper layer is between 3 and 4 km. If such a second cloud layer is present, then the radiative influxes of heat in the lower cloud layer are zero, and they cannot cause temperature deviations, not can they lead to the formation of an inversion above the lower cloud layer. Since, nevertheless, a temperature inversion does indeed frequently form above the lower cloud layer, and the correlation coefficients between the temperature at h_i' and h_i are positive, the decisive factor shaping the thermal regime of double-layer cloudiness is not radiation, but rather dynamic factors, such as vertical motions, advection, and turbulent mixing.

The cloud water content fluctuates considerably in both time and space [68]. In a field of downward motions, the cloudiness becomes dissipated primarily in places where the water content is a minimum: this is, in the present author's opinion, how the cloudless interlayers form. It is quite natural that in such an interlayer (above the lower cloud layer) temperature inversions are produced by downdrafts more rarely, and the temperature differential in them will be less than in the inversions over the upper cloud layer (it should be recalled that, before formation of the cloudless interlayer, the temperature here dropped with height, while the time of temperature increase according to the dry-adiabatic law in the interlayer was less than in the atmosphere above the upper cloud layer). These considerations help explain the difference in the mean values of γ, as well as the different frequencies of temperature inversions above the lower and upper cloud layers (see Table 7.8).

The autocorrelation coefficients r_{qq} for the specific humidity in the case of double-layer cloudiness are also positive, with high absolute values. For instance, at the boundaries of the first layer, r_{qq} is 0.88 during the cold season and 0.97 during the warm season, while at the boundaries of the second layer the corresponding values are 0.95 and 0.95. The correlation between the specific humidity at h_i, on the one hand, and h_i' , on the other, is naturally somewhat weaker than the correlation between the q values at the boundaries of a given cloud layer, but this correlation is still quite close. For instance, the autocorrelation coefficient between the q values at h_1 and h_1' is 0.69 during the cold season and 0.82 during the warm season; the corresponding values for h_2 and h_1' are 0.81 and 0.88, those for h_1 and h_2 are 0.64 and 0.76, and, finally, for h_2 and h_2' this coefficient is 0.76 for the cold season and 0.84 for the warm season.

The foregoing values of r_{qq} indicate that the humidity fields of both the first and second cloud layers are shaped by factors exerting the same (if only in sign) effect on the cloud layers. Vertical motions and advections may well be such factors.

The cross-correlation between the temperature and the specific humidity at levels h_i and h_i' is just as close for double- as for single-layer cloudiness. For instance, the coefficients of cross-correlation r_{Tq} between T and q at a given level during the cold season are 0.95 at h_1, 0.94 at h_2, 0.96 at h_1', and 0.95 at h_2'. During the warm season these coefficients at all four levels

(h_1, h_2, h_1', h_2') have the very same value, namely 0.96. The correlation between T and q at different levels is somewhat weaker. For instance, the correlation coefficients r_{Tq} between T at level h_1 and q at levels h_2, h_1', h_2', are, respectively, 0.84, 0.66, and 0.60 during the cold season and 0.97, 0.82, and 0.78 during the warm season. The cross-correlation coefficients r_{qT} between q at level h_1 and T at levels h_2, h_1', and h_2' are, respectively, 0.81, 0.61, and 0.59 during the cold season and 0.91, 0.77, and 0.78 during the warm season. All the other r_{Tq} and r_{qT} are positive as well and, as a rule, are not less than 0.7.

The close statistical correlation between the temperature and the specific humidity at the boundaries of the cloud layers can be explained physically as follows: close to the cloud boundaries the air is nearly saturated, so that any temperature rise there will be accompanied by an increase in the specific humidity. In conclusion, a statistical analysis of experimental material indicates the following: dynamic factors (vertical motions, advection, turbulence) play a determining role in the formation and evolution of St-Sc cloudiness.

7.2. Statistical methods for determining cloud boundaries and temperatures

The close correlations between the various quantities discussed in Section 7.1 can be used to solve various problems of practical interest. Since the height of the lower cloud boundary h_1 is measurable, the correlation between h_1 and h_2 can be used to find the height of the upper boundary h_2, a problem of considerable importance for aviation. A new stimulus to this line of thought was the advent of satellite determinations of the heights and temperatures of the upper cloud boundary. In view of the lack of information about the state of the atmosphere below h_2, two so-called inverse problems were posed: to reproduce the temperature and humidity profiles in the layer from the upper cloud boundary down to the ground, and also to determine the height of the lower cloud boundary h_1 from known values of h_2 and T_2. Two approaches to the solution of these inverse problems have been worked out.

The first approach is the determinate one [135, 137, 191], based on the equations describing the influx of heat and moisture, together with the equations of motion and continuity. For specified boundary conditions at h_2, the indicated equations are solved in order to reconstruct, for a given moment, the vertical profiles of T and q in the layer from h_2 down to the ground. This method is also used to reconstruct the profile of the cloud water content and to determine the height of the lower cloud boundary.

A statistical approach to reconstructing the temperature and humidity profiles of a cloud was developed in [143, 167, 257] and elsewhere. This method is based on the equation of linear regression

$$y_i' = a x_i' \quad (i = 1, 2, \ldots, N) \tag{7.2.1}$$

between the deviations of x and y: $x' = x - \bar{x}$ and $y' = y - \bar{y}$. Coefficient a in eqn. (7.2.1) is determined from the condition of minimum of the functional

$$\delta^2 = \sum_{i=1}^{N} (y_i' - a x_i')^2. \tag{7.2.2}$$

Using the rule for finding the minimum ($\partial\delta^2/\partial a = 0$) one obtains from (7.2.2) the following expression for coefficient a:

$$a = \frac{B_{xy}}{\sigma_x^2} = \frac{\sum_{i=1}^{N} x_i' y_i'}{\sum_{i=1}^{N} (x_i')^2}, \tag{7.2.3}$$

where B_{xy} is the covariation function.

By substituting the expression for a into relation (7.2.2), we find the minimum value of the error δ in determining quantity y with the aid of (7.2.1):

$$\delta_{\min}^2 = \sigma_y^2 - \frac{B_{xy}^2}{\sigma_x^2}. \tag{7.2.4}$$

Next the minimum relative error is introduced:

$$\varepsilon_{\min} = \frac{\delta_{\min}}{\sigma_y} = \sqrt{1 - r_{xy}^2}. \tag{7.2.5}$$

Here x and y can represent any pair of quantities, one of which (x) is measured or determined using some other method. Malkevich and Polovina [167] constructed an equation of linear regression for finding the height of the upper cloud boundary ($y = h_2$) from the measured height of the lower boundary ($x = h_1$). Although in some cases the errors in these determinations were quite large (as high as 0.75 km), most of the h_2 values obtained were nevertheless of satisfactory accuracy.

Devyatova [86] used the data of the statistical analysis described in Section 7.1 to construct regression equations for many pairs x and y. The calculated coefficients a and minumum errors $\delta_{\min}$ and $\varepsilon_{\min}$ are given below:

x	T_2	h_2	h_1	h_2	T_2
y	T_1	h_1	h_2	h_3	q_2
x	T_2	T_3	q_3	T_0	q_0
y	q_1	T_2	T_2	T_2	T_2
			Cold season		
a	1.01	0.86	0.86	0.93	0.14
δ_{min}	1.4 °C	0.20 km	—	0.14 km	0.2
ε_{min}	0.28	0.48	—	0.54	0.32
a	0.16	1.03	0.16	0.93	0.17
δ_{min}	0.3 g/kg	3.0 °C	0.6 g/kg	3.0 °C	0.5 g/kg
ε_{min}	0.39	0.53	0.61	0.57	0.51
			Warm season		
a	1.02	1.00	0.91	1.00	0.28
δ_{min}	1.0 °C	0.17 km	—	0.17 km	0.5
ε_{min}	0.20	0.28	—	0.31	0.32
a	0.29	0.94	0.27	—	0.22
δ_{min}	0.6 g/kg	2.7 °C	1.2 g/kg	—	1.0 g/kg
ε_{min}	0.32	0.48	0.61	—	0.65

Here the subscripts refer to the following levels: 1 is the lower cloud boundary, 2 is the upper cloud boundary, 3 is the upper boundary of the above-cloud inversion, and 0 is the level 2 m above the ground. According to the data of an independent sampling, comprising 152 aircraft soundings, the actual errors δ were found as the difference between the measured value of some quantity y and the value calculated using the method of optimum extrapolation. It was found that for T_1 the absolute value of δ did not exceed 1.5°C in 80% of the cases; for h_1 the values of $|\delta|$ did not exceed 200 m in 75% and 300 m in 91% of the cases. The aboslute value of the errors in determining q_2 and q_1 did not exceed 0.5 g/kg in 94 and 92% of the cases, respectively. The values of $|\delta|$ for T_0 and q_0 did not exceed, respectively, 4°C in 83% of the cases and 1.0 g/kg in 93% of the cases.

The method of optimum extrapolation with the accuracy specified here makes it possible to reconstruct the vertical temperature and humidity profiles below the upper boundary of single-layer cloudiness, provided the temperature and humidity of the upper cloud boundary are determined from satellite data.

7.3. The statistical correlation between temperature increments in time

It is of great interest to evaluate the roles played by various factors in cloud formation. On the basis of the numerical models discussed in other chapters of this monograph, it can be concluded that large-scale vertical motions, advection, and turbulent mixing are the factors determining the formation of stratiform (frontal) clouds such as nimbostratus (Ns), altostratus (As), and cirrostratus (Cs). There is, however, less agreement about the formation of such clouds as stratus (St), stratocumulus (Sc), altocumulus (Ac), and cirrocumulus (Cc). In scientific literature, and especially in textbooks, the opinion is frequently expressed (to be sure, it is generally not substantiated by any estimates) that radiation plays a decisive role in the formation and evolution of these clouds. However, a number of studies [107, 274] have shown that vertical motions and turbulent mixing are more important than radiation in forming St-Sc-Ac-Cs clouds as well.

In [195] an attempt was made to resolve the problem of the formation and evolution of St-Sc clouds, with the aid of a statistical analysis of experimental material. It is known that these factors cause a variation in time of not only the cloudiness (field of water content) but the temperature as well. Thus, there is a single-valued relationship between the temperature increment ΔT during a time τ and the specific heat influx ε due to various factors:

$$c_p \Delta T = \varepsilon\tau.$$

It will be recalled that the radiative influx of heat at the lower cloud boundary is positive, while that at the upper boundary is negative. This means that because of radiation the air temperature close to the lower boundary increases with time ($\Delta T > 0$), whereas close to the upper boundary it drops ($\Delta T < 0$). As an example, the temperature variations ΔT during 10 h, caused by the absorption and radiation of infrared radiation, calculated by Feigel'son [311] are given below:

z.km . . .	0	1	2	3	4	5	6	7	8	9
ΔT.°C. . .	0	1	0.3	—0.2	—0.9	—0.3	—0.3	—0.3	—0.3	—0.2

Close to the lower boundary, which in this example was at a height of 2 km, absorption of radiation raised the temperature by 0.3°C; close to the upper boundary, which was at a height of 3 km, and also above the cloud, radiation reduced the temperature ($\Delta T < 0$). Since, results similar with regard to the sign of ΔT were obtained for the other characteristics of clouds and the atmosphere, the correlation between the observed temporal variations of the temperature at the upper and lower cloud boundaries would have to be negative, if radiation played a decisive role in the formation of the temperature field and the clouds themselves. Taking into account the heating of the cloud by solar radiation, according to measurements [79] and according to calculations, does not alter this conclusion, since the resultant temperature variation close to the upper cloud boundary stays negative (that is, the rate of cooling is greater than the rate of heating).

Under the influence of vertical currents and turbulence, the air temperature varies in one direction (rises or drops), as a rule, throughout the troposphere. For example, according to the data of [52], the air-temperature increments ΔT over a 24-h period caused by the upward motion and turbulence, for specified values of the parameters of the problem, are the following:

z, km . . .	0	1	2	3	4	5	6	7	8	9	10
ΔT, °C. . .	0	−2.3	−3.5	−4.7	−6.0	−7.2	−8.0	−8.4	−8.0	−6.9	−4.0

For other values of the parameters, different values of ΔT are obtained, but in every case the signs of ΔT at the upper and lower cloud boundaries are the same, provided the temperature variation is due to vertical motions. Consequently, if vertical currents play a determining part in shaping the thermal regime of the clouds, the correlations between the temperature variations observed at the cloud boundaries must be positive.

In order to answer the questions posed above, the author turned to data from aircraft soundings of the atmosphere in different parts of the Soviet Union, carried out from 1951 to 1968. Instances of single-layer cloudiness, predominantly of the St-Sc type, were selected. The clouds remained over the observation point for at least 12 h and their boundary heights varied during this time by no more than 200 m. Table 7.10 gives the calculated correlation coefficients $r_{f\varphi}$ between the temperature increments $f_i = T_i(t + \tau) - T_i(t)$ during a 12-h period at the lower cloud boundary and the synchronous temperature variations $\varphi_i = T(t + \tau) - T(t)$ during this time at the upper cloud boundary. These coefficients pertain to different samplings, including observations during the cold and warm seasons at various points. These data indicate that the correlation coefficients for all samples are positive, with values between 0.41 and 0.72.

Table 7.10. Correlation coefficients between synchronous temperature variations during 12 h at upper and lower boundaries of St-Sc clouds.

Sample	Place and years of observations	Season	$r_{f\varphi}$	Sample volume
1	Moscow (1956-1968), Leningrad (1953-1966)	Cold	0,55	100
2	Arkhangelsk (1960-1961) Moscow (1951-1962), Omsk (1957-1962); Seven points in all Leningrad (1953-1964)	Warm	0,72	100
3	The same seven points (1951-1964)	Cold	0,41	100
4	Moscow (1951-1954)	Warm	0,51	50
5	Moscow (1951-1954)	Cold	0,53	68

Although the importance of advection is not to be underrated, the effect of advection alone cannot account for the formation of the temperature inversion above the clouds. Actually, synoptic experience has shown that advection of the same sign encompasses, as a rule, atmospheric layers of considerable depth (usually the bulk of the troposphere). Consequently, as a result of advection, the temperature varies (especially at levels that are close to one another) by practically the same amount. At the same time, in order to produce an inversion above the clouds, the depth of which is generally several hundred meters, the temperature above the clouds must vary considerably more rapidly than the temperature in the upper part of the cloud.

However, this difficulty can be circumvented by assuming that vertical motions are the decisive factor in shaping the thermal regime of St-Sc clouds. Actually, according to the formulas $\Delta T = w(\gamma - \gamma_a)\tau$ outside the cloud and $\Delta T = w(\gamma - \gamma_a')\tau$ inside the cloud, the temperature variation due to a downward motion ($w < 0$) will be positive ($\Delta T > 0$), both above the cloud, where $\gamma < \gamma_a'$, and inside the cloud, if $\gamma < \gamma_a'$; however, the temperature will rise more rapidly above the cloud than in the cloud, since $|\gamma - \gamma_a| > |\gamma - \gamma_a'|$. Obviously, such temperature increments in and above the cloud will produce a temperature inversion above St-Sc clouds.*)

The correlation between the temperature variations at the cloud boundaries and at the upper boundary of the inversion is also, as a rule, positive (Table 7.11). At each place in the table there are three values of r, one for each of samples 1, 2, and 3 in Table 7.10, respectively. The fact that one of the correlation coefficients between ΔT at heights h_1 and h_3 is negative (-0.12), while one of those between ΔT at h_2 and h_3 is close to zero (0.07), does not contradict the hypothesis that various factors shape the temperature field. Actually, if in the cloud $\gamma > \gamma_a'$ (moist-unstable stratification), then the foregoing formulas indicate that the temperature variations due to w will have opposite signs above the cloud and inside it. However, since the coefficients of the correlation between ΔT in the cloud and at the upper boundary of the inversion are generally positive, it follows from the foregoing considerations that a moist-stable stratification ($\gamma < \gamma_a'$) prevails in St-Sc clouds.

For the same places at which aircraft soundings were made to obtain the data of samples 1, 2, and 3 in Table 7.10, instances of double-layer cloudiness predominantly of the St-Sc-Ac type were selected; below the first cloud layer the air was clear in all these cases, while above the second layer an upper level of cloudiness of not more than 3 was present. Each of the three samples included 100 cases of variation during 12 h of the temperature and specific humidity at the boundaries of the cloud layers and the upper boundaries of the temperature inversions. Table 7.12 gives the calculated coefficients for variations of the temperature and specific humidity at various moving levels (a dash indicates an absence of calculations due to insufficient data).

According to theoretical estimates, radiation does not affect the lower cloud layer (which is in a state of radiative equilibrium), but it promotes cooling of the upper boundary of the second layer

*) It is generally accepted that downward motions are a determining factor in producing a temperature inversion above convective clouds (Cu cong., Cb).

Table 7.11. Correlation coefficients between synchronous variations during 12-h period of temperature (left of main diagonal) and specific humidity (right of main diagonal) at lower (h_1) and upper (h_2) boundaries of single-layer St-Sc cloudiness and at upper boundary of above-cloud inversion (h_3).

	h_1	h_2	h_3
h_1	1	0.42; 0.72; 0.60	0.05; 0.59; 0.32
h_2	0.55; 0.72; 0.41	1	0.40; 0.60; 0.50
h_3	0.48; 0.64; —0.12	0.50; 0.83; 0.07	1

and heating of the lower boundary. Consequently, if radiation plays a predominant role, the correlation coefficients between the temperature variations in the lower cloud layer would be close to zero, while in the second layer they would be negative.

Actually, however, the coefficients of the correlation between the ΔT values at the upper and lower boundaries of the first layer, and especially the second layer, are not only nonnegative but in fact are quite positive: in the second layer the coefficients for ΔT at the upper and lower boundaries fluctuate for different samples from 0.46 to 0.92, and in the first layer from 0.07 to 0.60. A very close correlation is observed between ΔT at the upper boundary of the inversion over the second cloud layer and ΔT at the upper as well as the lower boundary of this layer: the coefficients of the correlation between ΔT at h_3' and h_2' vary from 0.51 to 0.96, and between ΔT at h_3' and h_1' from 0.29 to 0.97. The correlation between ΔT at the boundaries of the first layer and at the boundaries of the second layer, although less close, is still, as a rule, positive (of 12 values of the coefficient, only 3 were negative) and quite significant. The correlation between the variations of the specific humidity at various moving boundaries of double-layer cloudiness is positive and quite close for practically all the samples (of 28 correlation coefficients, only 3 were less than zero).

The data of Table 7.12 and of [87] indicate that a temperature inversion is observed above the first cloud layer as well as above the second. As mentioned above, such a temperature distribution close to the upper boundary of the lower cloud layer cannot be attributed to radiation. However, this phenomenon can be explained quite simply on the basis of the hypothesis that vertical motions play a predominant role. The field of the cloud water content is nonuniform: zones of increased water content alternate with zones of lower water content. Therefore, for a downward motion, which is naturally accompanied by a gradual dissipation of the cloud (often taking tens of hours), some parts of the cloud are dispersed sooner than others, that is, a stratification of the cloud mass takes place. Reasoning as above, it can be concluded that in the cloud-free

Table 7.12. Correlation coefficients between synchronous variations during 12-h period of temperature (left of main diagonal) and specific humidity (right of main diagonal) at lower (h_1 and h'_1) and upper (h_2 and h'_2) boundaries of first and second layers of St–Sc clouds, and upper boundaries of inversions above first (h_3) and second (h'_3) layers.

	h'_1	h'_2	h'_3
h_1	1	0.52; 0.69; 0.44	0.28; —; 0.40
h_2	0.38; 0.60; 0.07	1	0.42; —; 0.24
h_3	0.24; —; —0.05	0.74; —; 0.39	1
h'_1	—0.15; 0.42; 0.08	0.29; 0.55; 0.19	— — —
h'_2	—0.16; 0.45; 0.02	0.32; 0.58; —0.04	— — —
h'_3	— — —	— — —	— — —

	h'_1	h'_2	h'_3
h_1	—0.23; 0.13; 0.29	—0.05; 0.21; 0.09	— — —
h_2	0.18; 0.31; 0.11	0.29; 0.39; 0.14	— — —
h_3	— — —	— — —	— — —
h'_1	1	0.79; 0.92; 0.73	0,31; 0.98; 0.01
h'_2	0.89; 0.92; 0.46	1	0.51; 0.97; —0.39
h'_3	0.83; 0.97; 0.29	0.94; 0.96; 0.51	1

interlayer right next to the upper boundary of the lower layer, a temperature inversion will be produced by the downward motion.

For samples 4 and 5 (see Table 7.10), the correlation coefficients for the temperature variations at the upper and lower boundaries of single-layer St–Sc cloudiness during 24 h were determined. They were 0.44 and 0.65 for samples 4 and 5, respectively. In accordance with the hypothesis of the determining role of vertical motions statistical correlations are observed between the variations of the heights of the upper and lower cloud boundaries: a downward motion,shifts the cloud boundaries, as a rule, in the same direction. For samples 4 and 5, the coefficients of correlation between the height increments of the upper and lower cloud boundaries are, respectively, 0.87 and 0.54 for τ = 12 h and 0.87 and 0.39 for τ = 24 h.

The correlation between the temperature variations at any cloud boundary and the variation in the height of this boundary is comparatively weak (for sample 4, for instance, the correlation coefficients at the upper boundary of single-layer cloudiness are 0.22 for τ = 12 h and -0.13 for τ = 24 h).

In addition to establishing correlations between variations of temperature and other quantities, the present author also calculated the matrices of the correlation coefficients for the temperatures themselves at various moving levels. These coefficients turned out to be just as high (0.8–0.9 or more) as those published in [86, 87]. For example, for samples 2 and 3, the coefficients of the correlation between the temperatures at the lower and upper boundaries of single-layer cloudiness are 0.95 and 0.86, respectively (for each sample, these coefficients were calculated for 200 cases).

It would be interesting to compare the correlation coefficients for the temperature and its variation in time at various levels. Certain simple considerations lead to the following relation, valid for a uniform isotropic random process:

$$r_{\Delta T^{(1)},\, \Delta T^{(2)}} = \frac{r_{T^{(1)}(t)\, T^{(2)}(t)} - r_{T^{(1)}(t),\, T^{(2)}(t+\tau)}}{1 - r_{T^{(1)}(t)\, T^{(1)}(t+\tau)}}. \qquad (7.3.1)$$

Here r are the following correlation coefficients: $r_{\Delta T^{(1)} \Delta T^{(2)}}$ between the synchronous temperature variations during time τ at levels 1 and 2; $r_{T^{(1)}(t) T^{(2)}(t)}$ between the temperatures at levels 1 and 2 at a given time t; $T^{(1)}(t) T^{(2)}(t+\tau)$ between the temperatures at level 1 and at time t and at level 2 at time $t + \tau$; $r_{T^{(1)}(t) T^{(1)}(t+\tau)}$ between the temperatures at level 1 at times t and $t + \tau$.

For sample 2, the values are $r_{T^{(1)}(t) T^{(2)}(t)} = 0.95$, $r_{T^{(1)}(t) T^{(2)}(t+\tau}$ $= 0.87$, $r_{T^{(1)}(t) T^{(1)}(t+\tau)} = 0.90$; for these values, formula (7.3.1) gives $r_{\Delta T^{(1)} \Delta T^{(2)}} = 0.80$, which does not differ overly much from the value of the correlation coefficient (0.72) found by means of direct calculations.

For sample 3, formula (7.3.1) gives $r_{T^{(1)} T^{(2)}(t)} = 0.86$, $r_{T^{(1)}(t) T^{(2)}(t+\tau)} = 0.81$, $r_{T^{(1)}(t) T^{(1)}(t+\tau)} = 0.89$ and $r_{\Delta T^{(1)} \Delta T^{(2)}} = 0.45$; direct calculations gave 0.41. Thus, the statistical correlations obtained by the author for the temperature variations show a satisfactory fit with data on the correlations between the temperatures themselves (these were established on the basis of a longer series of observations, since it was not necessary for cloudiness of a certain type to remain over the observation point for a time τ).

The data presented show that vertical motions, as well as advection and turbulent mixing, are indeed the factors determining the formation of St-Sc-Ac cloudiness and the temperature field, in particular, of above-cloud inversions.

7.4. Conditions preceding and accompanying cloud formation

Since low clouds, including St-Sc, are of great practical interest, especially with regard to aircraft flights, much attention has been given, and is being given, to their study. Pchelko [262], Zak [107], and Selezneva [247] were among the first to analyze the observational data and the synoptic-aerological material.

Pchelko [262] concluded that low clouds (300 m or lower) are observed during the cold season, primarily in the vicinity of warm fronts (that is, for heat advection); and during the warm season, immediately beyond cold, slowly moving fronts (for advection of cold). These conclusions agree with the results of Gogoleva 1/M who established that during the cold season, low clouds are observed for heat advection in 85% of the cases 1/M and also with Khrgian's results [317].

Zak's studies [107] are of great interest today although they were carried out some 40 years ago. These investigations were based on data of atmospheric soundings, including four series (in April 1939 and May 1940) of frequent soundings at points situated along the line Minsk-Orsha-Smolensk-Vyaz'ma-Moscow-Kovrov. During the winter half-years of 1938-39 and 1939-40, all the cases of an anticyclonic regime over the European USSR and West Siberia were selected (47 periods, with a total duration of 236 days).

Table 7.13 shows the frequencies of various weather conditions in anticyclones during the winter half-year.

Table 7.13. Frequency (%) of weather conditions in anticyclones.

Season	Clear	Fog	St-Sc	As-Ns	Ac
Prewinter	22	11	62	2	2
Winter	44	10	32	13	2
Prespring	62	11	17	5	5
Average	40	10	36	12	2

As is to be expected, the frequency of clear weather is quite high (40%, on the average) in an anticyclone. The frequency of St-Sc cloudiness is just about as high (36%, on the average). However, in different parts of an anticyclone, the frequency of weather conditions differs substantially. For instance, in the central and eastern parts of an anticyclone, clear weather predominates (72%), while St-Sc clouds have frequencies ranging from 14 to 20% there. At the same time, in the western and southern parts of an anticyclone, St-Sc clouds are observed most frequently (43 and 46%), as well as frontal As-Ns clouds. If days with clear weather and fog are excluded, then the frequencies of St-Sc, As-Ns, and Ac in an anticyclone are 70, 26, and 4%, respectively.

A determining condition for the existence of St-Sc clouds is, according to Zak, the presence of a temperature inversion, the lower boundary of which is situated above the condensation level, determined from data obtained at the ground assuming adiabatic lifting of an air parcel. An analysis of synoptic-aerological material led Zak to conclude that "a considerable portion of the observed inversions are of frontal origin, that is, they constitute horizontal, or nearly horizontal, parts of interfaces between different air masses. Inversions originating in a uniform air mass as a result of subsidence of this mass are observed comparatively rarely" [107, p. 118].

All the vertical sections plotted by Zak exhibited comparatively narrow inversion layers and regions, quite deep along the vertical, in which the temperature was either constant with height or else dropped gradually. Such regions are distributed along the horizontal over the whole anticyclone. Zak considers them to be frontal interfaces.

In this author's opinion, a sizable contribution to the creation of stable zones and inversion layers is made by the frontal cloudiness itself, which originated in a cyclone and then was transported by the airflow to the western and southern periphery of the anticyclone, where cloudiness is observed most frequently. In any case, most temperature inversions, according to Zak, originate in an anticyclone not because of radiation, but rather because of transport of air masses from the cyclone to the anticyclone.*) Radiation can cause only an inversion close to the ground, which, if fog appears, may become a lifted inversion, while the fog turns into low cloudiness.

The very formation of St-Sc cloudiness was attributed by Zak to convective transport of water vapor away from the ground near the inversion layer, but he ruled out the effect of a downward advection and diffusion of vapor. This conclusion is borne out by data on the lapse rate (the maximum frequency of 51% in the lower layer for St-Sc corresponds to a gradation of 0.61°C-1.00°C per 100 m) and on the air humidity near the ground (for St-Sc, the relative humidity exceeds 80% in 76% of the cases and the specific humidity is higher than 3 g/kg in 77% of the cases, whereas for clear weather the corresponding figures are 36 and 2%).

Although the importance of the turbulent influx of water vapor from the ground cannot be denied (cloud-formation models developed by the author and others accord a leading role to this mechanism), there is certain illogicality in Zak's arguments. On the one hand, advection is assigned a decisive role in producing a temperature inversion; on the other hand, the influx of water vapor in the warm mass is denied a role in the cloud formation.

*) Interestingly enough, in this lengthy paper, wholly devoted to a description of St-Sc clouds in anticyclones, Zak used the term "radiative" only once, and then with regard to temperature inversions close to the ground.

However, this is possible if it is assumed that the warm air mass is itself heated by a downdraft as it advances into the anticyclone. This is indicated by Zak's data: for an inversion depth $h < 1000$ m, there is no correlation between the St-Sc frequency and h, and only for $h > 1000$ m does the probability of clear weather increase (to 70%).

The temperature differential ΔT in the inversion, if $\Delta T < 10°C$, will not affect the St-Sc probability either, but for $\Delta T > 10°C$ the likelihood of clear weather is greater. These facts are hardly in accordance with the idea that advection plays a decisive part in producing the temperature inversion and that water-vapor transport from the ground is instrumental in the creation of clouds. Why, indeed, should the mechanism of water-vapor transport from the ground and vapor concentration under the inversion cease for high values of h and ΔT? However, the indicated facts agree with the hypothesis that downward vertical motions play a decisive part in producing temperature inversions close to the upper boundary of clouds traveling from a cyclone (trough) to the western or southern periphery of an anticyclone. In this case h and ΔT increase with time, and there will be a simultaneous slow subsidence of the cloud and inversion layers. It is quite natural that as it heats while descending, the cloudiness dissipates gradually (although the downflow of vapor tends to slow this process); thus, the likelihood of clear air will be greater, the higher the values of h and ΔT, since they increase as a result of the same factors which cause the clouds to dissipate.

Consequently, the data so carefully analyzed by Zak indicate clearly that dynamic factors (advection, turbulent mixing, and vertical motions) play a decisive role in creating temperature inversions and cloudiness in anticyclones.

Selezneva [274] presented two typical schemes for the development of St-Sc clouds during the autumn-winter season, based on an analysis of sounding data obtained over the northwestern part of the European USSR. According to the first scheme, the most important features of the clouds observed in the center of an anticyclone, for the advection of cold, are as follows. The air temperature drops with height under the cloud and inside it; above the cloud the temperature also initially drops with height. Later, under the influence of downward vertical motions (their velocity was estimated at the upper edge of the boundary layer using the method of the author of [176]), a temperature inversion intensifying with time, as well as a layer where γ is low, is formed above the cloud. The relative humidity, which is high near the ground, reaches 100% in the cloud, but in the inversion layer it rapidly drops to 30-40%, and often to 10-20%. The wind speed in the lower layer increases with height, and often reaches a maximum at some height close to the upper cloud boundary or above it, then remains practically constant further up. The wind direction does not vary substantially with height. Under the cloud and inside it, turbulent mixing develops, as indicated by the low values of the Richardson number there. The turbulent flux of water vapor is directed upward since the specific humidity under the cloud and inside it drops with height . The water vapor flux supplies the cloud with moisture and sustains its existence.

The second scheme describes the development of a cloud in the presence of heat advection. In this case, above the cloud an inversion of the temperature and specific humidity is observed, while above the inversion the temperature and specific humidity decrease linearly with height; there is no layer of low γ. The wind velocity, as a rule, is considerable and increases with height; the wind direction also varies greatly upon transition from the layer near the ground to the inversion layer. The turbulent mixing is intense, above the cloud as well as inside it.

The cloud and the inversion above it form because of heat advection (defined simply as the movement of warm air over a cooler underlying surface). Naturally, as the air mass moves over the surface, a temperature inversion is produced near the ground, the inversion depth increasing with time. However, since the turbulent mixing is intense because of the wind, while the heat flux in the inversion layer is directed downward, the air temperature close to the ground rises with time, and this temperature rise is accompanied by a disappearance of the inversion in the lower part of the layer and the creation of a lifted inversion. The accumulation of water vapor under the layer, together with the general lowering of the air temperature, is the main mechanism of cloud formation for heat advection.

Podgurskaya [252] recently made a study of the origin of St-Sc clouds, using data from network soundings of the atmosphere with radiosondes and aircraft, together with hourly observations on the ground and maps of the weather during the cold seasons of 1955-1963 in the Ukraine. A total of 1323 cases were selected, in which solid St-Sc cloudiness lasting at least 3 h was observed.

Table 7.14 gives the results of an analysis of the weather conditions preceding the advent of St-Sc clouds. In anticyclones and in a low γ field, St-Sc clouds appear after clear weather in a considerable number of cases (32-42%). However, most of the time the appearance of St-Sc is preceded by Ac-As and Ns cloudiness. In a cyclone this happens in more than 75% of the cases, and in an anticyclone or low γ fields in 45-60% of the cases. This indicates that there is a close correlation between the formation of St-Sc and the cloudiness of fronts, as well as of low pressure regions as a whole.

The appearance of St-Sc clouds after clear weather can be due to their arrival from some other region or to their formation right in the region. It is difficult to decide which, on the basis of observations at a single point or over a small region. However, if the clouds appear almost simultaneously over a considerable territory (for instance, over the entire Ukraine), then it is justified to assume that the cloudiness originated over this territory.

Table 7.14. Frequency (%) of weather conditions preceding appearance of St-Sc clouds.

Synoptic system	Weather conditions: clear	Ac-As cloudiness of 5-10	Ns cloudiness of 10	fog	No. of cases
Anticyclone					
northern part	41.9	43.6	1.6	12.9	62
eastern part	32.4	52.6	6.3	8.7	207
southern part	32.5	46.4	8.3	12.8	179
western part	36.3	39.1	8.5	16.1	323
Cyclone					
warm sector	11.6	60.4	14.7	13.3	68
rear part	19.8	53.8	24.5	1.9	106
Low γ field	34.2	45.3	6.8	13.7	225
Transition field from cyclone to anticyclone	19.4	55.9	16.3	8.4	153

Table 7.15 gives the results of an analysis of the synoptic conditions for the appearance of St-Sc clouds after clear weather, according to data of observations at Kiev, Kharkov, Odessa, and Lvov. It is seen that even after clear weather, St-Sc clouds are, on the whole, more frequently a result of migration than of simultaneous formation (235 cases versus 154).

The role of migration is especially great in the western and southern parts of an anticyclone, as well as in a transition field or in a cyclone as a whole; spontaneous formation of St-Sc is never observed in a cyclone after clear weather. In anticyclones, St-Sc clouds arriving from regions of lower pressure accounted for 235•0.734 = 172 cases, while the spontaneous formation of St-Sc

Table 7.15. Frequency (%) of synoptic conditions for appearance of solid St-Sc cloudiness after clear weather.

Synoptic system	Migration	Spontaneous formation
Anticyclone		
northern part	5.9	4.6
eastern part	14.4	17.9
southern part	16.2	12.6
western part	36.9	9.9
Cyclone		
warm sector	6.3	—
rear part	1.4	—
Low γ field	6.4	40.2
Transition field from cyclone to anticyclone	13.5	14.9
No. of cases	235	154

over most of the Ukraine accounted for only 154•0.45 = 69 cases, that is, in anticyclones St-Sc clouds arrive from regions of lower pressure 71% of the time, and only 29% of the time do they originate spontaneously after clear weather.

Since a comparatively small portion of the clouds in cyclones and other low-pressure regions are St-Sc, it can be inferred from the data presented that during their travel from a cyclone to an anticyclone, Ns-As stratiform clouds are transformed into St-Sc-Ac wave clouds. Only in a field with a low lapse rate (low γ field, see Table 7.15) is the frequency of simultaneous St-Sc formation after clear weather substantially higher than the frequency for migration from elsewhere (62 cases versus 15).

However, Table 7.15 does not give a clear indication of the probability that St-Sc clouds will appear over the Ukraine spontaneously or after arrival from elsewhere. Therefore, Table 7.16 was compiled from the data of Tables 7.14 and 7.15 in which clouds appearing spontaneoulsy over a large territory after clear weather were included in the category of newly formed St-Sc clouds. All the other St-Sc clouds that appeared in the territory in question can be placed in the same group (migration and transformation).

The latter category includes St-Sc clouds appearing after clear weather due to migration, as well as St-Sc that were preceded by other forms of clouds or fog. There is apparently good reason to assume that in the latter case, St-Sc cloudiness also appears

Table 7.16. Probability (%) of appearance of St-Sc clouds due to spontaneous formation and migration from elsewhere.

Synoptic system	Migration and transformation	Spontaneous formation	No. of cases
Anticyclone			
northern part	88.6	11.4	62
eastern part	88.6	13.4	207
southern part	89.4	10.6	179
western part	95.4	4.6	323
Cyclone			
warm sector	100	—	68
rear part	100	—	106
Low γ field	72.5	27.5	225
Transition field from cyclone to anticyclone	85.0	15.0	153

because of either the arrival of St-Sc from other regions or the migration of Ns-As clouds and fog, which, as they travel, are transformed into St-Sc. According to Table 7.16, in an overwhelming majority of cases St-Sc clouds appear in a given territory because of migration of the clouds from elsewhere, accompanied by a transformation of some of the clouds from one form to another. In all the synoptic systems considered, the spontaneous formation of St-Sc accounted for 5-15% of the cases, and only in a low γ field did the proportion increase to 27%. Since in other regions the probability of the spontaneous formation of St-Sc is also very slight, it can be concluded from the material considered that most St-Sc clouds originate via the transformation of other cloud forms (primarily Ns-As) or fog.

The St-Sc clouds have lifetimes of a few hours, as Table 7.17 indicates.

The relative humidity of the air close to the ground prior to the appearance of St-Sc after clear weather increased in 60% of the cases for migration and in 74% of the cases for spontaneous formation, remaining unchanged in 10 and 26% of the cases, respectively. For migration, the relative humidity decreased in 30% of the cases.

Table 7.17. Frequency (%) of lifetimes of St-Sc clouds after clear weather.

Process	Lifetime, h					No. of cases
	<3	4–6	7–9	10–12	>12	
Migration	50	25	11	7	7	235
Simultaneous formation	75	12	7	4	2	154

The appearance of St-Sc clouds after clear weather as a result of migration is of almost equal probability at any time of day, but from 1000 to 1500 h these clouds are 1/3-1/2 as frequent as from 1900 to 0900 h. At the same time, over a large territory St-Sc clouds form only from 2200 to 0900 h, with a maximum frequency (28%) in the morning (0700-0900 h). Clearly, this diurnal variation in the appearance of St-Sc cloudiness can be attributed to the diurnal temperature variation, since these clouds form in the lower part of the boundary layer: for simultaneous formation, the height of the lower boundary is less than 200, 400, 600, and 800 m in 30, 74, 90, and 100% of the cases, respectively, when the relative humidity increases, and in 20, 51, 80, and 96% of the cases when the relative humidity remains unchanged.

7.5. Frontal clouds; precipitation and cloud water content

In the early 1950s, Drozdov [96] pointed out that during the movement of fronts over the southeastern part of the European USSR in the summer, it frequently happens that no precipitation falls and the clouds gradually become dissipated. Mogila and Polovina [218] made a study of the fronts whose clouds do not give precipitation. Data of aircraft soundings of the atmosphere at four places in the Ukraine (Lvov, Kiev, Kharkov, and Odessa) from 1953 to 1963 were used.

According to Table 7.18, at all fronts without precipitation St-Sc clouds predominate: at cold fronts they have frequencies of 53-65% (at different points), and at warm or occlusion fronts 50-68%. The frequencies of As-Ac clouds are also quite high (above 30%, on the average). Those of nimbostratus at fronts without precipitation, on the other hand, are low: 3-7% at cold fronts, 2-8% at warm fronts, and 2-12% at occlusion fronts (note that these clouds really should not be called "nimbostratus", since no precipitation falls from them). In most instances (64-74% at cold fronts, 52-70% at warm fronts, and 58-74% at occlusion fronts), at fronts without precipitation single-layer cloudiness is observed; double-layer cloudiness is observed in 25-30% of the cases.

Table 7.18. Frequency (%) of various cloud forms at fronts without precipitation (Kharkov data in numerator, Kiev data in denominator), for period from 1953 to 1963 (October-March).

Front	Sc	St	As	Ac	Ns	Cs	No. of cases
Cold	43/53	23/11	10/24	18/10	3/2	3/—	252/202
Warm	34/37	20/18	17/28	20/12	7/6	2/—	243/265
Occlusion	31/48	19/18	19/19	19/13	12/2	—/—	48/62

In the present author's opinion, these experimental data are an indication that dynamic factors play a decisive part in the formation of St-Sc clouds. The absence of precipitation at the front probably shows that the front is breaking up (small temperature contrasts, vertical upcurrents weak or even becoming downcurrents, etc.). Naturally, the cloud system of the front then also starts to be transformed: the Ns-As cloudiness created at the front during its active period deteriorates into St-Sc cloudiness. 85-90% of the time, the depth of the latter does not exceed 1000 m.

The frequency of fronts without precipitation is very high. For instance, according to the data of [146], on the average, a total of 106 (Odessa region) to 137 (Kiev region) fronts pass over the Ukraine in a year. During the cold season (autumn-winter) the annual average passing over Kiev is 76 fronts; over Kharkov it is 74 fronts. According to Table 7.18, over an 11-year period, during the cold season (October-March) a total of 529 fronts without precipitation passed over Kiev, and 543 passed over Kharkov. Thus, the annual average number of fronts without precipitation but with St-Sc for the cold half-year in the Kiev region is $529 \div 11 = 48$ fronts out of 76, and in the Kharkov region it is 50 fronts out of 74. Consequently, it can be concluded that a considerable portion of the St-Sc cloudiness evolves out of Ns-As cloudiness.

The authors of [256] used observations of the water content and phase of frontal clouds, carried out during 700 aircraft network soundings from 1957 to 1963 and special flights from 1961 to

1963. Liquid-water Ns-As clouds are observed during the warm season in 55-77% of the cases, and during the cold season in 38-51% of the cases. For temperatures above -12°C, Ac, Sc, and Frnb clouds are predominantly liquid-water clouds (frequency higher than 60%). Ns-As and As liquid-water clouds predominate at a higher temperature (-8°C). At low temperatures, the frequencies of mixed Ns-As and crystalline As and Ac clouds are high. In the cases of mixed frontal and layer clouds, there is a high probability of precipitation reaching the ground; whereas the liquid-water clouds do not give precipitation in most cases.

Vertical profiles of the water contents of St and Sc clouds of various depths were constructed in [251]. The water content was determined over 100-m layers. In order to exclude the effect of temperature, only data pertaining to a narrow temperature interval (from -5.0 to 0.1°C) were included in the sample. The results are given in Table 7.19.

Table 7.19. Mean water content (g/m^3) of St and Sc clouds of various depths (number of cases shown in parentheses).

Cloud depth, m	Height of layer above cloud base, *m*				
	⩽ 100	100-200	200-300	300-400	> 400
100—200	0.11 (71)	0.16 (62)			
200—300	0.11 (67)	0.18 (224)	0.24 (128)		
300—400	0.10 (186)	0.19 (219)	0.25 (272)	0.30 (94)	
400—500	0.11 (74)	0.18 (142)	0.25 (149)	0.31 (130)	0.32 (60)
500—600	0.13 (35)	0.19 (75)	0.26 (82)	0.31 (79)	0.33 (96)
>600	0.12 (49)	0.19 (99)	0.24 (117)	0.31 (101)	0.32 (309)

An important conclusion can be drawn from the data of Table 7.19: at a given height above the cloud base, the water content is practically identical for clouds of different depths, that is, the vertical profiles of the water content in the lower parts of clouds of any depth are the same. The mean (over the cloud) water content increases only via a rise in the water content with height up to the maximum value level, which is found in the upper part of a cloud of any depth, rather than via a variation in the form of the vertical profile of the water content.

The dependence of the mean (over the cloud) water content on the depth of St and Sc clouds was established experimentally in [35, 50, 179, 214, 216]. Table 7.20 gives information about the mean (over the cloud) water contents for various cloud depths and temperatures. According to these data, an increase in cloud depth from 300 to 600 m is accompanied by a rise in the mean (over the whole cloud) water content by 80-100% at negative temperatures and by 60% at positive temperatures. For a specified cloud depth, the water content increases with a rise in termperature. It should be noted that the rise in the mean water content as a function of cloud depth is observed up to depths of about 600 m; for greater depths the mean water content may even decrease, because of the nonuniform structure of deep clouds and the presence of interlayers in them.

Table 7.20. Mean water content (g/m^3) of St-Sc clouds for various cloud depths and temperatures.

Cloud depth, m	Temperature, °C				
	-15...-10	-10...-5	-5...0	0...5	5...10
100	0.02	0.03	0.04	0.06	0.08
150	0.04	0.05	0.08	0.10	0.15
200	0.05	0.07	0.11	0.15	0.19
250	0.07	0.09	0.14	0.19	0.23
300	0.08	0.11	0.16	0.22	0.26
350	0.09	0.13	0.18	0.24	0.29
400	0.10	0.15	0.20	0.26	0.32
450	0.11	0.16	0.22	0.29	0.34
500	0.12	0.17	0.24	0.31	0.36
550	0.14	0.19	0.26	0.33	0.38
600	0.15	0.20	0.28	0.35	0.40

The level of maximum water content, according to the studies of St-Sc clouds over the Ukraine, fluctuates between the middle of the cloud and its upper boundary; the maximum frequency corresponds to heights equal to 0.8-0.9 of the cloud depth. With an increase in cloud depth, the distance between the level of maximum water content and the upper cloud boundary becomes greater: for a depth of 200 m this level is 30-40 m from the upper boundary, and for a depth of 600 m it is 80-90 m from it.

The largest vertical gradients of the water content are observed in the lower 100-m layer, and they decrease with increasing height. At lower temperatures the water-content gradients will be smaller than at higher temperatures.

All these experimental results are in accordance with the predictions of theory [179, 192].

The water-content distribution of Ns-As clouds are more complicated than those of St-Sc. In Ns-As, several levels of heightened water content are frequently observed. For cloud depths up to 1.5 km, there are often (60% of cases) two levels of heightened water content (close to the base and in the middle of the cloud); for depths of 2-3 km, there may be three levels (60-75% of cases).

The mean water content of Ns-As rises with an increase in cloud depth:

Cloud depth, km	<0.5	0.5—1.0	1.0—2.0	2.0—3.0	3.0—4.0
Mean water content, g/m^3	0.14	0.19	0.25	0.39	0.42

As a rule, the maximum frequency of water-content values shifts toward higher δ^* with an increase in the Ns-As depth (Table 7.21).

For shallow clouds (< 0.5 km in depth), values of δ^* not greater than 0.1 g/m^3 predominate; in clouds with depths from 0.5 to 2 km, the values of δ^* are generally between 0.1 and 0.2 g/m^3, while in clouds deeper than 2 km the maximum frequency for δ^* values lies in the interval from 0.3 to 0.5 g/m^3.

Table 7.21. Frequency (%) of water content of Ns-As clouds of various depths.

δ^*, g/m^3	Cloud depth, km				
	<0.5	0.5-1.0	1.0-2.0	2.0-3.0	3.0-4.0
<0.1	52.9	32.0	46.5	—	—
0.1—0.2	23.5	44.0	50.0	—	—
0.2—0.3	12.0	16.0	27.3	16.7	—
0.3—0.4	0.5	4.0	18.2	22.2	50.0
0.4—0.5	5.8	4.0	—	44.4	33.3
>0.5	—	—	—	16.7	17.7

The behavior of the lapse rate γ is also of great interest. In Ns clouds less than 1 km deep, $\gamma > \gamma_a'$ under the cloud, $\gamma \leqslant \gamma_a'$ in the cloud, and $\gamma \leqslant 0$ (isotherm, inversion) above the cloud. This is a deteriorating cloud, with a water content less than 0.2 g/m^3, the maximum value being in the central or top part of the cloud. Such clouds should really be called stratus or strato-cumulus, which, according to the hypothesis developed in this book, are formed from Ns. For developing Ns clouds more than 2 km deep, the lapse rate above the cloud is greater than under the cloud, where it is close to γ_a'. This experimental result also agrees with theoretical predictions: because of upward motions, the air temperature drops with time more rapidly above the cloud than in the cloud, so that γ also increases above the clouds.

It is much more difficult to measure the water contents of crystalline clouds than of droplet clouds. Thus, as yet only a few determinations of the water contents of crystalline clouds have been made. According to the data of Minervin [214], who developed a device for measuring the water content of crystalline clouds, the latter drops sharply with height and above 2 km it is about an order of magnitude less than the water content of droplet clouds (Table 7.22). The measurements for which results are reported in Table 7.22 pertain mainly to the upper parts of Ns-As clouds, and more rarely to As clouds that are changing into Cs. When snow fell, the water content of both cloud and precipitation was measured.

In Japan, a special sounding device was developed to measure the water content of snow and of crystalline clouds. The measurement results were analyzed, taking into account the dynamics of cloud development. In the early developmental stage, the number of crystals is very low and the whole cloud consists of supercooled droplets. For instance, on 22 January 1970, in such a cloud with a lower boundary at 1300 m and an upper boundary at 2100 m, the water content of the solid phase fluctuated between 0.02 and 0.04 g/m^3, whereas the water content of the liquid phase increased sharply with height in the lower part of the cloud (from 0.05 to 0.50 g/m^3), and then rose gradually to 0.85 g/m^3 in the upper part.

Table 7.22. Water content (g/m^3) of crystalline clouds.

Water content	Height above ground, km					
	<1	1-2	2-3	3-4	4-5	>5
Mean	0,17	0,088	0,032	0,026	0,016	—
Maximum	0,67	0,74	0,14	0,17	0,078	0,052
No.of measurements	21	47	28	34	17	1

In the mature stage, the crystals are distributed quite uniformly throughout the cloud, droplets being observed only in the lower half of the cloud. On 29 January 1970, in a cloud with a lower boundary at 500 m and an upper boundary at 1300 m, the contribution of droplets to the water content fluctuated from 0.02 to 0.20 g/m^3, while that of crystals fluctuated from 0.02 to 0.13 g/m^3. During the decay stage, the cloud consists only of ice crystals. On 25 January 1970, the water content of such a crystalline cloud increased with height from 0.01 g/m^3 close to the lower boundary to 0.10 g/m^3 in the upper part of the cloud, and it dropped to 0.02 g/m^3 close to the upper cloud boundary. The temperature in the cloud varied from -10 to -22°C in all three cases.

CHAPTER 8

STATISTICAL DATA OF CLOUDS AND CLOUD BEHAVIOR

8.1. Cloud boundaries and depth; amount of cloud cover

The statistical characteristics of clouds were obtained from data of aircraft soundings of the atmosphere at 32 sites in the Soviet Union [3]. The cloud-structure parameters considered below were calculated for most points from sounding data gathered from 1957 to 1964. The greatest number of soundings were made at Moscow (1386 in winter, 1431 in spring, 1372 in summer, 1370 in autumn), Leningrad (862, 854, 860, and 891, respectively), and Arkhangelsk (561, 409, 461, and 610, respectively).

Since during the cold half of the year (autumn and winter) the air is saturated with water vapor more often than during the warm seasons (spring and summer), the frequency of low-lying solid cloudiness (8-10) is considerably higher in winter and autumn than in spring and summer (Table 8.1).

Table 8.1. Frequency (%) of clear (0-2), partly cloudy (3-7), and overcast (8-10) skies over Moscow (numerator) and Leningrad (denominator).

Month	Cloud cover								
	low-lying			intermediate			high		
	0–2	3–7	8–10	0–2	3–7	8–10	0–2	3–7	8–10
January	14/17	12/9	74/74	59/69	11/7	37/24	65/71	21/12	13/17
April	45/37	22/18	33/45	55/56	17/14	28/30	60/57	27/24	13/19
July	35/52	42/25	23/23	49/59	30/25	21/16	48/58	42/24	10/18
October	16/19	18/13	66/68	58/61	15/10	27/29	63/61	22/19	15/20

For example, in Moscow and Leningrad in January, low-lying cloudiness of 8-10 is observed in 74% of the cases, while a cloud cover of 0-2 is observed in only 14 and 17%, respectively. In July, on the other hand, in both Moscow and Leningrad cloudiness of 0-2 and 3-7 predominates (frequency of 77%); in both cities overcast skies are observed in only 23% of the cases.

Statistical characteristics obtained on the basis of 5-7 years of aircraft soundings of the atmosphere can be considered quite reliable, as was shown by a comparison of data gathered over many years. However, the soundings were not carried out completely regularly: there was a predilection for days when cloudiness was present. For this reason, the frequency of a solid cloud cover

(8-10) according to the aircraft-sounding data proved to be 5-20% higher 1/M and, in individual instances, 1.5-2 times higher 1/M than the frequency according to daily ground-based observations.

The predominance of solid low-lying cloudiness in winter and of clear and partly cloudy skies in summer is found in practically the entire Soviet Union. The only exceptions are high-latitude regions and the Soviet Far East: at Arkhangelsk, for instance, the frequency of low-lying cloudiness of 8-10 is no less than 50% during the summer months; at Vladivostok the frequency of overcast skies is highest in June and July (60-64%), lowest in November (8%); and during the other months it varies from 18 (September) to 33% (March).

Although it is commonly thought that overcast skies have a high frequency of occurrence in Leningrad, the foregoing data do not bear out this opinion: overcast skies (low-lying, as well as total, cloudiness) have about the same frequency over Leningrad as over Moscow. Data gathered over many years gives a frequency difference between them which does not exceed 1%; aircraft soundings over Leningrad in July show the skies there to be even clearer than over Moscow, and only in April does overcast weather occur 12% more often in Leningrad than in Moscow.

The aircraft soundings made it possible to compare the vertical cloudiness distributions in the lower and middle (up to 7 km) troposphere (Table 8.2).

Inspection of the table shows that in winter, solid cloudiness, and in fact all cloudiness (1-10), forms most frequently in the layer from 0.5 to 1 km: in this layer the frequency of a cloud cover of 8-10 is 48% in both Moscow and Leningrad, while the probability of any cloudiness at all (1-10) is 54 or 55% in both cities; the mean cloud cover is 5.2 for Moscow and 5.1 for Leningrad. Similar (albeit slightly lower, especially for Leningrad) values of these characteristics are observed in the lower 1/2-km layer. Above 1 km, with increasing height the frequency of solid cloudiness, the probability of any cloudiness at all (1-10), and the mean cloud cover all decrease monotonically: in Moscow they drop from 40 and 46% and 4.5 in the layer from 1 to 2 km down to 10--15% and 1.3 in the layer from 5 to 6 km; in Leningrad they drop from 47 and 51% and 5.0 in the layer from 1 to 2 km down to 16 and 20% and 1.8 in the layer from 5 to 6 km.

The cloud distribution in summer is somewhat different. Since from winter to summer the air temperature increases most in the lower layers, the conditions there will be less suitable for water-vapor condensation in summer than in winter (air at a higher temperature is further away from saturation). Consequently, the frequency of a solid cloud cover and the probability of cloud formation in the lower layers (from the ground up to 1 km) will both be considerably lower in summer than in winter. For instance, for Moscow in summer the frequencies of a cloudiness of 8-10 in the 0-0.5 and 0.5-1.0 km layers are only 8 and 12%, respectively (for winter frequencies of 42 and 48%); the probabilities of any

Table 8.2. Frequency (%) of slight (0-2), variable (3-7), and solid (8-10) cloudiness in various layers of the troposphere over Moscow (numerator) and Leningrad (denominator).

Layer, km	Amount of cloud cover				Mean cloud cover
	0–2	3–7	8–10	1–10	
			Winter		
0–0.5	46/57	12/6	42/37	55/43	5.0/4.0
0.5–1	46/46	6/6	48/48	55/55	5.2/5.1
1–2	54/49	6/4	40/47	46/51	4.5/5.0
2–3	70/64	5/3	25/33	30/36	3.0/3.4
3–4	78/71	5/3	17/26	22/30	2.0/2.8
4–5	82/77	5/3	13/20	18/24	1.6/2.2
5–6	85/80	5/4	10/16	15/20	1.3/1.8
6–7	96/89	1/1	3/10	4/11	0.3/1.0
			Summer		
0–0.5	86/88	6/5	8/7	14/13	1.2/1.0
0.5–1	79/79	9/11	12/10	22/25	1.8/1.7
1–2	46/58	34/24	20/18	57/50	3.8/3.0
2–3	46/55	32/24	22/21	58/51	4.0/3.4
3–4	57/66	24/17	19/17	47/37	3.4/2.6
4–5	72/76	15/10	13/14	31/26	2.2/1.9
5–6	82/84	10/7	8/9	19/18	1.4/1.3
6–7	95/92	3/2	2/6	5/8	0.3/0.6

cloudiness at all (1-10) are in these same layers 14 and 22%, respectively (for winter probabilities of 55 and 55%).

However, in summer the turbulent mixing (including mesoscale convection) becomes intensified, and thus the water-vapor content in the middle troposphere increases and more favorable conditions for the formation of clouds, especially ones of convective origin, are created in spite of the temperature rise. Inspection of Table 8.2 shows that in the middle troposphere (at heights from 1 to 6 km) in summer, the frequency of variable cloudiness increases sharply (by as much as a factor of 5-8); the frequency of a solid cloud cover, on the other hand, is only about half as high in the layer from 1-2 km and remains practically unchanged in higher layers. Thus, the likelihood of the formation of any cloud cover at all (1-10) in summer is considerably higher than in winter, in all layers located above 1 km for Moscow and above 2 km for Leningrad. For instance, in Moscow the probabilities of cloud formation in the 1-2, 2-3, and 3-4 km layers are 57, 58, and 47%, respectively, in summer and 46, 30, and 22% in winter; in Leningrad in the 2-3, 3-4, and 4-5 km layers the corresponding probabilities are 51, 37, and 26% in summer and 36, 30, and 24% in winter. The most favorable conditions for cloud formation in summer are created in the 2-3 km layer: there all the characteristics (frequency of solid and variable cloud cover, probability of cloud formation, and mean cloud cover) reach a maximum; above and below the 2-3 km layer there is a monotonic decrease in almost all the cloudiness characteristics (the frequency of variable cloudiness is an exception, since in the 2-3 and 1-2 km layers it is practically the same - 32 and 34% for Moscow and about 24% for Leningrad).

It should be noted that in winter the frequency of a cloudiness of 1-2 is very low. Actually, in some layers the sum of the frequencies of variable and solid cloudiness equals the probability of cloud formation; in other layers this sum is 1% less than the probability (for instance, for Moscow in the 0-0.5 and 0.5-1 km layers and for Leningrad in the 3-4 and 4-5 km layers). In summer the frequency of a cloud cover of 1-2 reaches 2-3% in some cases (for instance, for Moscow in the 1-3 and 4-5 km layers, and for Leningrad in the 3-4 km layer), and even as much as 4-6% (for Moscow in the 2-3 and 3-4 km layers, and for Leningrad in the 0.5-1 and 2-3 km layers); only in one layer (the 1-2 km layer for Leningrad) does the frequency of a cloud cover of 1-2 reach 8%.

Frequencies of combinations of different amounts of cloud cover in the lower (up to 3 km), middle (3-6 km), and upper (above 6 km) troposphere. Atlas [3] includes information about the frequencies of combinations of 0-2, 3-7, and 8-10 cloud covers in three layers of the troposphere, relative to the total number of aircraft ascents during the season, separately for daytime (0800--1900 h local time) and nighttime (2000-0700 h). Inspection of the tables indicates that the frequencies of most of the 27 combinations do not exceed a few percents (they generally are from 1 to 2%). Only the frequency of a solid cloud cover below 3 km, for a cloudiness of 0-2 in the 3-6 km layer and above 6 km, reaches some tens of percents (as a rule, not more than 40%) in winter and autumn in the European USSR, West Siberia, and Western Kazakhstan. In spring and summer, the frequency of a combination of clear skies (cloudiness of 0-2) in all three layers is quite high, namely 20--30%. Strange as it may seem, this frequency is higher at night than in the daytime: for Moscow in springtime it is 32% at night and 12% during the day, and in summer the corresponding figures are 27 and 5%; for Leningrad in spring the frequencies are 30 and 25%, and in summer 32 and 24%, for night and day, respectively. In the European USSR, the combination of a solid cloud cover in all three layers has a frequency of 8-10% in winter and autumn and 6-8% in summer and spring.

The degree of cloud cover at different atmospheric levels for observations from fixed heights. Tables 8.3 and 8.4 give data on this cloudiness characteristic. Numbers at intersections between rows and columns with the same height value give the mean amount of cloud cover (in percent) at the given height.

With increasing height above the ground, the degree of cloud cover in winter increases sharply in the lower kilometer layer (in Moscow, to 71.4% in the daytime and 70.2% at night), and then increases gradually with height. Since in summer, as discussed above, clouds usually form in the 1-2 and 2-3 km layers, it is

Table 8.3. Degree of cloud cover (%), Moscow.

Level,km	Fixed height,km													
	0	1	2	3	4	5	6	0	1	2	3	4	5	6
	Winter, daytime(n=371)							Summer, daytime(n=385)						
6	91.7	68.2	47.1	34.4	24.8	17.6	8.7	76.2	74.6	65.0	45.8	29.6	17.2	8.1
5	90.5	65.3	43.3	29.5	18.7	9.8	17.6	75.4	73.7	63.9	43.5	26.7	12.3	17.1
4	88.6	61.6	38.5	23.3	9.9	18.7	24.8	73.7	71.7	61.2	39.7	19.6	26.7	29.6
3	86.2	58.0	32.3	12.7	23.3	29.5	34.4	69.1	66.4	54.4	27.5	39.7	43.5	45.8
2	80.8	44.9	15.7	32.3	38.5	43.3	47.1	57.7	54.1	36.5	54.4	61.2	63.9	65.0
1	71.4	24.7	44.9	57.0	61.6	65.0	62.8	23.0	16.8	54.1	66.4	71.7	73.7	74.6
0	00.0	71.4	80.8	86.2	88.6	90.5	91.7	00.0	23.0	57.7	69.1	73.7	75.4	76.2
	Winter, nighttime (n=349)							Summer, nighttime(n=267)						
6	94.6	70.8	45.3	29.5	20.4	12.1	6.3	72.3	68.5	59.9	47.1	28.0	15.5	7.1
5	84.0	69.2	42.5	26.4	16.7	7.7	12.1	70.2	66.3	57.2	43.8	23.6	7.7	15.5
4	92.9	66.1	38.0	20.5	10.1	16.7	20.4	65.6	61.5	51.6	36.6	10.6	23.6	28.0
3	90.2	60.6	30.7	8.9	20.5	26.4	29.5	54.1	48.4	36.2	16.3	36.6	43.8	47.2
2	85.7	52.3	15.0	30.7	38.0	42.5	45.3	38.1	30.0	11.4	36.2	51.6	57.2	59.9
1	70.2	23.5	52.3	60.6	66.1	69.2	70.8	22.0	9.5	30.0	48.4	61.5	66.3	68.5
0	00.0	70.2	85.7	90.2	92.9	94.0	94.6	00.0	22.0	38.1	54.1	65.6	70.2	72.3

Table 8.4. Degree of cloud cover (%), Leningrad.

Level, km	Fixed height, km													
	0	1	2	3	4	5	6	0	1	2	3	4	5	6
	Winter, daytime (n=233)							Summer, daytime (n=232)						
6	91.9	70.1	45.4	30.8	23.6	16.9	10.1	64.2	61.3	51.8	34.9	24.1	15.9	8.5
5	90.3	68.0	41.0	26.1	18.0	10.6	16.9	63.0	60.1	50.2	32.9	21.3	12.1	15.9
4	89.0	64.7	37.3	21.9	13.3	18.0	23.6	60.8	57.8	47.0	29.1	16.7	21.3	24.1
3	86.6	62.1	32.9	15.1	21.9	26.1	30.8	56.3	52.8	41.2	21.3	29.1	32.9	34.9
2	84.2	57.2	24.2	32.9	37.3	41.0	45.4	45.6	41.4	26.5	41.2	47.0	50.2	51.8
1	67.5	32.0	57.2	62.1	64.7	68.0	70.1	21.5	15.3	41.4	52.8	57.8	60.1	61.3
0	00.0	67.5	84.2	86.6	89.0	90.3	91.9	00.0	21.5	45.6	56.3	60.8	63.0	64.2
	Winter, nighttime (n=221)							Summer, nighttime (n=189)						
6	93.8	73.9	50.1	37.1	24.3	16.8	11.6	73.5	70.0	61.8	47.0	32.3	18.8	9.6
5	92.6	72.2	47.7	34.3	21.4	13.0	16.8	72.2	68.7	60.3	45.3	29.8	13.9	18.8
4	91.7	69.6	44.9	31.4	16.3	21.4	24.3	67.8	64.0	55.7	39.7	20.7	29.8	32.3
3	86.7	62.0	34.2	19.0	31.4	34.3	37.1	58.2	52.6	42.3	22.4	39.7	45.3	47.0
2	83.8	56.3	23.5	34.2	44.9	47.7	50.1	43.2	33.3	18.9	42.3	55.7	60.3	61.8
1	68.8	32.6	56.3	62.0	69.6	72.2	73.9	28.1	14.7	33.3	52.6	64.0	68.7	70.0
0	00.0	68.8	83.8	86.7	91.7	92.6	93.8	00.0	28.1	43.2	58.2	67.8	72.2	73.5

Table 8.5. Integral distribution function (%) $F(z_u < X)$ of height of upper cloud boundary.

Season	Time	X, km									$\bar{z}_u$	$z_{u,max}$	$z_{u,min}$	n
		0.5	1.0	1.5	.2.0	2.5	3.0	4.0	5.0	6.0				
							Moscow							
Winter	Day	5.4	30.0	47.4	62.1	72.4	76.9	86.3	93.4	98.3	2.06	6.58	0.18	224
	Night	5.2	24.2	43.5	57.3	68.7	76.3	84.9	93.8	98.6	2.15	6.34	0.19	290
Summer	Day	—	1.8	10.8	19.8	36.9	56.8	74.8	89.2	96.4	3.09	6.66	0.76	111
	Night	4.5	10.1	20.2	24.7	34.8	48.3	79.8	91.0	100	2.95	5.80	0.19	89
							Leningrad							
Winter	Day	5.7	26.8	43.8	50.8	62.2	69.0	79.8	86.6	90.0	2.01	5.68	0.23	176
	Night	6.4	26.3	43.1	56.3	66.3	74.0	86.3	93.1	98.6	2.17	6.51	0.22	220
Summer	Day	—	5.3	12.3	26.3	38.6	52.6	73.6	87.6	98.2	3.09	6.38	0.88	57
	Night	4.1	8.2	9.6	16.4	23.2	45.2	64.4	83.6	95.9	3.38	6.42	0.31	73

Note. Here $\bar{z}_u$, $z_{u,max}$, and $z_{u,min}$ are, respectuvely, the mean, maximum, and minimum values of the height of the upper cloud boundary z_u (km), and n is the number of observations.

precisely in these layers that the degree of cloud cover varies most. For example, in Leningrad during the daytime, ground observations show the 1-km level to have a cloud cover of 21.5%, observations at a height of 1 km give a 41.4% cloud cover at the 2-km level, and observations at a height of 2 km give a 41.2% cloud cover at the 3-km level.

Heights of upper cloud boundaries for lower and middle cloud levels. Table 8.5 gives some data on the integral distribution function $F(z_u < X)$ of the height of the upper cloud boundary z_u for the lower and middle cloud levels over Moscow and Leningrad. In summer the upper cloud boundaries are seen to be higher than in winter, by about 1 km on the average. For Moscow during the daytime, the upper boundary is located below 1 km in 30% of the cases in winter but only in 1.8% of the cases in summer; it is below 2 km in 62.1% of the cases in winter and in 19.8% of the cases in summer, etc. It should be kept in mind that dense clouds whose upper boundaries were above the ceiling of the Li-2 aircraft sounder (6-7 km) were not sampled. Clouds exhibiting the greatest vertical development (nimbostratus and convective) were not covered by observations either, since either the low height of the lower cloud boundaries or stormy conditions generally made atmospheric sounding impossible. Therefore, the z_u values in Table 8.5 are too low.

Total cloud depth. The total depth is determined for layers of different thicknesses. The table below gives the mean values of the total cloud depth $z_u - z_1$ and also the probability P of cloudiness in the different layers. This probability is defined as the ratio of the number of cases with clouds in the given layer to the total number of ascents during which clouds were observed (the numerator corresponds to winter, and the denominator to summer):

		Moscow	
Layer, km	0—0.5	0—1	0—2
$z_u - z_1$, km	0.22/0.20	0.39/0.36	0.56/0.57
P, %	59.6/16.7	76.0/30.2	88.0/74.9
Layer, km	0—3	0—4	0—5
$z_u - z_1$, km	0.68/0.99	0.78/1.16	0.88/1.33
P, %	93.5/88.9	96.2/96.4	99.0/98.9
		Leningrad	
Layer, km	0—0.5	0—1	0—2
$z_u - z_1$, km	0.19/0.17	0.37/0.32	0.64/0.61
P, %	50.3/16.2	74.5/35.4	90.9/71.4
Layer, km	0—3	0—4	0—5
$z_u - z_1$, km	0.85/0.90	1.01/1.10	1.14/1.27
P, %	93.4/87.6	96.8/95.9	98.5/98.7

In Moscow, the mean total cloud depth in the layer from the ground up to 6 km was 0.95 km in winter and 1.43 km in summer; the corresponding figures for Leningrad were 1.25 km in winter and 1.40 km in summer. The probability of cloudiness in the lower layers (0-0.5 and 0-1 km) in summer was 1/3-1/2 that in winter; however, the mean values of the total cloud depth in these layers were practically the same for summer and winter. Over the entire layer from 0 to 6 km, the total cloud depth was greater in summer than in winter: by 0.49 km for Moscow and by 0.15 km for Leningrad. For the European USSR, the mean total cloud depths in the 0-0.5 and 0-1 km layers were 0.2 and 0.4 km, that is, about 40% of the vertical extent of the layer.

Depths of clouds of various types. The cloud depth can vary over a wide range, from several tens of meters to several kilometers (Table 8.6).

Table 8.6. Integral distribution function (%) $F(z_u - z_l \leqslant X)$ of depths of clouds of various types (numerator corresponds to day, and denominator to night). Moscow.

Cloud type	X, km								n
	0.2	0.4	0.6	0.8	1.0	2.0	4.0	6.0	
Winter									
Sc	51/39	82/72	91/91	96/97	98/99	99/100			274/284
St	53/44	74/71	85/86	96/97	98/100	99/100	100/100		109/96
Ns—As—Cs	1/1	9/7	18/20	33/44	40/57	68/80	89/91	98/99	152/94
Ac	57/50	86/84	95/94	97/97	98/98	100/99			162/135
Summer									
Sc	36/51	63/80	80/91	90/94	93/96	100/100			89/150
Cu—Cb	10/22	20/30	31/34	37/43	43/52	68/69	90/95	100/100	313/23
Ns—As—Cs			9/21	21/31	35/38	85/66	100/93	100/100	34/29
Ac	56/57	83/84	93/93	95/96	96/98	99/100	100/100		260/291

Inspection of the table shows that in 40-50% of the cases the depths of St-Sc and Ac do not exceed 200 m, and in 70-80% of the cases they do not exceed 400 m. The mean depths of Sc clouds (day and night, respectively) are 270 and 310 m in winter and 370 and 270 m in summer; the maximum depths are 3.02 and 1.47 km in winter and 1.55 and 1.45 km in summer (day and night, respectively). Frontal Ns-As-Cs clouds are denser: only in 35-40% of the cases is the depth 1 km or less, while in 15-35% it is greater than 2 km (even though, as mentioned above, the depths of these clouds were underestimated due to the low ceiling of the aircraft ascents). The mean depths of frontal clouds are 1.76 and 1.46 km in winter and 1.35 and 1.74 km in summer (for day and night, respectively). The maximum depths of Ns-As-Cs clouds are 6.13 and 6.08 km in winter and 3.66 and 4.75 km in summer. In 40-50% of the cases, the depths of Cu-Cb do not exceed 1 km; in about 70%, they are less than 2 km. The mean depths of these clouds are 1.61 km in the daytime and 1.48 km at night; the maximum depths are 6.47 km in the daytime and 4.95 km at night. Data on Cu-Cb clouds from winter observations are not given, due to the paucity of these clouds (only five cases were observed).

The probability of a transition from one cloudiness state to another. According to observational data, a solid cloud cover and clear skies are the most stable configurations. The probabilities P that a solid cloud cover will persist over Moscow for a time τ are the following:

τ,h	1	2	3	6	12	24
P,% . . .	94/80	90/72	88/66	84/54	80/43	76/38

where numerator corresponds to winter, and denominator to summer.

The probabilities that clear skies, observed at a certain time t over Moscow, will persist until a time $t + \tau$ are given below for winter (numerator) and summer (denominator):

τ,h	1	2	3	6	12	24
P,% . . .	82/77	70/61	64/50	50/34	36/20	35/32

It is seen that both states are more stable in time in winter than in summer (the difference between the probabilities that clear or overcast weather will persist reaches, for τ = 6-24 h, several tens of percents). This is because a temperature variation of a given magnitude will, at higher temperatures (in summer), be accompanied by a greater change in water content than at lower temperatures (in winter). If, for instance, the temperature rises by 1°C, then the probability that the clouds will disperse (provided there were clouds at the observation point) is higher in summer than in winter. At lower temperatures, the probability of cloud formation (provided there was no cloudiness at the observation point) is also higher in summer than in winter. Thus, in the summertime it is more likely that the state of the atmosphere which existed at the initial time will change rather than persist.

The probability that weather other than overcast or clear will persist is lower. As an example, Table 8.7 gives the probabilities of a change from one cloudiness state to another for two time intervals.

For low τ there is considerable likelihood that the cloudiness observed at the initial time t will remain at a time $t + \tau$ (in any case, this probability will be higher than the probability of a transition to any other state). For τ = 1 h the probabilities that

Table 8.7. Probability (%) of transition of cloudiness from one state to another. Moscow (numerator, τ = 1 h; denominator, τ = = 24 h).

Cloudiness at time *t*	Cloudiness at time *t*+τ					*n*
	0	1–3	4–6	7–9	10	
			Winter			
0	82/35	8/6	4/7	3/10	3/42	1 742/1729
1–3	27/12	38/9	17/8	12/14	6/57	689/687
4–6	9/12	21/7	33/8	25/15	12/58	860/846
7–9	2/10	4/5	12/7	52/16	29/62	1 996/1969
10	—/6	—/3	1/4	5/11	94/76	11 920/11 789
			Summer			
0	77/32	16/20	5/17	1/18	1/13	2544/2485
1–3	17/18	56/21	20/19	6/27	1/15	3003/2982
4–6	2/10	18/16	49/34	28/34	3/19	3912/3862
7–9	—/6	2/10	16/19	70/41	12/24	7232/7191
10	—/7	1/9	2/14	17/32	80/38	5285/5216

cloud covers of 1-3, 4-6, or 7-9 will persist are, respectively, 38, 33, and 53%. For high τ values, the likelihood that the initial state will be preserved drops sharply, and the probability of a change to another state increases. For instance, the probabilities that cloud covers of 1-3, 4-6, and 7-9 will persist for 24 h are, respectively, only 9, 8, and 16%. However, the probabilities that these amounts of cloud cover will change into a cloudiness of 10 in 24 h in winter are 57, 58, and 62%. Even clear skies have after 24 h changed into a cloud cover of 7-10 with a probability of 52% (whereas the probability that clear weather will persist is only 35%). An analysis of data for other places indicates that during winter the greatest tendency is for all states of the atmosphere to change (in 24 h) into a solid cloud cover (the probabilities of these transitions reach a maximum). In Leningrad, for instance, the probabilities in winter of a change from cloud covers of 0, 1-3, 4-6, and 7-9 to completely overcast (10) in τ = 24 h are, respectively, 44, 50, 54, and 58%.

This evolution of cloudiness can be explained on the basis of the plausible assumption that in winter the air temperature more frequently drops from one day to the next than rises (so that clouds are more often formed than dispersed). In any case, a comparison of the data for winter and summer (see Table 8.7) indicates that in summer the above-mentioned predominance of transitions to a solid cloud cover is not observed. For instance, for τ = 24 h for Moscow, cloud covers of 0, 1-3, 4-6, and 7-9 will change to a solid cloud cover (10) with frequencies of only 13, 15, 19, and 24%, respectively. On the other hand, in 62% of the cases in summer, a solid cloud cover changes into some other state (that is,

Table 8.8. Integral distribution function (%) $F(t_{cont} < t^*)$, Moscow.

Cloud cover	t^*, h										$\bar{t}_{cont}$, h	$t_{cont,max}$	n
	1	3	5	8	11	15	24	42	60	96			
							Winter						
0	32.5	57.7	72.8	82.6	89.7	95.5	97.3	98.8	99.7	100	5.4	90	326
1—3	63.2	94.2	99.4	100	—	—	—	—	—	—	1.6	8	426
4—6	68.2	96.5	99.1	100	—	—	—	—	—	—	1.5	88	580
7—9	49.8	86.4	95.3	98,8	99.6	99.9	100	—	—	—	2.1	16	946
10	20.5	38.5	48.9	58,5	65.6	72.6	82.2	91.2	95.1	98.6	14.6	195	822
							Summer						
0	25.4	60.0	76.3	87.3	93.5	98.7	99.4	99.6	99.8	100	4.2	62	595
1—3	45.3	82.6	94.6	98.5	99.4	99.8	100	—	—	—	2.3	20	1316
4—6	54.4	88.4	96.9	99.3	100	—	—	—	—	—	1.9	15	2016
7—9	35.5	67.4	83.7	93.9	97.8	99.3	99.9	100	—	—	3.2	29	2227
10	35.4	61.1	71.2	82.0	88.8	93.4	97.5	99.5	99.8	100	5.0	63	1050

Note. Here $\bar{t}_{cont}$ and $t_{cont,max}$ are the mean and maximum values of t_{cont}; n is the number of observations.

partial or complete dispersal of the clouds), whereas in winter this process is observed in only 24% of the cases. A cloud cover of 7-9 persists or decreases in summer in 76% of the cases, and in winter only in 38% of the cases. These data show that the temperature variation (usually a drop in winter and a rise in summer) is a significant, albeit not the only, factor in the evolution of cloudiness.

Continuous lifetime of clouds (t_{cont}). Table 8.8 gives data on continuous cloud lifetimes. Inspection of the table shows that a solid cloud cover and clear skies (and, in summer, a cloud cover of 7-9) have the longest continuous lifetimes. The mean values of t_{cont} for a solid cloud cover are close to 15 h in winter and close to 5 h in summer; the mean values for clear skies range from 4 to 6 h. The maximum values of t_{cont} for a solid cloud cover are about 200 h in winter and about 60 h in summer; clear skies may persist for 60-90 h.

8.2. Air humidity in clouds

Estimates show that the relative air humidity f in droplet clouds can differ from the saturation humidity only by a fraction of a percent, and in the steady state it practically has to equal the saturation value. Conversely, if it is assumed that f = 90%, then a droplet 1 μm in radius will evaporate completely (according to the estimates of various investigators) in 0.2-0.5 s, while a 10-μm droplet will evaporate in no longer than 1 min. Nevertheless, measurements show that the relative humidity in clouds is, as a rule, considerably less than the saturation value. For example, Burkovskaya and Minervin [50] processed 27,175 humidity measurements in clouds, obtained by means of network aircraft soundings of the atmosphere over Moscow (Vnukovo) from 1951 to 1963, and they found that the mean values of f for air temperatures above -7.5°C lie between 95 and 99%, while at lower temperatures they are from 81 to 96% (Table 8.9).

For temperatures above -7.5°C, the mean values of f for a given type of cloud are practically independent of the cloud temperature and phase state. At lower temperatures, in all cases there is a drop in relative humidity, which is especially great in mixed Ns and St clouds *) (in mixed Sc clouds, the relative humidity at low temperatures is slightly higher than in droplet Sc clouds). However, even in the densest nimbostratus droplet clouds, at positive temperatures the raltive humidity is 1-2% below saturation, while in St-Sc clouds this difference reaches 2-4%. Since random measurement errors are eliminated during averaging, it is to be assumed that either a systematic measurement error exists or the relative humidity in clouds is actually less than the saturation value. The

*) A mixed cloud contains both water droplets and ice crystals (Translator).

Table 8.9. Mean relative air humidity (%) in clouds. Moscow, 1951--1963.

Temperature, °C	Sc		St		Ns	
	droplet	mixed	droplet	mixed	droplet	mixed
17·5... 12·6	98·9		95·8		98·5	
12·5... 7·6	96·5		97·6		99·1	
7·5... 2·6	96·8		98·2		99·0	
2·5... 0·0	95·3	97·4	97·7	96·9	98·6	96·9
—0·1... —2·5	96·0	95·5	98·0	96·2	98·6	97·2
—2·6... —7·5	95·1	95·2	96·4	95·7	96·7	94·8
—7·6... —12·5	93·9	94·1	95·4	95·6	93·4	92·6
—12·6... —17·5	90·6	91·7	93·2	90·9	90·0	88·9
—17·6... —22·5	82·4	84·3	95·6	84·2	88·4	86·5
—22·6... —27·5	84·3	88·0		87·5	89·2	84·9
—27·6... —32·5						83·8
—32·6... —37·5						81·0

latter assumption, in the opinion of the authors of [50], is highly unlikely.

One reason for the systematic underestimation of f is the instruction in the Manual to the effect that when a value of f higher than 100% is obtained during processing, f = 100% is to be entered in the corresponding tables (as a consequence, the distribution of measured values of f, as shown by an analysis, is far from normal). Another reason for the systematic understating of f is that the level where the horizon appears is taken as the upper cloud boundary. However, it frequently happens that the aircraft emerges from the cloud (in the wave trough), and the horizon is no longer visible. In such cases the measurement result is referred to the cloud, although in actuality the measurement is made outside of the cloud, where f is considerably lower than 100%. At low temperatures the underestimation of f is due to icing over of the screen of the meteorograph's ventilator shaft, which reduces the ventilation rate and leads to moisture losses as the air moves through the narrow, icy ducts.

In order to obtain more reliable data on the relative humidity in clouds, humidity measurements were carried out on expeditions mounted by the Central Aerological Observatory from 1962 to 1970. The temperature and humidity were measured with the aid of an electrometeorograph, the sensors of which were shielded to keep cloud droplets from hitting them. Cases in which the aircraft remained in the clouds for a long time were processed; random errors (positive as well as negative) for all values of f were retained. Table 8.10 gives the results of processing the measurement data.

For positive temperatures the relative humidity in St and Ns clouds is, according to the table, close to 100%. However, in Sc clouds f is considerably less than 100% (possibly because, when flying close to the upper cloud boundary, emergences from the clouds were not always noticed).

Table 8.10. Mean relative air humidity (%) in clouds of various types (according to data of expedition flights).

Temperature, °C	Sc	St	St fr.	Ns	Ns, Frnb	Ns	Ns-As	Ns-As
	droplet			droplet, rain	droplet	mixed		crystalline
12.5... 7.6		99.5	95.8	99.5	97.8			
7.5... 2.6	93.8	101.2	91.9	102.0	98.2			
2.5... 0.0	97.6	99.2	89.3	99.6	99.1	99.1		
—0.1... —2.5	96.9	92.3	88.2	97.0	101.1	100.0	93.6	94.6
—2.6... —7.5	96.5	90.8	98.6	95.0	94.5	94.6	88.9	96.4
—7.6... —12.5	102.3	102.0		102.0	93.0	92.9	89.3	95.1
—12.6... —17.5					96.2	107.0	87.6	97.7
—17.6... —22.5							99.0	98.2
—22.6... —27.5							90.0	98.4

Since St fr. and Frnb clouds have clear spaces in them, the relative humidity in these clouds is considerably lower than the saturation value.

At negative (sub freezing) temperatures, a decrease in relative humidity is observed. Here, the humidity of mixed clouds is practically the same as the humidity of supercooled droplet clouds. However, it is not completely clear why the humidity in crystalline Ns-As clouds is slightly higher than in mixed clouds of this same type, and higher than the saturation value over ice.

Table 8.10 includes some f values higher than 100%, because random errors of both signs were retained (the distribution of f according to the expedition data is close to normal). The rms errors in the measurements of f with the electrometeorograph range from 5 to 7%, while the measured f values fluctuate from 50 to 110%. On the whole, the data obtained in measurements of the humidity in supercooled clouds are, as noted in [50], very unreliable. The methods used to measure the humidity are in need of considerable improvement.

8.3. Cloud water content

Cloud water contents were measured from 1957 to 1968 at several places in the Soviet Union, via systematic aircraft soundings of the atmosphere. The data gathered during these measurements are given in generalized form in [3].*)

Table 8.11 gives information about the mean values $\bar{\delta}^*$ and standard deviations σ_{δ^*} of the absolute water contents of clouds of various kinds over the Soviet Union. The table shows that the mean water contents of clouds (primarily droplet and mixed clouds) vary from hundredths of a g/m^3 to 0.2-0.3 g/m^3, but the standard

*) The material on cloud water content and microstructure presented in [3] was analyzed by Borovikov and Mazin [37].

Table 8.11. Mean values $\bar{\delta}^*$ and standard deviations σ_{δ^*} of absolute water contents of clouds over the Soviet Union, averaged over year, summer, and winter (g/m^3)

Temperature, °C	$\bar{\delta}^*$	σ_{δ^*}	n	$\bar{\delta}^*$	σ_{δ^*}	n	$\bar{\delta}^*$	σ_{δ^*}	n
		Year			Summer			Winter	
15... 10	0,25	0,22	527	0,25	0,19	401	—	—	—
10... 5	0,28	0,23	1494	0,31	0,23	642	—	—	—
5... 0	0,26	0,21	3683	0,29	0,21	793	0,23	0,22	539
0... —5	0,21	0,19	7164	0,23	0,19	1003	0,19	0,15	1996
—5... —10	0,17	0,15	5310	0,17	0,11	442	0,16	0,12	2047
—10... —15	0,15	0,13	2734	0,16	0,12	141	0,15	0,13	1201
—15... —20	0,12	0,07	806	—	—	—	0,12	0,07	419
—20... —25	0,09	0,04	234	—	—	—	0,09	0,04	104

deviations are quite high in comparison with the mean value. This means that the cloud water content varies over a wide range: from values near zero up to 2-3 g/m^3 in layer clouds and wave clouds and up to 20-30 g/m^3 in cumulonimbus. Table 8.11 shows clearly the above-mentioned, theoretically predicted dependence of the mean water content on the air temperature (averaging makes it possible to exclude, or at any rate to diminish, the effects of other factors). The highest water contents correspond to positive air temperatures (the maximum being between 10 and 5°C), and with a drop in temperature the mean water content decreases monotonically. Such a dependence of the mean water content on the air temperature is also observed for individual types of clouds (Sc-St, Ns, Ac, As) and for different regions of the Soviet Union (northern and western European USSR, Transcaucasia, Siberia, the Arctic).

The integral distribution function $F(\delta^* < X)$ of the cloud water content is shown in Table 8.12. Inspection of this table shows that the lower the air temperature, the more likely will be the formation of clouds with low water contents (the interval of very high temperatures, from 15 to 10°C, is an exception). For instance, for air temperatures from 5 to 0°C, cloud water contents less than 0.05 g/m^3 are encountered in 4.7% of the cases; for temperatures from -10 to -15°C, in 12.1% of the cases; and for temperatures from -20 to -25°C, in as many as 26.9% of the cases. Clouds with water contents lower than 0.2 g/m^3 are observed in these same temperature intervals in 46.7, 75.2, and 93.5% of the cases, respectively. Cloud water contents higher than 0.5 g/m^3 are observed for positive air temperatures in 10-14% of the cases (at temperatures from 10 to 5°C, for example, they are observed in 13.8% of the cases), whereas at low negative temperatures, clouds having such high water contents are extremely rare (at -15 to -20°C, water contents $\delta^* \geqslant 0.5$ g/m^3 are observed in 1% of the cases; at -20 to -25°C, they are observed in only 0.4%). The drop in the mean water content and the increase in function $F(\delta^* < X)$ upon transition from temperatures of 10-5°C to temperatures of 15-10°C have not yet been explained. To be sure, if the water content is high enough ($X \geqslant 1$ g/m^3), then for any temperature (including 15-10°C) the function $F(\delta^* < X)$ will increase with a decrease in the water content.

It would be of interest to obtain an analytical form for the distribution function of the cloud water content. The data in Table 8.13 on the asymmetry and excess of the distribution indicate that the water content does not have a normal (Gaussian) distribution, since for a normal distribution the asymmetry and excess both go to zero.

Table 8.12. Integral distribution function (%) $F(\delta^* < X)$ of cloud water content (over one year).

Temperature, °C	X, g/m^3								
	0.05	0.1	0.15	0.2	0.3	0.4	0.5	0.6	0.7
15... 10	5.7	20.9	34.7	47.4	69.4	82.0	89.7	94.1	96.0
10... 5	4.7	19.3	31.7	42.7	63.7	77.4	86.2	90.9	94.8
5... 0	4.7	18.0	33.5	46.7	68.5	81.3	88.2	92.5	95.2
0... −5	6.6	26.1	44.4	58.5	77.5	88.3	93.4	95.9	97.3
−5... −10	10.6	35.7	57.0	70.7	87.3	94.0	97.1	98.4	99.2
−10... −15	12.1	38.5	61.4	75.2	90.4	95.6	98.0	98.8	99.4
−15... −20	16.9	52.1	73.7	87.3	94.4	98.4	99.0	99.7	100
−20... −25	26.9	62.0	85.0	93.6	97.9	99.2	99.6	99.6	99.6

Temperature, °C	X, g/m^3								
	0.8	1.0	1.2	1.4	1.6	1.8	2.0	2.5	3.0
15... 10	97.2	98.3	98.5	99.1	99.4	99.4	99.4	99.4	99.4
10... 5	96.8	98.8	99.5	99.5	99.7	99.8	99.9	100	
5... 0	96.8	98.7	99.5	99.8	99.9	100			
0... −5	98.4	99.5	99.8	99.9	99.9	99.9	100		
−5... −10	99.5	99.8	99.9	99.9	100				
−10... −15	99.8	100							
−15... −20									
−20... −25	99.6	100							

Mazin [37] proposed describing the water-content distribution with the aid of the exponential formula

$$F_{\text{M}}(\delta^* < X) = 1 - \exp\left(-\frac{X-a}{b}\right), \tag{8.3.1}$$

where a and b are parameters of the distribution.

Although, on the whole, formula (8.3.1) describes the distribution of δ^* satisfactorily, for low δ^* a comparison of experimental values of F with those calculated using (8.3.1) reveals a sizable discrepancy. Moreover, for $\delta^* < a$ (experimental data give a = = 0.04-0.06 g/m^3), the function F_M assumes negative values.

Table 8.13. Asymmetry (A), excess (E), and parameters of lognormal distribution of cloud water content over USSR (over one year).

	Temperature, °C			
	15... 10	10... 5	5... 0	0... −5
A	1.868	2.255	2.609	2.204
E	17.273	10.599	19.314	19.018
δ_0^*, g/m^3	0.19	0.22	0.20	0.16
σ	0.76	0.72	0.71	0.77

	Temperature, C			
	−5... −10	−10... −15	−15... −20	−20... −25
A	1.801	3.992	5.160	9.844
E	17.095	4.353	31.583	21.000
δ_0^*, g/m^3	0.13	0.11	0.10	0.08
σ	0.76	0.75	0.54	0.42

Note. Here A = μ_3/σ^3 , E = μ_4/σ^4 , and μ_3 and μ_4 are the third and fourth central moments.

Matveev [204] investigated the possibility of describing the distribution of δ^* with the aid of a lognormal law. The probability density of such a distribution has the form

$$f(\delta^*) = \frac{1}{\delta^* \sigma \sqrt{2\pi}} \exp\left[-\frac{(\ln \delta^* - \ln \delta_0^*)^2}{2\sigma^2}\right], \quad (8.3.2)$$

where $\delta_0^* = (\delta_1^* \delta_2^* \cdot \ldots \cdot \delta_n^*)^{1/n}$ is the geometric mean of the water content, and $\sigma^2 = \overline{(\ln \delta^* - \ln \delta_0^*)^2}$ is the variance of the logarithm of the water content.

The parameters of distribution (8.3.2), namely δ_0^* and σ, are related as follows to the arithmetic mean ($\bar{\delta}^*$) and variance ($\sigma_{\delta^*}^2$) of the water content:

$$\delta_0^* = \frac{\bar{\delta}^*}{(1 + \sigma_{\delta^*}^2/\bar{\delta}^{*2})^{1/2}}, \quad \sigma^2 = \ln(1 + \sigma_{\delta^*}^2/\bar{\delta}^{*2}). \quad (8.3.3)$$

The integral function of the lognormal distribution

$$F(\delta^* < X) = \int_0^X f(\delta^*)\, d\delta^*$$

after the new independent variable

$$\tau_\delta = \frac{\ln \delta^* - \ln \delta_0^*}{\sigma}$$

is introduced, becomes

$$F(\delta^* < X) = \frac{1}{2} + \Phi(\tau_X), \qquad (8.3.4)$$

where

$$\Phi(\tau_X) = \frac{1}{2\pi} \int_0^{\tau_X} \exp\left(-\frac{\tau_\delta^2}{2}\right) d\tau_\delta$$

is the probability integral, for which detailed tables exist.

Table 8.13 gives values of the parameters of lognormal distribution (8.3.2) and (8.3.4). A comparison of the function $F(\delta^* < X)$ calculated using formula (8.3.4) with the experimental values in Table 8.12 revealed that the lognormal law describes the experimental data very accurately: the difference between the calculated and experimental values of F remains, as a rule, between 1 and 2%.

The criteria for the fit between the experimental and theoretical distributions were taken to be the familiar criteria χ^* and λ, plus the so-called "series criterion". An analysis indicated that the best fit with the experimental data is given by a lognormal distribution, where, in all temperature intervals for distribution (8.3.4), the null hypothesis is valid for a 2.5% significance level.

8.4. Phase states of clouds

Clouds can consist of water droplets, ice crystals, or a mixture of droplets and crystals. Clouds of the first category are called droplet (or liquid-water) clouds; the second, crystalline (or ice) clouds; and the third, mixed clouds. At positive air temperatures (> 0°C), only droplet clouds are observed. At temperatures below freezing, the droplets are either supercooled in a metastable state or frozen. Freezing is a random process.

Table 8.14 shows how often clouds assume the various phase states at different temperatures. The table, which was compiled on the basis of 61,580 observations in 41,500 clouds, shows that supercooled droplets are common in clouds, down to very low

Table 8.14. Frequency (%) of phase states of clouds at various temperatures over the USSR (yearly average).

Cloud phase	Temperature, °C					
	0 ... −2	−4 ... −6	−8 ... −10	−12 ... −14	−16 ... −18	−20 ... −22
Droplet	83.6	69.0	53.7	36.7	22.7	16.8
Mixed	14.1	25.9	35.4	41.8	39.9	35.2
Crystalline	2.3	5.1	10.9	21.5	37.4	48.0
No. of observations	10 312	12 286	11 317	8450	5890	4814

Cloud phase	Temperature, °C					
	−24 ... −26	−28 ... −30	−32 ... −34	−36 ... −38	−40 ... −42	−44 ... −46
Droplet	10.3	5.8	3.2	2.3	1.6	0.0
Mixed	30.9	25.3	17.2	11.1	6.8	6.4
Crystalline	58.8	68.9	79.6	86.6	93.2	93.6
No. of observations	3592	2435	1395	687	309	93

temperatures (-42°C). With a reduction of temperature, the frequency of droplet clouds decreases monotonically, while that of crystalline clouds increases: above -10°C the frequency of droplet clouds is higher than 50%, and that of crystalline clouds is lower than 10%; below -20°C the opposite is true: the frequency of droplet clouds is lower than 20%, and that of crystalline clouds is higher than 50%. If, however, the liquid phase in both droplet and mixed clouds is taken into account, then the probability of liquid water being present is quite high: at temperatures from 0 to -10°C it is higher than 85%, at -12 to -22°C it is 50-75%, and even as low as -24 to -34°C the probability of the liquid phase is 20-40%. The nature of the dependence of the phase state of a cloud on the temperature remains the same, regardless of the season or the geographical region of the Soviet Union. For instance, at temperatures of -8 to -10°C, the frequency of droplet clouds is 54.7% in summer and 55.3% in winter, while for crystalline clouds it is 10.4% in summer and 10.3% in winter. At -20 to -22°C, droplet clouds are observed in 11.7% of the cases in summer and in 17.5% of the cases in winter; the corresponding frequencies of crystalline clouds are 49.8 and 45.3%. It should be noted that the number of instances of temperatures of -20 to -22°C varies greatly from summer (247) to winter (2029); naturally, this difference can affect the errors in calculating the data presented above.

The air temperature is the main factor determining the phase of a cloud particle. The most important of the other, secondary, factors is the type of cloud, which is, in turn, dependent on the cloud-formation processes. Table 8.15 gives the frequencies of droplet and crystalline clouds of various types. As the temperature decreases, there is seen to be a monotonic decrease in the frequency of droplet clouds, together with a rise in the frequency of crystalline clouds, for all cloud forms. However, there is a significant difference in the frequencies of the liquid and solid phases for St, Sc, and Ac, on the one hand, and for Ns and As, on the other. At practically all temperatures, the frequency of the liquid phase is some tens of a percent higher in the first group (St, Sc, and Ac) than in the second group (Ns and As), and the frequency of the solid phase is several times less.

For instance, if in the first group droplet clouds are observed in 65-75% of the cases at temperatures of -8 to -10°C, in the second group the frequency of droplet clouds at these same temperatures will be only 28-35%. The probability of crystalline clouds at -8 to -10°C is 2-4% in the first group and 22-26% in the second. The ratio between the frequencies of the phases in clouds of the first and second groups remains about the same at other temperatures. Sc clouds at very low temperatures (-24 to -34°C) are an exception, but, since the frequencies of the phases in these clouds were determined on the basis of very few observations (for example, for the interval from -28 to -30°C there were only 11 observations, and for the interval from -32 to -34°C there were only 4), therefore the error is high.

The probability of the freezing of supercooled droplets (and thus of the formation of ice crystals) is, according to the experimental data of Bigg, influenced most by the droplet temperature (degree of supercooling) and the rate of cooling, as well as by the droplet size and the time a droplet stays at the given temperature. With an increase in cooling rate from 0.05 to 0.5°C/min, the freezing point of a supercooled droplet rises 2°C. Even though it is difficult to estimate quantitatively the cooling rates in clouds of different types, still it can be stated with certainty that the rates of droplet cooling in Ns-As clouds are considerably higher than in St-Sc-Ac.

Table 8.15. Frequency (%) of phase states of various cloud types (yearly averages over entire USSR).

Cloud type	0 . . . −2	−4 . . . −6	−8 . . . −10	−12 . . . −14	−16 . . . −18	−20 . . . −22	−24 . . . −26	−28 . . . −30	−32 . . . −34
	Droplet								
St	95.5	88.6	75.8	60.8	17.6	12.5			
Sc	94.8	85.4	75.2	59.0	42.1	42.0	28.6		
Ac	90.6	81.1	65.8	52.6	43.8	37.2	31.3	24.0	14.4
As	64.9	47.9	35.1	24.6	16.1	11.8	6.9	3.6	2.2
Ns	66.9	43.4	27.8	17.4	10.9	5.3	3.6	3.8	4.3
	Crystalline								
St	0.1	0.6	1.9	5.9	23.5	37.5			
Sc	0.4	1.1	2.0	3.9	9.5	17.5	40.0	72.7	100
Ac	0.9	2.3	4.2	7.9	12.3	13.9	15.3	20.2	35.2
As	10.0	13.7	22.6	36.3	49.5	58.0	67.0	75.2	83.5
Ns	5.8	13.5	26.0	40.5	54.7	63.6	72.3	77.9	82.4
	Mixed								
St	4.0	10.8	22.3	33.3	58.9	50			
Sc	4.8	13.5	22.8	37.2	48.4	40.5	31.4	27.3	0.0
Ac	8.5	16.6	30.0	39.5	43.9	48.9	53.5	55.8	50.5
As	25.1	38.4	42.3	39.1	34.4	30.2	26.1	21.2	14.3
Ns	27.3	43.1	46.2	42.1	34.4	31.1	24.1	18.3	13.3

Column heads: Temperature, °C

8.5. Horizontal dimensions and fluctuations of cloud fields

Mazurin [166] and Romanov [266] were among the first to evaluate the horizontal dimensions of cloud fields. Mazurin determined areas of stratiform clouds and precipitation zones on the basis of synoptic charts of the Northern Hemisphere for 1958–1960. The area of a cloud field fluctuates over a wide range: from hundreds of thousands of square kilometers to as much as 20 million km^2. During every season, cloud fields 0.25–4 million km^2 in area are usually observed (frequency exceeds 70%). The highest frequency of occurrence (about 50%) corresponds to areas from 0.6 to 3 million km^2 during the cold half-year and from 0.6 to 2 million km^2 during the warm half-year.

During the course of a day, a cloud field with an area greater than 5 million km^2 varies in extent by 10-20%, a field 1-5 million km^2 in area varies by 30-40%, and a 1 million km^2 field varies by 100-300%. In most cases (especially during the warm half-year), precipitation zones are patchy. For cloud fields up to 5 million km^2, the area of the precipitation zones increases with an increase in the field area. In 70% of the cases, the ratio of precipitation area to cloud-field area is from 0.1 to 0.5. The reduced cloud depth (total depth of all cloud layers above observation point) in an intensifying cyclone is a maximum (about 4.5 km, on the average) in the central part, and, as it diminishes toward the periphery, has practically the same values at equidistant points from the cyclone center.

Romanov [266] analyzed data of observations (more than 20,000 from 1958 to 1961) of clouds in the upper level, made from a Tu-104 aircraft. In the layer from 7 to 11 km, 10-23% of the total flight time was in the clouds or close to their boundaries (along the Moscow-Tashkent, Tbilisi-Tashkent, and Tashkent-Novosibirsk routes). Clouds with horizontal (predominantly transverse) dimensions of 100-200 km had the highest frequency (about 38%). It should be noted that warm fronts that are poorly delineated in the lower layers are, in Soviet Central Asia, very frequently accompanied by high (up to 14 km), very dense cirrus and cirrostratus. Romanov also considered the ratio between the heights of the clouds and the troposphere, a subject which was further investigated by Baranov [14].

Istomina [114] made use of cloud photos taken from an aircraft to estimate cloud sizes. For a quantitative description of the cloud-field structure, she proposed constructing autocorrelation matrices of the random brightness vectors. The data of three photos gave the principal dimensions of the microirregularities of St and St-Sc clouds as 0.8-1.4 km and 0.6-1.5 km.

Hayden (referred to in [280]) used data from *ESSA* satellite observations of clouds for four months in 1967 and 1968 to estimate the average sizes of cloud clusters in the intertropical convergence zone [ITC] of the Pacific. The cloud clusters are usually about 275 km wide in winter and 450 km wide in summer and the frequency maxima for the distances between clusters lie in two intervals: 6-8° and 10-12° (without any seasonal difference). It is noted that cloud clusters in the tropics are smaller than cloud clusters in temperate zones.

Spiridonova [280] made use of television pictures of clouds obtained with *Cosmos* and *Meteor* satellites from 1967 to 1971 to find the horizontal sizes of clouds and cloud clusters in a region bounded by latitudes of 160°E and 70°W and by 60°N and the equator (North America and the adjacent part of the Pacific, about 1/3 of the area of the hemisphere). The smallest cloud size that could be determined was, for technical reasons, 3.7 km. The observations correspond to the daylight hours (from 0400 to 1700 h local time). Table 8.16 shows the values obtained for the mean and maximum sizes of the clouds and the spaces between them.

The mean cloud size is seen to increase with an increase in the total amount of cloud cover: for slightly cloudy skies (2-5) it may be only 1/10-1/20 of the size for a solid cloud cover (10). The sizes of the clear spaces decrease with an increase in the amount of cloudiness, but only by a factor of 2 or 3 for a variation in cloud cover from 2-5 up to 10.

The ratio between the mean sizes of clouds and the intervening spaces varies over a wide range. For a cloud cover of 2-5, a space is, as a rule, 3-5 times as large as a cloud, while in October (over land) it can be 20 times as large. For a cloudiness of 5-8, the mean cloud size is close to the mean size of an intervening space, while in April and July it is 1/2-1/3 as great. The mean cloud sizes in January over the ocean are smaller than over the land; in April the opposite is true; in July and October the clouds are about the same size over land and sea.

The maximum cloud sizes fluctuate from tens of kilometers for a cloud cover of 2-5 to 1000-2000 km for a solid cloud cover (in April, over the ocean, a maximum cloud size of 3817 km was observed).

Table 8.16. Mean horizontal sizes (km) of clouds (numerator) and intervening spaces (denominator)

Cloud cover	Ocean				Land			
	I	IV	VII	IX	I	IV	VII	IX
2–5	12.9/40.3	18.7/34.1	7.9/78.2	10.5/26.6	16.1/46.8	11.6/53.8	10.2/64.0	7.5/150.7
5–8	28.7/20.5	31.6/48.7	16.5/60.3	26.7/21.8	29.3/31.2	17.0/45.0	9.4/26.9	19.0/14.0
8–[10]	46.4/18.7	25.6/32.3	49.1/30.0	69.2/23.9	140.3/57.8	38.2/28.6	21.2/20.5	75.6/18.5
10	192.6	87.2	75.8	210.5	410.7	101.7	63.2	115.2

Note. [10] denotes a solid cloud cover with gaps.

The standard deviations (km) of the cloud sizes are the following:

Cloud cover	Ocean				Land			
	I	IV	VII	IX	I	IV	VII	IX
2–5	9.8	15.2	8.8	9.9	14.7	13.5	8.6	6.5
5–8	26.8	32.6	28.8	24.9	26.4	31.2	19.4	18.8

An analysis of the differential curves of the cloud-size distribution (probability density) indicated that all clouds can be placed in one of two groups, on the basis of this characteristic. Clouds in the first group have a unimodal size distribution, the distribution mode corresponding to a minimum size (3.7 km), and the probability of the modal value being from 50 to 70% (higher in July than in January). Such cloud-size distributions are observed for a slight (2-5) or considerable (5-8) cloud cover. In the former case the frequency drops rapidly with an increase in size, going to zero at 70 km; in the latter case the maximum cloud size reaches 350 km.

The second group includes cloud covers of 8-10. For these clouds the size distribution is bimodal, and the range of possible cloud sizes is considerably wider. The first frequency maximum, corresponding to a size of less than 70 km, is apparently due to an imprecise visual deciphering of the photos (boundary and internal regions, which should be in the first group, are mistakenly placed in the second). The second mode, which in July corresponds to a size of 100 km, serves as a real characteristic of the second group of clouds (solid cloud cover, with or without clear spaces). For this mode, the maximum size of a cloud cover of 8-10 is about 560 km, and in the case of a cloudiness of 10 the maximum size is about 750 km. In January the distribution curve is similar in form to the July curve, except that it is shifted toward higher values. The probability of the second mode (close to 100 km) is about 20% in both January and July.

Kislov and Semenov [123] used observations of the cloud cover with *ESSA-9* satellites from 1 April 1971 to 29 March 1972 to investigate the variations in time of the amount of cloud cover n over the tropical part of the Indian Ocean (30°S-30°N and 30-110°E). Data on the cloud cover in five-degree spherical squares were used to plot the time-dependent spectral functions of n for the warm season (1 April to 31 September 1971) and the cool season (1 October 1971 to 29 March 1972). After a cosine filter was used to filter out oscillations with periods longer than 32 days, in each square the length of the series was $128 = 2^7$ observations.

An analysis of the spectral density of n showed that in the tropical part of the Indian Ocean, two kinds of fluctuations of the cloudiness are most typical: one with a period of 14-21 days (global) and one with a period of 3-6 days (synoptic).

The variances of these two fluctuations are of the same order of magnitude, and summed together they amount to 90-95% of the total variance. The fluctuations of n with a period of 14 to 21 days in the latitude zone from 20°S to 25°N are synchronous: an increase or decrease in n takes place simultaneously practically everywhere in the territory being considered. South of 20°S, cloud systems travel from west to east. Together with the synchronous fluctuations, fluctuations that are in antiphase are observed as well (especially east of 75°E): when monsoon cloudiness is building up over northern Hindustan, the cloudiness close to the equator (in the southern arm of the intertropical convergence zone) is almost completely dissipated. Conversely, breaks in the monsoon weather over Hindustan are contemporaneous with maximum activity of the cloud-formation processes in the southern arm of the ITC.

CHAPTER 9

THE GLOBAL CLOUDINESS FIELD*)

The statistical analysis of the field of cloudiness has been the subject of many studies [3, 12, 23, 24, 97, 123, 295]. The majority of these works contain cloud information obtained on the basis of observations carried out on the ground, primarily on continents and islands. The many uses to which information about the cloud field can be put have become apparent during the last 10 or 15 years, once satellites were provided with equipment operating in the visible and infrared parts of the spectrum. Using satellites, a volume of cloud data has already been amassed which is considerably greater than that obtained from observations on the ground (this is especially true for marine regions and sparsely populated land regions, as well as for the entire Southern Hemisphere).

Concerning cloud observations made on the ground and with satellites, it should be noted that since the distances from clouds to an observer on the ground are not great, the observer's field of view will encompass not only the lower surfaces (the areas of which must be used to determine the amount of cloudiness), but the lateral cloud surfaces as well, especially if the clouds are located in the lower half of the celestial sphere. For this reason the amount of cloud cover given by observations on the ground is, as a rule, too high. Another cause of the overestimation of the mean values (per day, month, season, or year) is that night observations of clouds (the amount of cloudiness being less at night than in the daytime) are not carried out in most countries. In order to ensure uniformity of the reference material used to determine the mean values (those given in [23, 24], in particular), night observations tend to be excluded, even in those few countries where they are carried out.

On the other hand, very thin ice clouds, especially in the upper level, as well as clouds situated over a snow or ice surface, especially at low temperatures, are not always detected by satellites. Consequently, the amount of cloudiness is somewhat underestimated (especially in the high latitudes of both hemispheres) in satellite observations.

The material to be analyzed below includes cloud data obtained with American *Tiros* satellites from 1966 to 1970 and with Soviet *Cosmos* and *Meteor* satellites from 1971 to 1975. The amounts of cloud cover were determined from nephanalysis maps and maps of cloud photomontage, many of which were compiled by co-workers

*) This chapter was written by Yu. L. Matveev.

at the USSR Hydrometeorological Center under the direction of I. P. Vetlov, T. P. Popova, L. S. Minina, G. N. Isaeva, Sh. A. Musaelyan, A. I. Burtsev, G. M. Ioffe, A. V. Leont'eva, and others.

A climatological generalization of the satellite data has also been begun at the Hydrometeorological Center [266]. L. T. Matveev and V. I. Titov supervised the culling of data on the amount of cloudiness from nephanalysis and photomontage maps, the tabulation and processing of these data, and their subsequent recording on magnetic tape (data for the Southern Hemisphere pertained to 1966-1975; data for the Northern Hemisphere, to 1971-1975). The statistical characteristics of the cloud field were calculated and analyzed with the aid of a computer by Yu. L. Matveev.

The amount of cloudiness was determined from nephanalysis and photomontage maps as the cloud-covered portion of a spherical rectangle 5° latitude and 10° longitude in size. Twice daily, 70-80% of the Earth's surface was examined by satellite observations.

For each 5° latitude zone, a total of 25,000-28,000 observations in the Northern Hemisphere and 15,000-18,000 observations in the Southern Hemisphere were processed. Such a large difference in the number of observations is due to the very limited amount of cloud information collected for the Southern Hemisphere from 1966 to 1970 (only 10-20% more data were collected for the Southern Hemisphere during this whole ten-year period than for the Northern Hemisphere for 1971-1975). However, for the Northern Hemisphere the volumes of cloud information were quite comparable for the first and second five-year periods. The amount of data is distributed quite uniformly over the easons, as well as (at a given latitude) over the grid points.

9.1. Seasonal average of the zonal cloudiness field

Zonal (averaged over latitude circles) cloud characteristics will now be considered. Table I of the Appendix gives separate seasonal averages of the zonal amounts of cloud cover f for oceans, continents, and the entire hemisphere, within zones 5° latitude in size.

For the Northern Hemisphere, during all seasons and over oceans as well as continents, two maxima and two minima of the amount of cloud cover are observed along the meridian: maxima in the middle and equatorial latitudes, and minima in the Arctic and the subtropics. Over the oceans, with decreasing latitude f starts to increase in all seasons and goes through a maximum at 50° latitude (cloud cover of 7.5 in winter, 7.2 in spring, 7.6 in summer, and 7.3 in autumn). Then the amount of cloudiness decreases to a minimum at a latitude of 15-25°; the winter minimum $f = 4.7$ is located closest to the equator (at 15°), and the summer minimum $f = 5.2$ is furthest from the equator (at 25°). The spring and autumn minima of 4.5 and 5 are located at latitudes of 15 and 20°, respectively. A second maximum of f is observed during all seasons at a latitude of 5°, but in summer and autumn close values of f are also noted at 10°. At the equator the amount of cloudiness is less than at the indicated latitudes. Clearly, the positions and the shifts of the extrema of f in the tropics conform to the movements, during the course of the year, of the intertropical convergence zone (ITC): in summer and autumn the ITC is at higher latitudes than in winter and spring. The extrema of f assume corresponding positions on the map.

The difference between the principal maximum (in the middle latitudes) and the deepest minimum of the amount of cloud cover over the oceans fluctuates between 4.3 in winter and 2.4 in summer.

Over the continents the latitude dependence of f is the same as over the oceans: moving along a meridian toward the equator, the amount of cloudiness goes through a minimum at 80°, and then increases until it reaches a maximum at a somewhat higher latitude (50–60°) than over the oceans. With a further southward shift (from the middle latitudes), the cloud cover decreases, and over the continents during all seasons it is a minimum at 20° latitude. The minimum of f is especially low in winter (1.8), and in spring and autumn, too (2.1 and 2.5, respectively). A second maximum of f is observed over the continents in the equatorial zone (at 0 and 5°S).

In contrast to the oceans, where the principal maximum of is observed in the middle latitudes and the deepest minimum is in the Central Arctic, over the continents the maxima of f in the equatorial region (at 0 and 5°S) and the middle latitudes are equal in summer and autumn, whereas in winter and spring the equatorial maximum is greater. The minimum of f over the continents is deeper in the tropics (at 20°N).

The difference between the maximum of f close to the equator and the minimum of f in the tropics over the continents is 4.7 in winter and 3.1 in summer.

During all seasons and at all latitudes (with the exception of the equator and 5°N during some seasons),the amount of cloudiness is much greater over the oceans than over the continents. For instance, in the latitude zone from 25 to 50°, the difference between the values of f over the oceans and continents during all seasons either exceeds or is close to 2.

The Southern Hemisphere is predominantly oceanic: there the land accounts for only 19% of the area of the hemisphere. Consequently, the zonal values of f on a latitude circle are close (from 40 to 65°S, where the land portion at any latitude is less than 5%, they are practically identical) to the values of f over the oceans.

The latitude distribution of the cloudiness in the Southern Hemisphere is the same as in the Northern Hemisphere: maxima of f in the low (0–10°) and middle (55–60°) latitudes, and minima in the subtropics and Antarctica. In the Southern Hemisphere the principal maximum of f in the middle zone is 0.5–1.0 higher than in the Northern Hemisphere, while the minimum in Antarctica is considerably lower than in the Arctic. The annual variation over the oceans and over the hemisphere is not very pronounced, but over the continents it is more marked, especially in the tropics. There, from 10 to 30°, the highest values of f are observed from December to February (summer), when the ITC lies the furthest south; accordingly, the lowest f values are in June-August (winter), when the ITC is in the Northern Hemisphere and convection over the land is weakly developed.

The equatorial maximum and the subtropical minimum in the Southern Hemisphere are considerably less pronounced than in the Northern Hemisphere.

The cloudiness pattern in the Antarctic should be considered separately. Although observations on the ground also indicate [105] a decrease in f in Antarctica, compared to the ocean around it (for instance, according to [12, 23], in the latitude belt from 80 to 90°S, f is 3-4 lower than in the belt from 60 to 70°S), satellite data make this decrease even greater: upon transition from the 60-65°S belt to the 80-85°S belt, the cloud cover decreases by as much as 5 or 6. According to Table I of the Appendix, the amount of cloudiness in Antarctica is small: about 5 close to the coast (70°S), and 1-3 inside the continent. There are two reasons for this: first, when very low temperatures (and thus a low water-vapor content) prevail over an ice surface and the pressure is high, conditions are unfavorable for cloud formation; second, if clouds do form, as a rule they will be very thin ice clouds with a low water content. Such clouds are frequently undetectable with equipment carried aboard satellites: in order to discern clouds at a temperature of -40 to -50°C, the difference between the temperatures at the upper cloud boundary and close to the ground (or water) cannot be less than 7-8°C, whereas at positive (above-freezing) temperatures a difference of 2-3°C suffices.

The ice surface itself, provided it is sufficiently elevated above sea level, has a definite, possibly significant, effect, since when water vapor sublimes it is deposited onto such a surface, drying out the air. According to [105], in Central Antarctica the precipitation is mainly in the form of needles and rime falling from a cloudless sky (sublimation precipitation). For instance, at *Vostok Station* there were 207 and 103 days per year, respectively, with needles and rime, and only 36 days on which snow fell. The probability of precipitation from clouds in Central Antarctica does not exceed 5-10%.

Thus it can be concluded that the amounts of cloudiness in the Arctic and Antarctic determined using satellites are only slightly less than the actually observed amounts.

Table II of the Appendix gives the mean values, over the seasons and the latitude circles, of the standard deviation σ, asymmetry A_f, and excess E_f of the amount of cloudiness. These data are based on the information about the density P of the distribution of f, which will be analyzed in Section 9.2.

Inspection of Table II shows that the standard deviation σ_f

varies from 2.4 to 3.8 in the Northern Hemisphere and from 1.5 to 3.4 in the Southern Hemisphere. In the low latitudes of the Northern Hemisphere, the variability of f is somewhat less than in the middle and high latitudes. However, in the Southern Hemisphere the minimum variability of f is observed in the middle latitudes (from 40 to 60°S).

Over most of the planet the asymmetry of the amount of cloud cover is negative. Exceptions are the high latitudes of both the Southern (all seasons) and Northern (winter and spring) Hemispheres. The absolute values of A_f in the Northern Hemisphere are, as a rule, less than the values at corresponding latitudes in the Southern Hemisphere. Especially low (close to zero) values of $|A_f|$ are observed in the region from 10 to 35°N. The excess of the amount of cloud cover in the Northern Hemisphere is everywhere negative (except for three values in summer), and its absolute value is generally higher than 0.5; in the Arctic and the subtropics it is greater than 1.0. In the Southern Hemisphere the excess is negative in the low latitudes (to 25°S) and positive in the middle latitudes (from 35 to 65°S). In the latter region the values of E_f are especially high (greater than 2.0). Since A_f and E_f differ considerably from zero, it is clear that the distribution of f is not a normal (Gaussian) distribution.

9.2. Cloud-field averages over large areas

It is of both theoretical and practical interest to obtain information about the amount of cloud cover, averaged over large areas (oceans, continents, hemispheres, even the entire planet). For such averaging, it is necessary to have data on the proportion of land and sea in each latitude zone, as well as on the variation with latitude of the area of the spherical rectangle (10° longitude by 5° latitude in size).

Rectangles including both land and sea are placed in the ocean (continent) category, if the water (land) in them occupies more than half the area. Table 9.1 shows the portions of land and water in each latitude zone (5° latitude in extent).

The portions of land and water (in percents of total area of hemisphere), evaluated according to the number of rectangles (5 x x 10° in size), are, respectively: 38 and 62% in the Northern Hemisphere and 20 and 80% in the Southern Hemisphere (the corresponding figures quoted in geographical handbooks are 39 and 61% (NH) and 19 and 81% (SH)). Here, in each hemisphere a region stretching from the equator to 2.5° into the other hemisphere was included.

The area of a spherical rectangle (of the size indicated above) is proportional to cos φ. Thus the amount of cloud cover,

Table 9.1. Portions (%) of land (L) and water (W) in rectangular latitude regions of the Earth

Latitude, deg.	85	80	75	70	65	60	55	50	45	40	35	30	25	20	15	10	5	0
Northern Hemisphere																		
L	0	19	22	50	78	61	56	58	53	42	42	47	36	28	25	19	17	28
W	100	81	78	50	22	39	44	42	47	58	58	53	64	72	75	81	83	72
Southern Hemisphere																		
L	100	100	81	44	0	0	3	3	6	3	11	25	28	25	25	19	25	
W	0	0	19	56	100	100	97	97	94	97	89	75	72	75	75	81	75	

averaged over the area, is given by the formula

$$\bar{f} = \frac{\sum_{(i)} f_i \cos \varphi_i}{\sum_{(i)} \cos \varphi_i},$$

where f_i is the cloud cover in the ith rectangle, the center of which is located at latitude φ_i; the summation is over all the rectangles within the surface being considered.

Table 9.2 shows the mean monthly, seasonal, and annual values of the cloud cover, averaged over the oceans, continents, hemispheres, and the planet as a whole.

The annual variation of the cloud cover is most pronounced for the continents. In the Northern Hemisphere, during the warm season (from May to October) the mean monthly values of f for the continents are 0.4–0.7 higher than during the cool season (from November to April). For the oceans of the Northern Hemisphere, March and April are the least cloudy months. In the Southern Hemisphere, for the continents the seasonal values of f are 0.2–0.4 less for autumn and winter than for spring and summer.

A comparison with the data of [23] shows that the satellite values of the cloud cover are several tenths less (and in some cases 1.0–1.5 less) than the values from observations on the ground. For example, the mean annual amounts of cloud cover in the Northern Hemisphere, for continents and oceans respectively, are 4.6 and 5.7 according to satellite data and 5.4 and 6.2 according to ground-based data; for the Southern Hemisphere the corresponding values of f are 4.2 and 6.3 (satellite data) and 5.2 and 6.4 (ground-based data).

The mean annual amounts of cloudiness for the Northern Hemisphere, the Southern Hemisphere, and the Earth as a whole are, respectively, 5.2, 6.0, and 5.6 (satellite data), and 5.9, 6.2, and 6.0 (ground-based data).

Table 9.2. Amounts of cloud cover for 1966-1975, averaged over oceans, continents, hemispheres, and entire Earth. Mean annual values according to observations on the ground shown in parentheses [23, 24].

Averaging period	Northern Hemisphere			Southern Hemisphere			Earth		
	Oceans	Continents	As a whole	Oceans	Continents	As a whole	Oceans	Continents	As a whole
I	5,7	4,3	5,1	6,4	4,3	6,0	6,1	4,3	5,6
II	5,6	4,4	5,2	6,2	4,7	6,0	6,0	4,5	5,6
III	5,2	4,1	4,8	6,2	4,4	5,8	5,8	4,2	5,3
IV	5,2	4,4	4,9	6,2	4,3	5,8	5,8	4,4	5,4
V	5,6	4,7	5,3	6,0	3,7	5,6	5,8	4,4	5,5
VI	5,9	4,7	5,4	6,3	3,9	5,8	6,1	4,4	5,6
VII	5,9	4,8	5,5	6,3	3,8	5,8	6,1	4,4	5,6
VIII	5,8	4,9	5,5	6,3	4,2	5,9	6,1	4,6	5,7
IX	5,9	4,7	5,4	6,6	4,2	6,1	6,2	4,5	5,7
X	5,8	4,6	5,3	6,3	4,6	6,0	6,0	4,6	5,6
XI	5,7	4,4	5,1	6,4	4,1	5,9	6,0	4,3	5,5
XII	5,9	4,6	5,4	6,2	4,3	5,8	6,1	4,5	5,6
XII-II	5,7	4,4	5,2	6,3	4,4	5,9	6,1	4,4	5,6
III-V	5,3	4,4	5,0	6,1	4,1	5,7	5,8	4,3	5,4
VI-VIII	5,9	4,8	5,5	6,3	4,0	5,8	6,1	4,5	5,6
IX-XI	5,8	4,6	5,3	6,4	4,3	6,0	6,1	4,5	5,6
Year	5,7 (6,2)	4,6 (5,4)	5,2 (5,9)	6,3 (6,4)	4,2 (5,2)	6,0 (6,2)	6,0 (6,3)	4,4 (5,3)	5,6 (6,0)

Whereas the values of f according to satellite and ground-based observations differ appreciably (by as much as 1.5-2 for the monthly means), the annual fluctuations corresponding to these sets of data are seen to agree satisfactorily (and often well) with one another, especially for the Northern Hemisphere. According to both ground-based and satellite data, the mean monthly zonal maximum of f in the middle latitudes of the Northern Hemisphere from January (when it is at 45°N, its southernmost point) to September (when it is at 62.5°N) shifts toward higher latitudes, while from September to January it shifts toward lower latitudes. The subtropical zonal minimum of f is located between 12.5 and 22.5°N from November to May and between 27.5 and 32.5°N from June to September. The tropical (equatorial) maximum of f lies in the Southern Hemisphere (between 7.5 and 15°S) from December to March and in the Northern Hemisphere (between 2.5 and 10°N) from May to November.

The zonal minimum of f in the middle latitudes of the Southern Hemisphere has no well defined direction to shift: during the whole year it lies between 55 and 60°S.

A comparison indicated that in the Northern Hemisphere the zonal monthly mean values of f go through a maximum and a minimum according to satellite and ground-based observations during the same months in 12 out of 18 latitude zones; in 4 zones the times of extrema of f differ by less than a month, and in 2 zones by more than a month.

The worst agreement is observed for the Southern Hemisphere, where only in 5 latitude zones do the times of extrema of f fall in the same months for both sets of data, while in 5 zones they are in adjacent months. This is primarily due to the small number of ground-based observations in the Southern Hemisphere. According to satellite data, throughout the latitude zone from 22.5 to 72.5°S, the mean monthly maximum occurs during the last winter month, August (according to ground-based data, from May to January), while the minimum is in January or February (according to ground-based data, from January to September). The satellite data are seen to agree best with existing ideas about the conditions of cloud formation. At all latitudes from 22.5 to 72.5°S, the area occupied by oceans is considerably greater than the continental area (Table 9.1). In winter the water temperature in the surface layer of the ocean is higher than the temperature of the air layer over it, a phenomenon that promotes the appearance of convection and the formation of cumuliform cloudiness. The cyclonic activity in the temperate zone is also more intense in winter than in summer. For both of these reasons, the largest amount of cloud cover is observed in the indicated zone in winter, and the least in summer

In the region from 2.5 to 22.5°S, both satellite and ground-based data place the maximum of f in January and the minimum in July, because of the shift in the location of the intertropical convergence zone: the ITC is south of the equator in January and considerably north of it in July.

9.3. Cloud-cover distribution density and distribution functions and their approximation

Distribution functions are one of the most informative statistical characteristics of the cloud field (as well as of fields of other meteorological quantities). Distribution functions are widely used in both theoretical studies and practical activity. If distribution functions are approximated with the aid of certain analytical expressions, the volume of information can be reduced considerably.

9.3.1. Effect of the averaging area on the cloud-cover distribution density

The distribution density and the distribution functions of the cloud cover f exhibit a number of peculiarities. For instance, for most meteorological quantities (temperature, pressure, humidity)

the distribution is close to normal. The distribution density of the cloud cover, on the other hand, is usually found from ground--based observations to be U-shaped [127]. Satellite observations, however, almost never indicate a U-shaped distribution [205, 296]; they generally give dome-shaped distributions.

As the present author has pointed out, this is because the distribution of f, more than any other meteorological quantity, depends on the size of the area over which the amount of cloudiness is determined. That such a strong dependence exists will be clear from the following considerations. If f is determined at a point (the zenith from the ground and the nadir from a satellite), then there are only two possibilities: overcast ($f = 10$) or clear ($f = 0$). On the other hand, if the area covered by an observation is large enough (for example, an entire continent or hemisphere), then in all cases the amount of cloud cover will be close to the average value (4-6), and the probability of low (0-2) or high (8-10) values of f will be very slight.

In order to obtain factual data on the distribution density and distribution functions of f, high-resolution pictures of the cloud cover were analyzed. The amount of cloud cover was defined as the portion of a spherical square covered by clouds. The sizes of squares with a common center varied from zero (square center) and 0.5° latitude x 0.5° longitude to as much as 10° x 10°.

Table 9.3. Empirical distribution density (%) of cloud cover during different seasons.

Square size	Season	Cloud cover										
		0	1	2	3	4	5	6	7	8	9	10
0	Wi	33	–	–	–	–	–	–	–	–	–	67
	Sp	51	–	–	–	–	–	–	–	–	–	49
	Su	74	–	–	–	–	–	–	–	–	–	26
	Au	51	–	–	–	–	–	–	–	–	–	49
0.5°x0.5°	Wi	17	6	5	2	3	2	4	4	4	16	38
	Sp	34	7	7	2	3	2	2	3	4	12	24
	Su	50	12	10	2	1	0	1	2	2	9	11
	Au	35	7	7	2	2	1	3	3	2	14	24
1°x1°	Wi	11	7	5	5	4	2	4	6	8	18	29
	Sp	26	10	8	5	5	2	3	5	5	13	18
	Su	39	18	9	6	2	1	2	2	3	9	9
	Au	26	10	7	5	4	2	4	4	5	17	16
10°x10°	Wi	1	2	3	9	10	8	12	19	18	17	2
	Sp	5	5	7	14	14	9	14	14	10	8	0
	Su	2	16	19	23	12	8	6	6	7	2	0
	Au	2	4	8	16	15	10	12	15	11	5	0

Table 9.3 gives (for only 4 square sizes for the sake of brevity) information about the empirical distribution density (to be called the "frequency") of the amount of cloud cover during all seasons, obtained from satellite observations from 1979 to 1981 over the European part of the USSR (total number of observations in each square: 2105 in winter (Wi), 2430 in spring (Sp), 1512 in summer (Su), and 2598 in autumn (Au)). These data will be described briefly.

A characteristic common to all seasons may be noted: when f is determined using small squares (0, 0.5° x 0.5°, 1° x 1°, 2° x x 2°), the distribution of f is found to be U-shaped, with maxima of the distribution density at low (0-2) and high (8-10) values of f and a minimum at 4-6; for large squares (8° x 8°, 10° x 10°), on the other hand, the distribution of f is close to dome-shaped, with the maximum of f shifted 3-7 points and minimum distribution densities at 0 and 10.

During *summer* in the European USSR, as shown by Table 9.3, the frequency of slightly cloudy weather is high. For small squares the frequency of a cloudiness of 0-2 drops steadily from 74% for observations at a nadir to 60% for a 2° x 2° square, while the frequency of a cloudiness of 8-10 drops from 26 to 18%. As the square size increases, the frequency maxima decrease and gradually become shifted, the first (principal) maximum toward higher f and the second maximum toward lower f. Since the second maximum differs little from the values around it, for large squares (beginning with 6° x 6°) the distribution can be considered essentially dome-shaped, although in general, the confluence of the maxima begins for the somewhat large square sizes used to evaluate f.

During *winter* the cloud-formation conditions on land differ considerably from the summer conditions. Whereas on the continents conditions are favorable for the formation of convective clouds in the summer, in the middle latitudes a solid stratiform cloud cover prevails in winter. This cover appears in low-pressure regions (cyclones and troughs) that have a higher frequency in winter than in summer.

According to Table 9.3, the principal frequency maximum of $f(P)$ corresponds to greater amounts of cloud cover, and the second maximum to smaller amounts. With a further increase in square size, the principal maximum of P, continuing to decrease, first shifts to 9 and then (for a 10° x 10° square) to 7 (the frequency values for cloud covers of 8 and 9 lie close by). The second maximum of P, beginning with a 2° x 2° square, exhibits a tendency to shift toward higher f but is not very pronounced. The frequency minimum for small squares (up to 1° x 1°) corresponds to a cloud cover of 5. Beginning with a 2° x 2° square, together with the remaining minimum at 5, increasingly deep minima appear at 0 and 10; the distribution-density curve acquires a two-humped character.

As Table 9.4 shows, the mean values $\bar{f}$ of the cloud cover are about twice as great in winter as in summer; the mean values over all the squares are 6.4 and 3.1, respectively. No definite tendency of variation of $\bar{f}$ with an increase in square size is noticeable: during one season (summer, for instance) $\bar{f}$ increases, and during another (winter, for instance) $\bar{f}$ decreases. From the physical viewpoint, there is a distinct likelihood that the mean cloud cover is independent of the square size.

Table 9.4. Moments of distribution of cloud cover.

Moment	Season	Square size							
		0	0,5°x0,5°	1°x1°	2°x2°	4°x4°	6°x6°	8°x8°	10°x10°
$\bar{f}$	Wi	6,7	6,4	6,5	6,5	6,4	6,4	6,3	6,3
	Sp	4,9	4,6	4,6	4,7	4,8	4,8	4,8	4,9
	Su	2,6	2,7	2,9	3,0	3,2	3,4	3,5	3,6
	Au	4,9	4,6	4,7	4,8	4,9	4,9	5,0	5,0
$\sigma_{f'}$	Wi	4,7	4,0	3,7	3,4	3,0	2,6	2,4	2,2
	Sp	5,0	4,3	4,0	3,7	3,2	2,9	2,6	2,5
	Su	4,4	3,8	3,6	3,3	2,9	2,6	2,4	2,2
	Au	5,0	4,4	4,1	3,6	3,2	2,8	2,6	2,3

A practical recommendation is that during cloudiness observations at stations, in addition to determining the amount of cloud cover over the entire sky the observer should note the presence or absence of clouds at the zenith. A mean value $\bar{f}$ estimated on the basis of the latter criterion will be much more accurate than an $\bar{f}$ value found from data of a survey of the entire sky, since in the latter case each value of f is determined with considerable error.

During the transition seasons (spring and autumn), the cloud-cover distributions for a given period rarely turned out to be very similar, being intermediate between the summer and winter distributions.

During all seasons, the standard deviations σ_f of the cloud cover, for estimates of f at a point and over small squares, are comparable to the mean values $\bar{f}$ (Table 9.4). With an increase in square size, σ_f decreases (to about half, upon transition from 0.5° x 0.5° squares to 10° x 10° squares). Thus, $\bar{f}$ can be evaluated somewhat more accurately using large squares rather than small ones (the standard deviation $\sigma_{\bar{f}}$ of the mean value, defined as $\sigma_{\bar{f}} = \sigma_f / \sqrt{N}$, is only about 0.1 for small squares and about 0.05 for large ones). Mean values of the cloud cover evaluated for different squares are practically identical, since the difference between the mean values are less than the errors in evaluating them ($\pm\ \sigma_{\bar{f}}$).

9.3.2. Approximating cloud-cover distributions

When solving problems in cloud dynamics and constructing models of the general circulation of the atmosphere and the Earth's climate, it becomes necessary to approximate the distribution functions of the cloudiness with the aid of analytical expressions. Such an approximation also serves to minimize the effect of errors in frequency calculations, which may be especially large when the distribution density P is low.

In order to parameterize (smooth out) the empirical distribution density and distribution function of f, several forms of the functions are introduced: a generalized lognormal function, a gamma function, and linear, power, exponential, Gramme-Charlier and other functions. The best fit with the experimental data was obtained by approximating the distribution as a generalized lognormal (Johnson) distribution, the density and function of which are

$$P(f) = \frac{1}{x\sigma_\eta\sqrt{2\pi}} \exp\left[-\frac{(\eta-\bar{\eta})^2}{2\sigma_\eta^2}\right], \qquad (9.3.1)$$

$$F(f) = \frac{1}{\sqrt{2\pi}} \int_{-\infty}^{y_f} \exp\left(-\frac{t^2}{2}\right) dt, \qquad (9.3.2)$$

where

$$\eta = \ln\frac{f+1}{10-f}; \quad y_f = \frac{\eta-\bar{\eta}}{\sigma_\eta} = a + b\ln\frac{f+1}{10-f}, \qquad (9.3.3)$$

$\bar{\eta}$ and σ_η^2 are the mean variance of a new variable η, and a and b are parameters of the given distribution, related to $\bar{\eta}$ and σ_η as follows:

$$\sigma_\eta = \frac{1}{b}; \qquad \bar{\eta} = -\frac{a}{b}. \qquad (9.3.4)$$

The parameters $\bar{\eta}$ and σ_η^2 are related to the mean value $\bar{x}$ and variance σ_x^2 of variable $x=(f+1)/(10-f)$ as follows:

$$\bar{\eta} = \ln\left[\frac{\bar{x}}{\left(1+\frac{\sigma_x^2}{\bar{x}^2}\right)^{1/2}}\right]; \qquad \sigma_\eta^2 = \ln\left(1+\frac{\sigma_x^2}{\bar{x}^2}\right). \qquad (9.3.5)$$

According to formula (9.3.2), the distribution of the cloud

cover is described by a normal law, on condition that the quantities η and y_f are used as the new variable and the upper limit of integration. The variable y_f is found from the table of the inverse function of the normal distribution, using the experimental values of function $F(f)$. Once the variable y_f has been determined for all f, the constants a and b in (9.3.3) are calculated using least squares.

A few remarks are in order regarding the meaning of parameters $\underset{\sim}{a}$ and $\underset{\sim}{b}$ of the lognormal distribution. It is clear that parameter $\underset{\sim}{a}$ gives the value of the distribution function for a cloud cover $\tilde{f}^* = 4.5$. For this value of f, the ratio $(f+1)/(10-f)=(4.5+1)/(10-4.5)=1$, variable $y_{f^*} = a$, and the distribution function

$$F(y_{f^*}) = \frac{1}{\sqrt{2\pi}} \int_{-\infty}^{a} \exp\left(-\frac{t^2}{2}\right) dt. \tag{9.3.6}$$

If $a = 0$, then $F(y_{f^*}) = 0.5 = 50\%$.

Consequently, if, for a sample, $a < 0$, then the frequency of slight cloudiness ($f < f^* = 4.5$) will be low ($< 50\%$), while the frequency of considerable cloudiness ($f > f^* = 4.5$) will be high ($> 50\%$). For $a > 0$, on the other hand, the frequency of slightly cloudy weather ($f < f^*$) will be high ($> 50\%$) and that of cloudy weather ($f > f^*$) will be low ($< 50\%$).

The meaning of parameter b can be clarified by the introduction of the value of the cloud cover f'' for which variable y_f is equal to $a+b$, that is, $y_{f''} = a+b$. For this, $(f'' + 1)/(10 - f'')$ must be equal to $e = 2.72...$, from which $f'' = 7.04 \approx 7$. The distribution function for $f = f''$

$$F(y_{f''}) = \frac{1}{\sqrt{2\pi}} \int_{-\infty}^{a+b} \exp\left(-\frac{t^2}{2}\right) dt$$

or

$$F(y_{f''}) = F(y_{f^*}) + \frac{1}{\sqrt{2\pi}} \int_{a}^{a+b} \exp\left(-\frac{t^2}{2}\right) dt, \tag{9.3.7}$$

where $F(y_{f^*})$ is the value of the distribution function for $f = f^*$.

Consequently, for a given $\underset{\sim}{a}$, parameter $\underset{\sim}{b}$ characterizes the difference between the values of function F for $f'' = 7$ and $f^* = 4.5$. Similarly, the value f' can be introduced, where $(f' + 1)/(10-f') = e^{-1}$ (hence $f' = 1.96 \approx 2$), along with the variable $y_{f'} = a - b$, and the distribution function

$$F(y_{f'}) = F(y_{f^*}) - \int_{a-b}^{a} \exp\left(-\frac{t^2}{2}\right) dt. \qquad (9.3.8)$$

The difference $F(y_{f''}) - F(y_{f*})$ for a nonzero value of a will not be equal to the difference $F(y_{f*}) - F(y_{f'})$, since for $a < 0$ the former will be less than the latter. For $a > 0$, on the other hand, the former will be greater than the latter. Here it is assumed, in accordance with data to be presented below, that parameter $b > 0$.

9.3.3. Approximating cloud-cover distribution functions using squares of different sizes

The method for approximating distributions that were described briefly in Section 9.3.2 will now be applied to smooth the empirical distribution densities and distribution functions of the cloud cover, which were discussed in Section 9.3.1.

Table 9.5 gives values of parameters $\underset{\sim}{a}$ and $\underset{\sim}{b}$ of the lognormal distribution for all seasons and for all square sizes. Parameter $\underset{\sim}{b}$ is positive during all seasons, and with increasing square area it always increases (by a factor of 4 or 5 upon transition from 0.5° x 0.5° to 10° x 10°). Parameter a, on the other hand, is positive in summer and negative in winter, and during the transition seasons its absolute value is low. The dependence of a on the

Table 9.5. Parameters of lognormal distribution of cloud cover over European USSR.

Parameter	Season	Square size						
		0.5°x0.5°	1°x1°	2°x2°	4°x4°	6°x6°	8°x8°	10°x10°
$10^3 a$	Wi	−376	−413	−428	−464	−475	−464	−472
	Sp	131	103	70	100	110	113	114
	Su	640	594	598	597	580	528	558
	Au	112	86	71	67	70	76	
$10^3 b$	Wi	258	343	469	616	745	866	959
	Sp	245	306	439	612	756	887	944
	Su	218	302	459	657	815	924	1056
	Au	206	306	449	634	792	939	1008

square size is weak and not single-valued. In summer the likelihood of slightly cloudy weather is high (since a is positive and quite high): both experimental data and calculations give a probability close to 75% that $f \leq f^* = 4.5$ for all squares. With an increase in averaging area, the differences between the distribution functions $F(7) - F(4.5)$ and $F(4.5) - F(2)$ increases, since b increases. The former difference is less than the latter for all squares, since $a > 0$.

In winter, the values of the distribution function for $f = f^* = 4.5$ are less than 50% (since $a < 0$): they are slightly below 30% for small squares and slightly above 30% for large squares.

In spring and autumn, parameter $\underset{\sim}{a}$ has a low absolute value. This means that the distribution function for $f = f^*$ is close to 50%, while the differences $F(y_{f^*}) - F(y_{f'})$ and $F(y_{f''}) - F(y_{f^*})$ for all squares are practically equal, the differences increasing in value with an increase in square size.

Since distribution parameters $\underset{\sim}{a}$ and $\underset{\sim}{b}$ were known, it was possible to calculate first the smoothed theoretical values F'' of the distribution function, and then the difference between F'' and the empirical values F' of function F: $\Delta = F'' - F'$.

During all seasons the mean values of the absolute value of this difference was found to be close to 3%. In 80-90% of the cases, the absolute value of Δ does not exceed 5%, there being an equal probability of positive or negative Δ. Similar results were obtained by evaluating the difference between the theoretical (P'') and empirical (P') values of the distribution density.

Several criteria (the Kolmogorov, Smirnov, Wilcoxson, and series criteria) have been employed to check the hypothesis that the theoretical distribution of f corresponds to the empirical distribution. All these criteria indicated the null hypothesis to be valid, meaning that the empirical and theoretical distributions belong to the same general set. Consequently, a generalized log-normal function approximates the empirical cloud-cover distributions with fairly high accuracy for both theory and applications.

9.3.4. Cloud-cover distribution functions averaged over latitude circles and seasons, and their approximation

Data of daily satellite observations of cloudiness from 1966 to 1975 were used to calculate empirical values of the distribution density and distribution functions of f at each point on a regular grid; these values were then averaged in various ways over time and space.

The zonal frequencies of f, averaged over latitude circles for all four seasons, will be analyzed below. First the frequencies of the f values, broken up into five gradations (Tables III and IV of the Appendix) will be considered.

In the middle latitudes of the Northern Hemisphere (35-65°N), the highest frequency of f (as a rule, above 30%) is observed for values from 7 to 8 (Table III). The frequency for $f = 9\text{-}10$ is close to this, generally being not less than 25% during all seasons.

Clear weather and a cloudiness of 4-6 in winter, spring, and autumn are of approximately equal probability; their frequency in the middle latitudes ranges from 15 to 20%. In summer a cloud cover of 4-6 is more probable (frequency of 20-25%), and clear weather is observed in 10-15% of the cases. A cloudiness of 1-3 is the least likely (frequency less than 10%) during all seasons.

In the tropics (from 30°N to the equator) a cloudiness of 4-6 prevails, its frequency being, as a rule, from 30 to 35%; the probability of a cloud cover of 7-8 is also considerable (25-30%). Moreover, in summer and autumn the frequency of a cloudiness of 7-8 in the region from 10 to 5°N is slightly higher than the frequency of a cloudiness of 4-6. From 30 to 15°N, the frequency of clear weather (20-25%) is comparable to that of a cloud cover of 7-8, as is a cloudiness of 9-10, especially in winter and spring. A cloud cover of 1-3 has the lowest frequency in both the tropics and the middle latitudes.

In the polar regions (70-85°N), clear weather has the highest frequency (30-45%) in winter and spring. During these seasons, a cloudiness of 4-6 or 9-10 is observed with a probability of 15-18%. In summer, a cloud cover of 4-6 or 7-8 is about equally probable in the polar regions (20-30%), and a cloudiness of 9-10 has a frequency of about 20%. In autumn, clear skies and a cloud cover of 7-8 are most likely (frequency of 25-30%). A cloudiness of 4-6 is observed 20% of the time.

In the polar regions of the Northern Hemisphere, a cloud cover of 1-3 is a little more likely than in the middle latitudes and the tropics, especially in winter or spring.

Passing to the Southern Hemisphere (Table IV), it may be seen that in the middle latitudes (45-65°) a cloudiness of 7-8 has the highest frequency. Its probability there is more than 50%, while from 50 to 60°S during all seasons it approaches 60%. The frequency of a cloud cover of 4-7 is quite high (20-30%) in the middle latitudes. A cloudiness of 9-10 in the middle latitudes occupies third place (10-20%), and the likelihood of clear weather in this region is negligible, being, as a rule, less than 1%.

In the subtropics and tropics (40°S of the equator), according to current thinking concerning the processes occurring there, a cloudiness of 4-6 prevails (frequency of 40-45% in the tropics and 30-35% in the subtropics). The likelihood of a cloud cover of 7-8 is somewhat lower, though still relatively high (30-35% in the tropics and 40-50% in the subtropics). A cloud cover of 1-3 occupies third place in the tropical region (frequency of 10-15%). The frequency of clear weather and a cloudiness of 9-10 in the southern tropics varies from 3 to 10%.

In the high latitudes of the Southern Hemisphere (70-85°S), the likelihood of clear skies is especially great (in comparison not only with other latitude zones, but even with the polar regions of the Northern Hemisphere), and the frequency of a cloud cover of 9-10 (overcast) is low. For instance, at 80 and 85°S clear skies have a probability of 60-80%; at 75°S, of only 45-50%.

The likelihood of overcast skies (9-10) at 75-85°S does not exceed 5%. Such a high frequency of slight cloudiness in the high latitudes is due both to the physical conditions of cloud formation and to methodical errors in cloud detection with the aid of satellite-borne equipment. A decrease in the temperature difference between the Earth's surface and the upper cloud boundary (which, for reliable detection, should not be less than 7-8°C at low temperatures is promoted by an inversion-type temperature stratification, like the one which, as a rule, exists over an icy surface in the Arctic or the Antarctic.

Using the method described in Section 9.3.2, the empirical functions of the cloudiness distribution (determined for each point on the cloudiness scale) were approximated. Table V of the Appendix gives parameters a and b of the generalized lognormal distribution of f. With the aid of this table and tables of the Gaussian function, it is easy to reproduce quite accurately the distribution of f at all latitudes of both hemispheres.

An evaluation of the sign of parameter a (Table V) shows that in the middle latitudes (60-40°) and the equatorial zone (0-5°) of the Northern Hemisphere, the frequency of slightly cloudy weather ($f < f^* = 4.5$) in all seasons is less than 50% (since $a < 0$ there), whereas in the subtropics (20-30°N) it is greater than 50% (since $a > 0$). In the subpolar and polar regions ($\varphi > 65°$), in winter and spring a cloud cover of less than 4.5 prevails, while in summer it is greater than 4.5. In summer, the region with a predominance of cloudiness greater than 4.5 is more extensive (up to 17.5°N) than in winter (up to 7.5°N), because of the northward shift of the intertropical convergence zone.

The other parameter of the lognormal distribution (b) generally increases toward lower latitudes. The considerations in Section 9.3.2 imply that in the low latitudes there is a greater difference between the values of the distribution function of f for f'' f 7 and $f^* = 4.5$, as well as for $f^* \approx 2$ and $f' \approx 2$ (and, consequently, the frequency of the cloud-cover gradations from ≈' to f^* and from f^* to f''), in comparison with the values for the middle and high latitudes.

In the Southern (oceanic) Hemisphere, the frequency of considerable cloudiness (more than 4.5) is higher than in the Northern Hemisphere. Over most of the hemisphere, except for the high latitudes (south of 67.5°S), parameter a is less than zero, so that the frequency of a cloud cover less than 4.5 is greater than 50%. Whereas in the tropics (2.5-22.5°S) the frequency of cloudiness of less than 4.5 is higher in winter (since $a > 0$ there), in summer a cloud cover of more than 4.5 prevails over the entire region from the equator to 67.5°S.

In the high latitudes (south of 67.5°S), a cloudiness of less than 4.5 prevails the year round, mainly because in this region (especially south of 72.5°S) the frequency of clear weather is high.

In the Southern Hemisphere, the second parameter of the

lognormal distribution (b) varies with latitude just as in the Northern Hemisphere: with the approach to the equator, the parameter b generally increases. This means that the frequency of gradations of f from 2 to 4.5 and from 4.5 to 7 is higher in the low latitudes than in the middle and high latitudes.

A comparison of the theoretical (F'') and empirical (F') values of the distribution function showed that the lognormal function approximates the observational data quite accurately: the mean (over the latitudes) values of the absolute value of the difference $F'' - F'$ (as well as of the difference $P'' - P'$) is 4-6%. In 75-80% of the cases these differences were from -7 to +7%.

An evaluation using the Wilcoxson and series criteria showed that at all latitudes of both hemispheres and during all seasons the null hypothesis is valid (correspondence of the lognormal distribution of f to the empirical distribution), for both a 5% and a 1% significance level.

9.4. Cloud-cover correlation functions

In Chapter 7, the autocorrelation functions and cross-correlation functions characterizing the statistical relationship between different quantities were considered. Now the autocorrelation functions (also known as the correlation coefficients) of the cloud covers, which gave an idea of the space-time structure of the cloudiness field, and in particular of the horizontal sizes of cloud fields, will be discussed.

In order to exclude the spatial nonuniformity and nonstatic nature of the field (due, in particular, to the annual variation of f), the cloud-cover values were normalized. For the mean monthly values the normalizing function is written as

$$\varphi_{ij}^{mn} = \frac{f_{ij}^{mn} - f_{ij}^{n}}{\sigma_{ij}^{n}},$$

where f_{ij}^{mn} is the mean monthly cloud cover at a point ij during the nth month of the mth year, and f_{ij}^{n} and σ_{ij}^{n} are, respectively, the mean value (from daily observations) and the standard deviations of f at point ij during the nth month.

A sample made up of φ_{ij}^{mn} was considered to be a set of random values uniform in space and static in time.

The correlation coefficients r were calculated for normalized mean monthly values of f, for various distances between points in the meridional and latitudinal directions. Table 9.6 gives the correlation coefficients, averaged over all the meridians $\bar{r}(y)$ and all the parallels $\bar{r}(x)$. In the calculations of $r(x)$, the variation of the distance between grid points with the latitude was taken into account. To do this, first the coefficients $r(x)$ were calculated for fixed values of the longitude difference, and then

Table 9.6. Correlation coefficients (%) for normalized mean monthly values of f along parallels $\bar{r}(x)$ and meridians $\bar{r}(y)$, 1971-1975.

$\bar{r}$	Distance, km									
	555	1110	1665	2220	2775	3330	3885	4440	4995	5550
Northern Hemisphere										
$\bar{r}(x)$	75	72	60	56	44	40	32	32	19	17
$\bar{r}(y)$	81	53	25	2	−16	−24	−23	−17	−6	–
Southern Hemisphere										
$\bar{r}(x)$	78	67	59	52	45	50	34	49	36	38
$\bar{r}(y)$	71	36	4	−19	−31	−38	−39	−37	−25	–

the dependence of r on x was plotted to get the values of $\bar{r}(x)$.

An analysis of Table 9.6 shows that the correlation between values of f in the latitudinal direction (from west to east) is much closer than in the meridional direction (from north to south): the coefficients $\bar{r}(x)$ at equal distances are considerably higher than the coefficients $\bar{r}(y)$. In both hemispheres, the correlation coefficients decrease with increasing distance. However, in the Southern Hemisphere, the coefficients $\bar{r}(x)$ diminish with an increase in x more slowly than in the Northern Hemisphere.

In the Southern Hemisphere, the coefficients $\bar{r}(y)$ decrease very rapidly with an increase in y, passing through zero at $y \approx 1700$ km and then becoming negative. Therefore, the difference between $\bar{r}(x)$ and $\bar{r}(y)$ in the Southern Hemisphere is even greater than in the Northern Hemisphere.

Clearly, the fact that $\bar{r}(x) > \bar{r}(y)$ for $x = y$ is due to a preponderance of zonal (latitudinal) transport over meridional transport. Thus the cloudiness fields will be more uniform along the prallels and more variable along the meridians. This is why in the Southern Hemisphere (which is more uniform), where latitudinal flows prevail over meridional flows even more, the correlation between the cloudiness along the parallels is stronger, and along the meridians weaker, than in the Northern Hemisphere.

The spatial scales of the correlation

$$x^* = \int_0^\infty \bar{r}(x)dx; \qquad y^* = \int_0^\infty \bar{r}(y)dy.$$

will now be evaluated. For a dependence of $\bar{r}$ on x and y which is close to exponential, the latter relations are used to obtain the following expressions for x^* and y^*: $\bar{r}(x) = 0.37$ and $\bar{r}(y^*) = 0.37$.

From the data of Table 9.6, we have $x_N^* = 3540$ km and $y_N^* = 1430$ km for the Northern Hemisphere and $x_S^* = 4950$ km and $y_S^* = 1094$ km for the Southern Hemisphere, that is, the correlation scales characterizing the horizontal sizes of the cloudiness fields along the parallels (x_N^* and x_S^*) are considerably greater than the scales along the meridians (y_N^* and y_S^*). Moreover, $x_N^* < x_S^*$ and $y_N^* > y_S^*$, because of the pronounced zonal (latitudinal) character of the cloudiness field in the Southern Hemisphere compared to the Northern Hemisphere.

Table 9.7. Averaged (over longitude or latitude) correlation coefficients (%) of normalized mean values of f over 10 days ($\bar{r}_{10}$), 5 days ($\bar{r}_5$), and 1 day ($\bar{r}_1$), for various time intervals (Δt), Northern Hemisphere, January, 1971-1975.

	Distance, km								
	0	555	1110	1665	2220	2775	3330	3885	4440
	$\Delta t = 0$ (synchronous)								
$\bar{r}_{10}$(x)	100	76	55	40	28	21	19	7	8
$\bar{r}_{10}$(y)	100	64	36	18	8	2	1	2	6
$\bar{r}_5$(x)	100	68	39	28	21	15	9	7	5
$\bar{r}_5$(y)	100	55	29	15	—	6	—	3	—
$\bar{r}_1$(x)	100	56	29	15	10	6	6	2	3
$\bar{r}_1$(y)	100	47	25	12	7	—	—	—	—
	$\Delta t = 1$ day								
$\bar{r}_{10}$(x)	90	70	48	37	29	20	16	7	8
$\bar{r}_{10}$(y)	90	38	33	17	7	2	1	—	—
$\bar{r}_5$(x)	86	63	37	27	20	12	9	6	2
$\bar{r}_5$(y)	86	50	26	14	—	5	—	3	—
	$\Delta t = 3$ days								
$\bar{r}_{10}$(x)	71	58	40	32	25	16	17	6	8
$\bar{r}_{10}$(y)	71	47	28	14	6	1	—	—	—
$\bar{r}_5$(x)	51	48	28	23	18	11	10	6	3
$\bar{r}_5$(y)	53	34	20	12	—	4	—	2	—
	$\Delta t = 5$ days								
$\bar{r}_{10}$(x)	53	46	32	27	21	15	14	4	7
$\bar{r}_{10}$(y)	52	36	22	11	5	0	—	—	—
$\bar{r}_5$(x)	21	34	18	18	14	9	7	3	3
$\bar{r}_5$(y)	24	19	14	9	—	—	—	—	—

In addition to the monthly averages, the average values of over periods of 10 days, 5 days, and 1 day were also analyzed. Table 9.7 gives the calculated space-time correlation coefficients, averaged over all the parallels ($\bar{r}(x)$) and all the meridians ($\bar{r}(y)$). The normalized moving averages for the months of January 1971-1975 were used.

A comparison of Tables 9.6 and 9.7 reveals the following important (and physically apparent) conclusion: the shorter the averaging period, the weaker the correlation (that is, the smaller the correlation coefficients) between values of f at points separated by equal distances. This is because as the averaging period becomes longer, the cloud field becomes more uniform, with an attendant enhancement of the correlation.

Just as for the monthly averages, the coefficients of correlation along the parallels ($\bar{r}_{10}(x)$, $\bar{r}_5(x)$, $\bar{r}_1(x)$) are higher than the corresponding value of r along the meridians ($\bar{r}_{10}(y)$, $\bar{r}_5(y)$, $\bar{r}_1(y)$). This means that the cloudiness field is markedly anisotropic.

However, with a shortening of the averaging period, the anisotropic nature of the cloudiness field becomes less pronounced, although it remains. The shorter the averaging period, the smaller will be the difference between coefficients $\bar{r}(x)$ and $\bar{r}(y)$ for fixed distances and times Δt, and the smaller will be the difference between the rates of change of $\bar{r}(x)$ and $\bar{r}(y)$ as the distance and Δt increase. It is to be expected that the correlations at fixed distances will become weaker with a lengthening of the time interval Δt between observations.

9.5. Parametrization of the global cloudiness field using complete systems of spectral functions

During recent decades, beginning with the basic studies of Blinova [28-30], spectral methods of solving problems in atmospheric dynamics and the theory of climate have become greatly used. These methods are based on a series expansion (parametrization) of the fields of meteorological quantities (pressure, wind speed, temperature), using complete systems of orthogonal spectral functions. Generally, spherical functions have been employed as such functions, and in recent years trigonometric functions (single and double Fourier series) have been used as well.

Parametrization of the fields of cloudiness and of other meteorological quantities makes it possible to represent a great deal of information in a more compact form with the necessary accuracy.

Due to the lack of data, all the expansions of fields of pressure, temperature, and wind velocity developed so far have been based on the forced assumption of a symmetrical or antisymmetrical distribution of the meteorological quantities in the Southern Hemisphere compared to those in the Northern Hemisphere. The tchnique of expansion in systems of orthogonal functions for the whole world first found application to the cloudiness field (since during satellite observations both hemispheres are examined almost uniformly), and it became possible to evaluate the error introduced by the hypothesis of symmetrical fields in the two hemispheres.

The parametrization of global fields with the aid of systems of spectral functions is applied when numerical forecasting methods are used to determine the derivatives with respect to the coordinates at the grid points, by differentiating the appropriate series of orthogonal functions. This makes the calculation of the derivatives more accurate than evaluations using finite differences.

The expansion of the amount of cloud cover $f(\lambda, \varphi)$ at a given latitude φ in a single trigonometric Fourier series has the form

$$f(\lambda,\varphi) = c_0(\varphi) + \sum_{m=1}^{M} c_m(\varphi)(m\lambda - \psi_m), \qquad (9.5.1)$$

where λ (which varies from 0 to 2π) is the longitude; m is the wave number, indicating how many waves of length z fit along a parallel at latitude φ (so that $m = 2\pi a \cos\varphi/z$, where a is the Earth's radius); $c_m(\varphi)$ and $\psi_m(\varphi)$ are the amplitude and phase of the mth harmonic ($c_0(\varphi)$ is the zonal value of the cloud cover); and M is the number of terms in the series.

The largest amplitudes (aside from c_0) are those of the first through fourth harmonics. The second harmonic, with two maxima and two minima at each parallel, reflects the influence of the continents and oceans on the cloudiness field. Without further analyzing the amplitudes and harmonics of series (9.5.1), Table 9.8 presents the estimated errors in parametrizing the seasonal cloudiness fields with the aid of series (9.5.1), for 6 terms (M = 5) and 10 terms (M = 9). In this table, $|\bar{\delta}|$ is the average (over all points of a grid 5° latitude by 10° longitude in scale) absolute value of the difference between the parametrized (reduced) value of the mean seasonal cloud cover and the actual value, and σ is the standard deviation of this difference. As was to be expected, with an

Table 9.8. Mean absolute deviations $|\bar{\delta}|$ and standard deviations σ of calculated (parametrized) cloud cover f from actual values (seasonal fields, 1966–1975).

	M	$10^2\lvert\bar{\delta}\rvert$				$10^2\sigma$			
		Wi	Sp	Su	Au	Wi	Sp	Su	Au
Northern Hemisphere	5 9	23 14	23 14	24 14	23 12	32 20	33 21	36 21	32 17
Southern Hemisphere	5 9	22 10	20 11	18 11	18 9	36 17	30 17	29 19	27 14
Earth	5 9	19 12	19 11	22 11	20 11	29 18	29 17	35 19	30 17

increase in the number of series terms, the parametrization errors ($|\bar{\delta}|$ and σ) are reduced, and for $M = 9$ they are less than 0.1-0.2.

If the averaging period is shortened, the cloud field becomes increasingly nonuniform. For parametrization with a specified accuracy, the more nonuniform the field is, the greater are the number of harmonics that must be considered. Conversely, for a given number of harmonics, the accuracy of the parametrization of the field will decrease in proportion to the shortening of the averaging period. For example, when the field $f(\lambda,\varphi)$ was averaged over 10 days, the absolute deviations $|\delta|$ increased to 0.4-0.5, for averaging over 5 days they increased to 0.60-0.65, and for averaging over 1 day to 1.0-1.1 (according to the data of a table not given here).

Another important quantity that is often used to evaluate the accuracy of parametrizing the field $f(\lambda,\varphi)$ is the ratio

$$d(\varphi)=\frac{1}{D(\varphi)}\sum_{m=1}^{M} c_m^2/2,$$

characterizing the contribution of the first M harmonics of series (9.5.1) to the dispersion $D(\varphi)$ of the cloud-cover values at latitude φ (this ratio approaches unity as $M \to \infty$).

When the cloud field is averaged over a season, the ratio $d(\varphi)$ for $M = 9$ exceeds 90% in all latitude regions (and in some cases is even greater than 95%).

The foregoing data show that expansion of the cloudiness field in single Fourier series ensures a fairly satisfactory parametrization accuracy, even for a comparatively small number of series terms ($M = 6 - 9$). However, this method has a serious shortcoming: the expansion in series is carried out for each latitude, making the number of expansion parameters (c_m and ψ_m) very large (for instance, for expansion of the global field for every 5° latitude and $M = 9$, the total number of amplitudes c_m and phases ψ_m is 665, and this is only half the number of actual values of $f(\lambda,\varphi)$ recorded at intervals of 5° latitude and 10° longitude).

Another shortcoming of the series (9.5.1) is the impossibility of using these series to evaluate the derivatives with respect to latitude, which enter into the equations of the dynamics of the atmosphere along with the derivatives with respect to the longitude.

Consequently, for parametrization of the field of cloudiness (and the fields of other meteorological quantities), double trigonometric Fourier series

$$f(\lambda,\varphi)=A_0^0+\sum_{m=1}^{M}\sum_{n=1}^{N}(A_n^m\cos m\lambda+B_n^m\sin m\lambda)\cos n\theta \quad (9.5.2)$$

and series of spherical functions

$$f(\lambda,\varphi)=\sum_{n=0}^{N}\left\{a_n^0P_n^0(x)+\sum_{m=0}^{n}(a_n^m\cos m\lambda+b_n^m\sin m\lambda)P_n^m(x)\right\}, \tag{9.5.3}$$

are used. Here, A_n^m, B_n^m and a_n^m, b_n^m are expansion coefficients, $P_n^m(x)$ are Legendre polynomials, and $x = \sin\varphi$.

Without further discussion of the method and results of calculating the coefficients, the errors involved in parametrizing the global cloudiness field with the aid of the series (9.5.2) and (9.5.3) will be considered. An analysis of the theoretical results indicates that the absolute value of the difference δ between the values calculated using the double Fourier series (9.5.2) for $M = N = 9$ and the actual mean seasonal values of $f(\lambda,\varphi)$ during all seasons do not exceed 1 in 95% of the cases, positive and negative values of δ being equally likely; only in summer in the Northern Hemisphere is the probability that $|\delta| < 1$ slightly less than about 90%. The mean (over the hemispheres and the entire Earth) values of $|\delta|$ for seasonal averages of the fields are about 0.4, for monthly averages they are 0.4-0.7, for 10 days about 0.8, for 5 days about 0.9, and for averages over 1 day 1.2-1.4. The assumption of symmetry of the fields in the Southern and Northern Hemispheres increases the mean (over the hemisphere) values of $|\bar{\delta}|$ to 0.45-0.50 for averaging over the season, 0.5-0.8 over a month, 0.9-1.0 over 10 days, 0.95-1.1 over 5 days, and 1.70--1.85 over 1 day.

A comparison indicated that the parametrization errors for highly smoothed fields (averaged over 5-10 years) are practically identical for double Fourier series or series of spherical functions (for the same number of series terms). However, these errors are substantially lower if double Fourier series are used to parametrize fields averaged over shorter periods (a month, 10 days, and especially 5 days or 1 day).

The following general conclusion can be drawn: the spectral expansion of global cloudiness fields in series of trigonometric and spherical functions provides, for a comparatively small number of series terms, the accuracy required for applications. The volume of stored information is thereby reduced to about 1/10 of the inital amount.

CHAPTER 10

THE STATISTICAL STRUCTURE OF THE HUMIDITY FIELD IN THE ATMOSPHERE

During recent decades, several studies (for instance [33, 257, 290, 291, and 294]) of the structure of the humidity field have been made using the methods of mathematical statistics. Komarov [140, 141] made use of a considerably larger volume of data to obtain the statistical characteristics of the humidity field in the atmosphere. He included material from radiosonde observations at many points in the Northern Hemisphere from 1961 to 1970. For each point, the data of 200-500 separate soundings of the atmosphere were used, ensuring an accurate calculation of the statistical quantities (the statistical error ε in determining, for instance, the mean value $\bar{q}$ is from 3 to 7% of $\bar{q}$, that is, $\varepsilon/\bar{q} = 0.03\text{-}0.07$). During processing, information about the air temperature and humidity at the principal isobaric surfaces was included as well as data for so-called separate points (the upper and lower boundaries of inversion and isothermal layers, the tropopause, and levels of sudden change in a vertical gradient). To provide a more detailed description of the field structure, the data were interpolated, so as to give a high height resolution (25 mb in the layer from the ground up to the 100-mb level, and 10 mb above the 100-mb level).

10.1. Expansion in natural orthogonal functions

In meteorology, wide use is made of the expansion of fields in terms of natural orthogonal functions (nof):

$$q'(p_i) = \sum_{j=1}^{n} a_j \varphi_j^i, \qquad i = 1, 2, \ldots, n,$$

where φ_j^i are nof, $q'(p_i)$ is the deviation of the specific humidity at level p_i from the mean value, and a_j are the expansion coefficients, related to the random values of $q'(p_j)$ by the linear combinations

$$a_i = \sum_{j=1}^{n} q'(p_j)\varphi_j^i, \qquad i = 1, 2, \ldots, n.$$

The natural orthogonal functions φ_j^i serve as eigenvectors of the symmetrical covariation matrix $\boldsymbol{r}(p_i, p_j)$ between the values of the specific humidity at levels p_i and p_j. They satisfy the system of equations

$$\sum_{j=1}^{n} r(p_i, p_j)\varphi_j^i = \lambda_i \varphi^i, \qquad i = 1, 2, \ldots, n,$$

where λ_i are the eigenvalues of the covariation matrix.

For the layer from the ground up to the 700-mb level, within which most of the water-vapor mass is concentrated, covariation matrix $r(p_i, p_j)$ and functions φ_i are calculated using data on the humidity at 13 levels (n = 13).

Fig. 10.1 a. Natural orthogonal functions (eigenvectors) of specific humidity at Beograd (1), Kaliningrad (2), and Barentsburg (3): (*a*) January

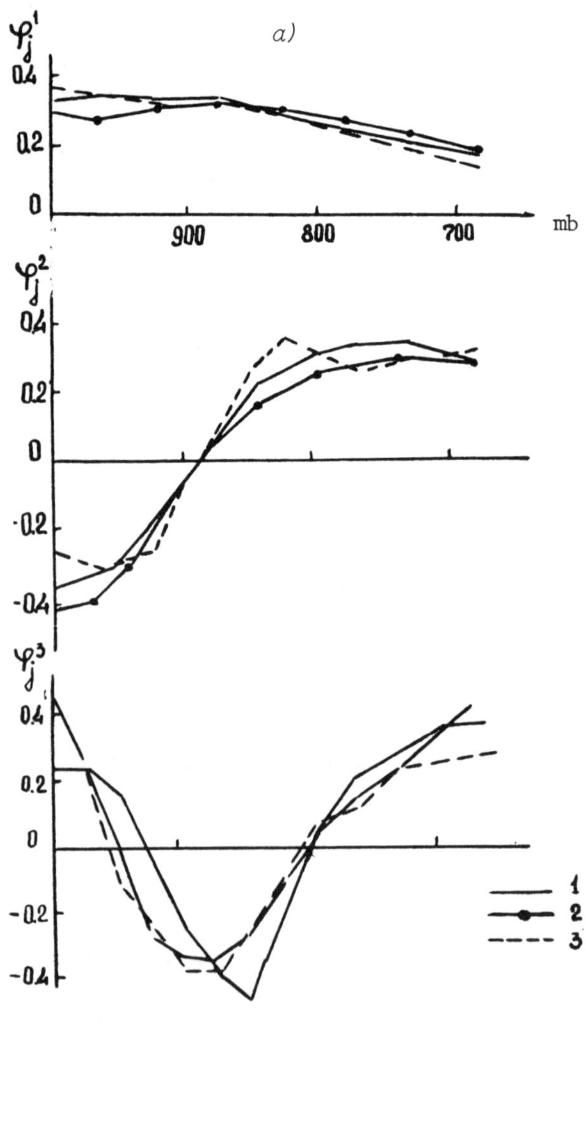

Fig. 10.1 b. Natural orthogonal functions (eigenvectors) of specific humidity at Beograd (1), Kaliningrad (2), and Barentsburg (3): (*b*) July

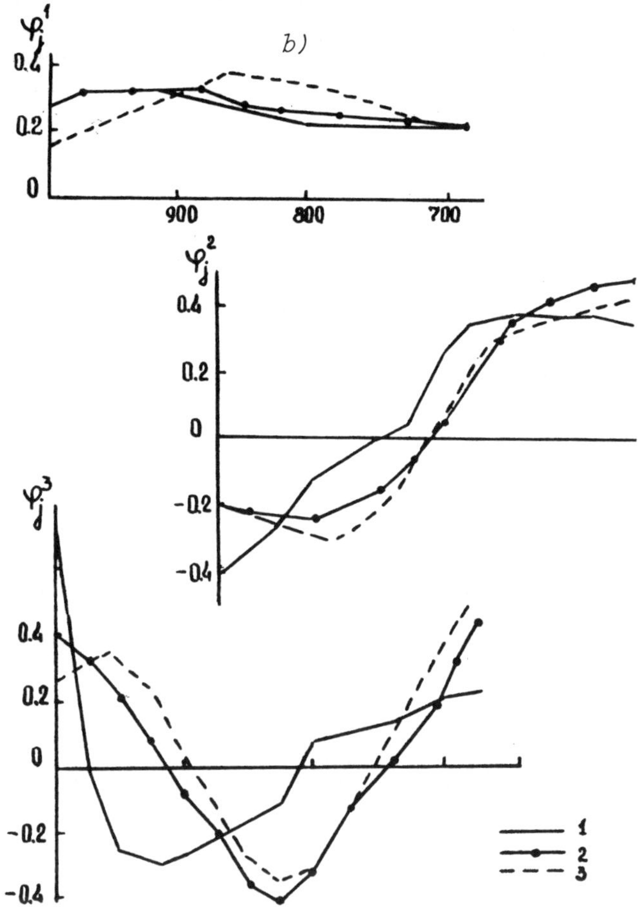

Figures 10.1a and 10.1b give the distribution with height of the first three eigenvectors of the specific humidity for stations at Beograd, Kaliningrad, and Barentsburg. The eigenvectors indicate the contributions of those processes of different scale which influence the formation of the humidity field. For instance, the first eigenvector characterizes the contribution of the processes of largest scale (planetary processes), to which can be attributed the positive correlation between the deviations of the specific humidity (as well as of the temperature and the other meteorological quantities) at all levels within the troposphere. Accordingly, the components of the first eigenvector φ^1_j at all levels are positive ($\varphi^1_j > 0$). All the rest of the eigenvectors (φ^2_j, φ^3_j etc.) reflect the influence which processes of smaller scale (than φ^1_j) exert on the structure of the humidity field.

According to Figs. 10.1 a and b, the components of the second eigenvector (φ_j^2) are negative in the lower part (up to 880-840 mb) and positive in the upper part of the layer being considered. The third component (φ_j^3) changes sign twice.

An analysis of the eigenvectors of the specific humidity at various points showed that the territory in question can be divided into several regions, within each of which the eigenvectors have close values. If we impose the condition that the basis system of the eigenvectors must provide an approximation of the vertical profiles of the temperature and specific humidity with an error not exceeding 5%, then the entire territory can be divided into only two regions: a northern region (region I) and a southern region (region II) from 60°N down.

Theory tells us that an expression in natural orthogonal functions ensures a rapid convergence of the series. Experimental data indicate that just the sum of the first three eigenvectors provides 93-97% of the information about the vertical profiles of the specific humidity in the lower troposphere.

10.2. The effect of cloudiness on the vertical profiles of temperature and humidity

The fields of temperature, humidity, and cloudiness have a marked effect on one another: cloud formation is intimately related to the atmospheric temperature and humidity, but once it has formed, the cloud cover starts to have considerable influence on the fields of temperature and humidity as well. This important topic has been considered in several works [140, 141, 291].

Table 10.1 gives data on the vertical profiles of the air temperature at various points for slight cloudiness (0-2) and overcast skies (8-10). In addition to the general trends (a drop in temperature and specific humidity with height in the troposphere, and a more gradual temperature drop near the tropopause), for slightly cloudy and overcast weather the distributions of temperature and specific humidity exhibit a number of important features. First, at high (Thule), middle (Buffalo), and even subtropical (Algiers) latitudes, the air temperature in winter for overcast weather is substantially higher than for slight cloudiness. At high and middle latitudes, a positive temperature difference ($\Delta T = T_{OC} - T_{SC}$) is observed throughout almost the whole troposphere (up to the 300-mb level). For instance, at Buffalo ΔT is 7.9°C at the ground and 7.4°C at 975 mb. With an increase in height, ΔT decreases to 0.4°C at 400 mb, until at 300 mb it goes through zero; above this level it is negative; and at the 100-mb level, $\Delta T = -1.5$°C. In the subtropics (Algiers), ΔT is positive close to the ground ($\Delta T = 0.8$°C), negative in the layer from 975 to 300 mb ($T_{OC} < T_{SC}$), and then positive again above 300 mb. In the tropics (Miami), the difference ΔT is negative throughout almost the entire troposphere. The absolute value of ΔT in the tropics and subtropics

	Level, mb															
	Ground		975		950		925		900		875		850		825	
	SC	OC	SC	OC	SC	OC	SC	OC	SC	OC	SC	OC	SC	OC	SC	OC
	Buffalo															
I	−10,4	−2,5	−10,1	−2,7	−10,3	−3,7	−10,3	−4,8	−10,8	−5,7	−11,3	−6,3	−11,5	−6,8	−11,8	−7
VII	21,5	20,7	21,4	20,4	21,0	19,4	19,5	18,1	17,8	16,6	15,9	15,1	14,1	13,4	12,4	11
	Thule															
I	−24,7	−17,6	−22,8	−18,1	−21,9	−17,8	−21,7	−17,8	−22,1	−18,0	−22,6	−18,8	−23,3	−19,5	−24,0	−20
VII	4,4	2,7	4,6	1,8	4,9	0,9	4,4	0,0	3,2	−1,1	2,0	−2,1	1,1	−2,9	0,1	−4
	Algiers															
I	10,5	11,3	11,4	10,3	10,4	8,8	9,1	7,3	7,7	5,9	6,2	4,3	4,7	3,0	3,7	2
VII	26,7	24,9	26,7	24,0	26,5	23,5	25,5	23,1	24,3	22,5	23,1	21,5	21,9	20,6	20,3	19
	Miami															
I	19,0	18,8	18,1	17,7	15,7	15,1	14,4	13,7	13,0	12,5	11,5	11,3	10,0	10,3	8,7	9
VII	28,1	26,9	25,7	25,4	24,0	23,9	22,7	22,3	21,3	20,8	19,7	19,2	18,1	17,5	16,7	16

Table 10.1. Height distribution for January (I) and July (VII) of air temperature for slightly cloudy (SC) and overcast (OC) weather

	Level, mb																	
	800		750		700		600		400		300		250		225		200	
	SC	OC	SC	OC	SC	OC	SC	OC	SC	OC	SC	OC	SC	OC	SC	OC	SC	OC
Buffalo																		
I	−12,5	−7,5	−13,6	−8,4	−15,1	−10,2	−18,9	−16,2	−34,8	−34,4	−48,0	−48,0	−52,6	−53,5	−54,5	−55,7	−55,2	−56
VII	11,0	10,5	7,9	7,5	5,1	4,9	−1,7	−1,9	−21,8	−21,0	−36,9	−35,7	−45,3	−43,9	−50,1	−48,6	−54,8	−53
Thule																		
I	−24,7	−21,2	−26,3	−22,8	−28,5	−25,4	−34,7	−31,5	−51,0	−48,4	−58,5	−58,4	−59,7	−61,4	−59,7	−62,1	−59,8	−62
VII	−0.9	−5,0	−3,5	−7,0	−6,6	−9,5	−13,4	−16,4	−34,3	−36,1	−48,2	−46,6	−49,2	−47,0	−46,7	−44,9	−43,6	−41
Algiers																		
I	2,6	1,0	0,2	−1,5	−2,3	−3,6	−10,0	−11,3	−32,0	−33,2	−47,3	−47,7	−54,6	−54,2	−58,2	−57,2	−60,7	−59
VII	18,6	17,6	14,8	14,2	10,7	10,3	1,8	1,0	−20,0	−20,3	−34,5	−35,5	−41,6	−42,9	−45,7	−47,2	−50,2	−51
Miami																		
I	9,6	8,1	7,7	5,5	5,4	4,1	−1,8	−2,4	−23,0	−23,0	−38,5	−38,2	−46,2	−46,0	−50,7	−50,5	−55,7	−55
VII	15,2	14,6	12,1	11,5	8,7	8,3	1,6	1,0	−18,5	−18,2	−33,2	−33,5	−42,3	−43,3	−47,5	−49,0	−53,4	−55

Table 10.2. The effect of cloudiness on the vertical profiles of temperature and humidity

is considerably lower than at high latitudes. Over the oceans, for $\varphi > 60°$, the difference ΔT is positive up to the 300-mb level; it is a maximum in the middle troposphere, where $\Delta T \approx 2 + 2.5°C$, whereas at the ground $\Delta T = 1.5°C$. At lower latitudes, $\Delta T < 0$ from the ground up to 900 mb, and $\Delta T > 0$ from 900 to 250 mb, where $\Delta T = 1 + 2.5°C$.

It is typical of the vertical temperature distribution in the high latitudes that inversions exist near the ground during clear weather and are absent during overcast weather. For instance, for clear skies at Thule, the mean January temperature increases with height from -24.7°C near the ground to -21.7°C at 925 mb; above this level the temperature drops, until at 800 mb it is the same as the ground temperature (-24.7°C). For clear weather at Buffalo, the temperature distribution is practically isothermal up to the 925-mb level, whereas for overcast skies the temperature drops with height throughout the troposphere. Over the oceans, the mean January temperature decreases with height during both clear and overcast weather.

In summer the effect of cloudiness is the opposite of that in winter: on cloudy days the air temperature in the high and middle latitudes is lower than on clear days. For example, at Thule the difference $\Delta T < 0$ from the ground up to 400 mb, while above this $\Delta T > 0$; at Buffalo $\Delta T < 0$ from the ground up to 500 mb, and further up $\Delta T > 0$. The difference ΔT is negative in summer in the subtropics and tropics as well. At Little Rock $\Delta T = -6°C$ near the ground, -4.4°C at 925 mb, and -2.4°C at the 800-mb level.

For the specific-humidity field, the values typically increase at practically all latitudes and heights for overcast weather, in comparison with clear or slightly cloudy weather. In January the difference $\Delta q = q_{OC} - q_{SC}$ at Buffalo is about 1.5 g/kg at the ground, about 2 g/kg close to the 850-mb level, 1.25 g/kg at 750 mb, and 0.3 g/kg at 500 mb. At lower latitudes (Algiers, Little Rock, Miami), the difference Δq varies similarly: at Miami it is 0.23, 1.52, 2.41, and 0.47 g/kg at the ground and at the 900-, 800-, and 500-mb levels, respectively.

In July in the high and middle latitudes, the difference Δq at all heights is positive, and it varies with height just as in January: first(in the lower troposphere) it increases (for instance, at Thule from 0.48 g/kg at the ground to 1.21 g/kg at 750 mb; at Buffalo from 0.17 g/kg at the ground to 1.83 g/kg at 800 mb), and then decreases (at Thule to 0.53 and 0.05 g/kg at 650 and 500 mb, respectively; at Buffalo to 1.43, 0.86, and 0.38 g/kg at the 600-, 500-, and 400-mb levels, respectively). In the tropics (Little Rock) only at the ground is the difference $\Delta q = -0.9$ g/kg less than zero; at all other levels, Δq is positive and varies with height just as at other latitudes.

The data presented are in accordance with theoretical models of the humidity distribution with height [192]. Here, the cloudiness in the given case serves as an indicator of the humidity profile. Cloudiness is observed for upward motions, when the profile of the specific humidity is characterized by a more gradual humidity drop with height in the lower and middle troposphere. Clearly, even if its values near the ground are lower, the specific humidity will, beginning at a certain height, be higher in cloudy weather than in clear weather (for which downward motions are typical), and thus the drop in q with height will be more rapid.

10.3. Correlations between variations of specific humidity

As in the case of the temperature, a close correlation is observed between the specific humidities at different levels of the troposphere. Figure 10.2 shows curves of the calculated coefficients $r_{qq}(0,z)$ between the specific-humidity values at the ground and an arbitrary height z. These coefficients were calculated for various stations in the Northern Hemisphere, according to data pertaining to the isobaric surfaces P_0 (ground), 975, 950, 925, 900, 875, 825, 800, 750, 700, 650, 600, 500, and 400 mb. The meridional sections in Fig. 10.2 were plotted from the calculated $r_{qq}(0,z)$ values for Panama, Swan Island, Miami, Buffalo, Coral Harbor, and Thule (Western Hemisphere); Barentsburg, Murmansk, Kaliningrad, Odessa, and Aden (Eastern Hemisphere).

An analysis of Fig. 10.2 shows that there is a sizable difference in the correlations between winter and summer, on the one hand, and different latitude regions and hemispheres, on the other. These correlations are closer in places where processes of the advective type predominate, since the latter spread throughout most of the troposphere. For instance, close correlations are observed in winter in the middle and high latitudes of the Western Hemisphere: there $r_{qq}(0,z)$ exceeds 0.8 up to a height of about 2 km and 0.6 at heights of 3-4 km. In the Eastern Hemisphere in January, the correlation becomes lower with an increase in height more rapidly than in the Western Hemisphere. In the middle latitudes there, only up to a height of about 1 km does $r_{qq}(0,z)$ exceed 0.8, and at a height of 2 km the correlation coefficients are already less than 0.6. This is because of the influence of the Siberian anticyclone, a spur of which extends into Eastern Europe in winter. Since there is a consequent high frequency of an inversion-type distribution of temperature and specific humidity in the lower troposphere, the interrelationships between the upward and downward transport of air masses will be weaker, and so will the correlations between the values of the humidity (at Yakutsk $r_{qq}(0,z)$ does not exceed 0.6 even at the 900-mb level).

In July the correlation is considerably weaker in the high latitudes: $r_{qq}(0,z)$ decreases rapidly with height there, since at this time of year the polar anticyclone forms, causing a transition to easterly transport in the middle and upper troposphere, while the transport remains westerly in the lower troposphere. In the middle latitudes, however, the correlation coefficients are about the same in summer as in winter; in the Eastern Hemisphere,

Fig. 10.2. Distribution of correlation coefficients $r_{qq}(0,z)$ over continents of Northern Hemisphere: (a) January, (b) July.

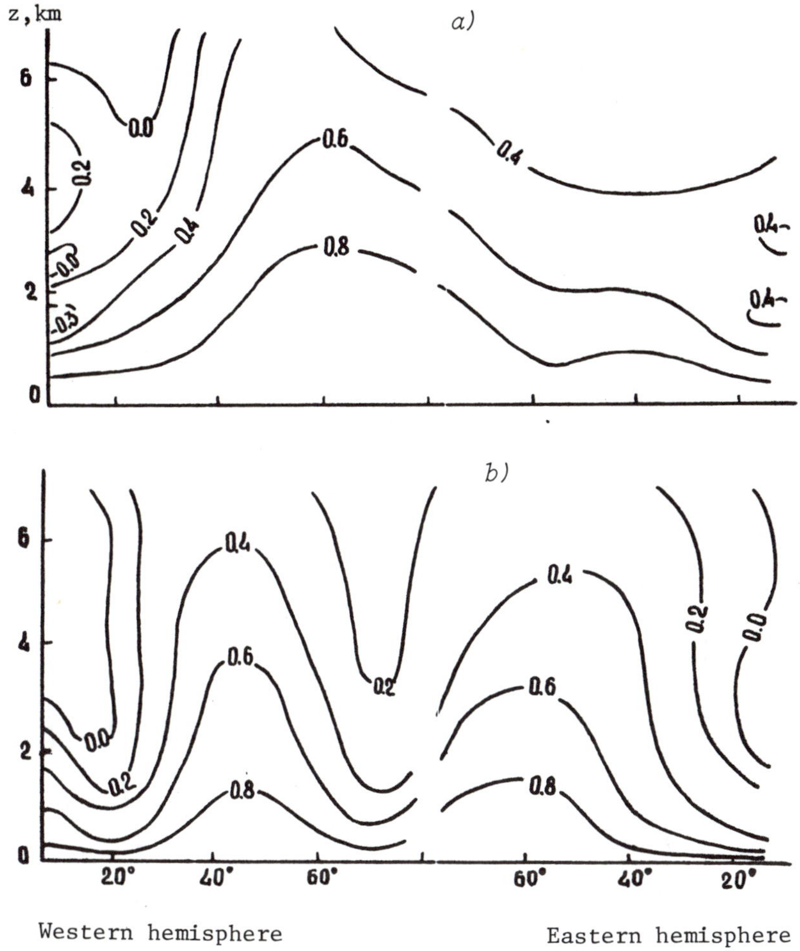

they are slightly higher.

In the tropics the correlations are quite weak in both winter and summer, and at a height of about 0.5 km (960 mb), $r_{qq}(0,z)$ is less than 0.3-0.5 as a rule. This is explained by the nature of the air currents: in the lower troposphere the prevailing winds are the northeasterly trades, while in the middle and upper troposphere the southwesterly antitrades prevail.

Finally, the cross correlations between the temperature and the specific humidity will be discussed. These correlations are of interest, first of all, because they are useful for purposes of weather forecasting and climatology [160-162]. Data on the air humidity is often lacking; and humidity measurements, especially at subfreezing temperatures, are less accurate than temperature measurements.

As Fig. 10.3 shows, in January the coefficients $rT_q(z,z)$ of the correlation between the temperature and specific humidity at a given height are high in the middle and high latitudes: generally, more than 0.8 from the ground up to 850 mb and not less than 0.6–0.7 throughout the entire troposphere.

Fig. 10.3. Distribution of cross-correlation coefficients $rT_q(\bar{z},\bar{z})$ over continents of Northern Hemisphere: (a) January, (b) July.

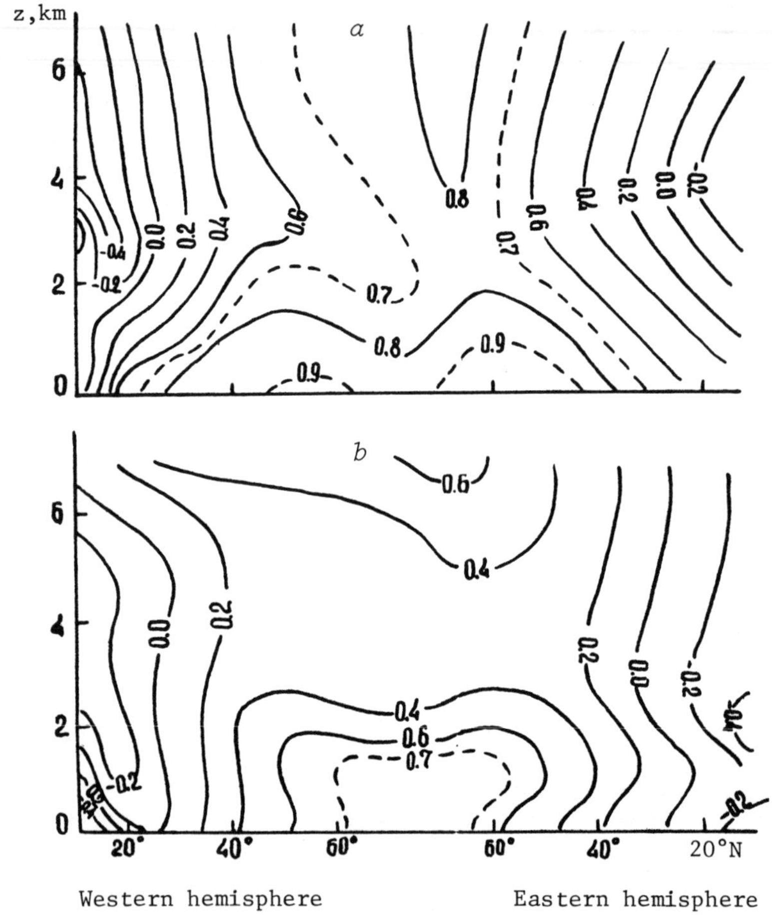

In July, the correlation between T and q is considerably weaker: in the high and middle latitudes, coefficients $rT_q(z,z)$ are mostly less than 0.6 (at Thule and Odessa they are less than 0.3–0.5 throughout the entire troposphere). In the tropics ($\varphi <$ $< 25°$), the correlation between T and q is very weak (rT_q is less than 0.2); moreover, in this region there is an inverse correlation between T and q ($rT_q < 0$) in the middle and upper troposphere: the specific humidity drops with a rise in temperature.

In order to construct their statistical models of the atmosphere, Vasishcheva and Shchukin [55] used radiosonde data obtained over Leningrad during the daytime from 1959 to 1974. Vertical profiles of the monthly averages (for January, April, July, and October) of the pressure, temperature, and relative and absolute humidity, as well as of their standard deviations were plotted for St, Sc, Ns, Cb, Cu, Ac, and Ci cloudiness of 8-10 and for clear weather. For each of the cloud forms, the number of observations varied from 20-30 to 80-90; only in the case of Sc clouds in October was it as high as 175.

Since clouds form mainly in low-pressure regions, negative deviations of the pressure from the standard value are observed. These deviations are especially great for Ns and Cb clouds: at small heights (up to 0.5 km) in January they range from -4 to -9 mb, and in the middle troposphere they are as high as -18 to -23 mb.

In January, April, and October the clouds have a warming effect on the underlying surface and the lower layer of the atmosphere. In January the mean air temperature for lower-level clouds (Sc, St, Ns) and Cb clouds near the ground is approximately 10°C higher than in clear weather (about -5°C for cloudy weather and about -15°C for clear weather). The temperature difference decreases with height, but in January it remains positive throughout the entire troposphere (at a height of 3 km this difference is about 5°C, and at 7 km it is about 3°C).

In July the air temperature close to the ground for Sc and Cb clouds is approximately 5°C lower than for clear weather. With increasing height, this difference decreases in absolute value, but its sign remains the same throughout the troposphere.

In the presence of clouds, the absolute humidity a during all seasons is higher (close to the ground by about 1 g/m^3) than in clear weather. The standard deviations $\underset{\sim}{a}$ in the lower troposphere in January are about 1 g/m^3 for lower-level clouds and Cb and they vary from 0.6 to 0.8 g/m^3 for clear weather; in July they are 1.6-2.4 g/m^3 for both overcast and clear weather.

CONCLUSION

During recent decades, a new branch of the physics of the atmosphere - cloud dynamics - has emerged. In it, the conditions of formation and development of individual clouds, cloud clusters, and cloud systems are investigated.

Prominent experts on cloud physics, such as Mason and Khrgian, have repeatedly pointed out the need for a monograph devoted to this new branch of atmospheric physics. Two collections of papers entitled *Cloud Dynamics* (one in 1960, edited by Anderson; the other in 1975, edited by Pruppacher) considered only the dynamics of convective clouds, mostly individual cumuli.

The present monograph is the first to describe in detail the state-of-the-art in cloud dynamics. The main emphasis is on the modeling of large-scale (mainly frontal) and mesoscale wave clouds, and of clusters of convective clouds. It is precisely these cloud types which most affect the influx of energy to the Earth's surface, the thermal regime, and the weather as a whole. Parametrization of the large-scale cloudiness is of great importance to the theory of the general circulation of the atmosphere and to the climate of the Earth.

This monograph also discusses models of fog formation, the humidity field, and the cloud field.

In addition to modeling problems, numerical hydrodynamic methods of predicting clouds and precipitation, as well as air temperature and humidity, are analyzed here. The predictability of stratiform cloudiness and precipitation using these methods was close to 80-85%.

Some results that have been established by means of numerical modeling of clouds and analysis of experimental data will now be given. Large-scale vertical motions play a decisive role in cloud formation. Turbulent mixing (not only in the boundary layer, but throughout the troposphere) and the advection of heat, moisture, and the energy of phase transitions of water have an effect on cloud formation comparable to that of vertical motions. The field of upward vertical motions, the frontal cloudiness, and the the frontal zone itself are formed largely by the convergence of air currents in regions of low pressure (troughs and cyclones), because of the presence there of surface friction and eddy viscosity. For such regions, especially during the stage of their development, stratiform (frontal) cloudiness of type Ns-As-Cs developing during the course of some time is typical. The formation time of the system fluctuates from several hours to tens of hours depending on the vertical velocity, temperature and initial humidity of the air on the intensity of turbulent mixing, and on other factors. The heights of cloud boundaries, the depths and water-content profiles of clouds, and the precipitation intensities all vary over a wide range.

Large-scale vertical motions (the velocities of which increase with height in the lower troposphere) not only are responsible for the formation of systems of frontal clouds (Ns-As-Cs); they also promote the increase with time of the lapse rate in the lower and middle troposphere, that is, they create conditions favorable for the appearance of mesoscale vertical motions of cellular structure under the influence of the buoyancy force. The latter, in turn, give rise to systems (clusters) of convective clouds outside the frontal cloud systems and associated with them. The presence of ascending motions causes the moisture content of the vertical air column and the cloud water reserves to increase.

In regions of large-scale ascending motions (of the order of 10^{-3} - 10^{-2} m/s), a field of steady precipitation of intensity more or less constant in time is produced. The heaviest showery precipitation falls from convective clouds, in which, around the time of the precipitation, the ascending motions (of the order of 10^{-1} - 10^{-1} m/s) either are greatly attenuated or else become descending motions. The very dynamics of convective-cloud development furthers such a change: in the cloud, air rises and is cooled moist-adiabatically, while outside the cloud, in adjacent cells, descending air is heated dry-adiabatically; therefore, in time the cloud becomes cooler than its surroundings and the positive buoyancy force becomes negative.

Since the velocity of the air motion in the middle and upper troposphere is greater than the velocity of the trough and the front, some cloudiness is seen to pass from the trough and the cyclone to the periphery of the anticyclone or ridge. Since the cloudiness is located in a low-pressure field there, the stratiform clouds of type Ns-As-Cs are transformed into St-Sc-Ac clouds. As soon as the cyclone starts to mature, a transformation of the cloudiness takes place in it as well. The transformation and dissipation of the cloudiness due to the downward motions take several hours, while the temperature inversion that forms above the clouds and the turbulent influx of water vapor may well prolong this process by several tens, or even hundreds, of hours. St-Sc clouds can also form from fogs of any provenance. The traditionally used term "internal clouds" should be applied to most (more than 80%) of St-Sc-Ac clouds only with extreme caution.

Conclusion

In this monograph, models of the formation of fogs of different provenance, including urban fogs, have been constructed and discussed at length.

The formation of temperature inversions (lifted, subcloud, above-cloud) is related to clouds and fogs. Vertical motions and condensation play a major role in their formation, and also in the stratification of cloudiness.

No theory can be considered adequate if it is not based on experimental results. Consequently, in this monograph considerable attention has been focused on the systematization of experimental material obtained with the aid of aircraft, satellites, radar, and radiometry. Most of the experimental data are in satisfactory and often excellent agreement with the theoretical conclusions.

APPENDIX

Table I: Seasonal mean values of zonal amounts of cloud cover, 1966-1975

f	Oceans				Continents				Hemispheres			
	XII-II	III-V	VI-VIII	IX-XI	XII-II	III-V	VI-VIII	IX-XI	XII-II	III-V	VI-VIII	IX-XI
85°N	3,2	3,4	5,6	4,9	–	–	–	–	3,2	3,4	5,6	4,9
80	3,7	3,9	5,9	5,4	2,4	3,0	4,5	4,0	3,4	3,7	5,6	5,1
75	4,0	4,2	6,0	5,7	3,4	3,6	4,8	4,6	3,9	4,1	5,7	5,4
70	4,6	4,7	6,1	6,2	3,9	4,2	5,6	5,7	4,3	4,4	5,8	5,9
65	6,1	5,8	5,9	6,9	4,5	4,8	6,2	6,1	4,9	5,0	6,1	6,3
60	6,4	6,2	6,7	7,1	5,2	5,4	6,2	6,4	5,6	5,7	6,4	6,7
55	7,1	6,9	7,2	7,2	5,4	5,5	6,0	6,3	6,1	6,1	6,5	6,7
50	7,5	7,2	7,6	7,3	5,6	5,3	5,4	5,4	6,4	6,1	6,3	6,2
45	7,3	6,8	7,0	6,9	5,3	5,0	4,6	4,6	6,3	5,9	5,8	5,7
40	6,9	6,3	6,0	6,2	4,9	4,7	3,9	3,7	6,1	5,6	5,1	5,2
35	6,5	5,9	5,3	5,8	4,4	4,3	3,6	3,4	5,6	5,2	4,6	4,8
30	6,3	5,7	5,3	5,7	3,2	3,4	3,5	2,8	4,9	4,7	4,5	4,4
25	5,6	5,3	5,2	5,1	2,3	2,5	3,3	2,6	4,4	4,3	4,5	4,2
20	4,9	4,7	5,4	5,0	1,8	2,1	3,1	2,5	4,1	3,9	4,8	4,3
15	4,7	4,5	5,5	5,1	2,2	2,5	4,6	3,6	4,1	4,0	5,3	4,7
10	5,0	5,0	6,2	5,9	2,4	3,8	4,9	4,5	4,5	4,8	6,0	5,6
5°N	5,9	5,6	6,3	6,0	4,5	6,0	5,6	6,0	5,6	5,7	6,2	6,0
0	5,4	5,1	5,4	5,4	5,9	6,2	6,2	6,2	5,5	5,4	5,6	5,6
5°S	5,6	5,2	5,1	5,4	6,5	6,2	5,4	6,4	5,8	5,4	5,2	5,6
10	5,7	5,2	5,4	5,6	6,4	5,6	3,5	5,3	5,9	5,3	5,0	5,5
15	5,8	5,4	5,6	5,8	6,1	4,8	3,2	4,6	5,9	5,3	5,1	5,5
20	5,8	5,6	5,9	6,0	5,4	4,2	3,1	4,3	5,7	5,3	5,2	5,6
25	5,8	5,8	6,2	6,3	4,9	4,2	3,7	4,2	5,6	5,4	5,5	5,7
30	5,7	6,1	6,4	6,5	4,3	4,2	4,1	4,3	5,4	5,6	5,8	6,0
35	6,1	6,5	6,6	6,8	4,4	4,8	5,5	4,6	5,9	6,3	6,5	6,5
40	6,5	6,8	6,8	7,0	4,7	5,1	5,9	4,7	6,4	6,8	6,8	7,0
45	6,9	7,2	7,0	7,3	6,0	6,3	6,6	6,1	6,8	7,1	7,0	7,2
50	7,1	7,3	7,1	7,4	6,4	6,2	6,5	6,1	7,1	7,3	7,1	7,4
55	7,3	7,4	7,1	7,6	7,1	7,0	6,7	6,9	7,3	7,4	7,1	7,6
60	7,3	7,5	7,0	7,6	–	–	–	–	7,3	7,5	7,0	7,6
65	6,8	7,1	6,5	6,9	–	–	–	–	6,8	7,1	6,5	6,9
70	5,9	6,5	5,7	5,9	3,1	2,7	3,0	2,8	4,8	4,9	4,6	4,6
75	4,7	5,2	4,2	4,4	2,5	2,4	2,3	2,2	2,9	3,0	2,7	2,6
80	–	–	–	–	1,8	1,9	1,7	1,6	1,8	1,9	1,7	1,6
85°S	–	–	–	–	1,1	1,1	1,0	0,9	1,1	1,1	1,0	0,9

Table II: Average values (over season and latitude circle) of moments of cloudiness distribution, 1966-1975

	σ_f				10^2A_f				10^2E_f			
f	XII-II	III-V	VI-VIII	IX-XI	XII-II	III-V	VI-VIII	IX-XI	XII-II	III-V	VI-VIII	IX-XI
85°N	3,5	3,6	3,5	3,6	68	58	−27	−11	−102	−114	−132	−140
80	3,7	3,6	3,3	3,5	54	44	−36	−19	−124	−114	−97	−137
75	3,7	3,6	3,4	3,5	38	24	−42	−34	−142	−144	−111	−142
70	3,8	3,6	3,3	3,4	18	8	−30	−58	−153	−148	−109	−96
65	3,8	3,6	3,1	3,2	7	−21	−35	−78	−157	−138	−80	−56
60	3,6	3,4	3,0	3,1	−40	−47	−76	−94	−131	−108	−34	−8
55	3,4	3,2	3,0	3,0	−67	−65	−80	−95	−85	−74	−26	0
50	3,3	3,2	3,2	3,1	−80	−65	−70	−60	−46	−72	−48	−53
45	3,2	3,3	3,2	3,2	−75	−55	−47	−65	−56	−88	−95	−73
40	3,3	3,2	3,3	3,3	−59	−41	−28	−30	−89	−101	−107	−116
35	3,2	3,3	3,4	3,3	−46	−27	−13	−16	−95	−112	−134	−121
30	3,4	3,3	3,2	3,3	−19	−5	−5	1	−126	−122	−116	−122
25	3,3	3,2	3,1	3,0	−2	6	−9	3	−126	−105	−104	−86
20	3,1	3,0	3,0	3,0	6	16	−22	−5	−112	−99	−90	−102
15	3,0	2,9	2,7	2,9	−1	10	−37	−18	−113	−88	−47	−81
10	2,9	2,7	2,5	2,7	−7	−13	−68	−48	−82	−61	11	−30
5°N	2,6	2,5	2,4	2,4	−45	−15	−64	−55	−20	−75	32	19
0	2,4	2,5	2,4	2,5	−40	−22	−37	−40	−2	−30	28	−5
5°S	2,3	2,3	2,3	2,3	−60	−62	−54	−64	−17	−17	−23	1
10	2,2	2,3	2,4	2,4	−55	−60	−51	−60	−4	−22	−44	−1
15	2,2	2,3	2,6	2,5	−53	−59	−56	−56	−4	−21	−54	−30
20	2,2	2,4	2,6	2,5	−53	−54	−65	−67	−14	−30	−42	−27
25	2,4	2,4	2,5	2,6	−50	−59	−67	−80	−37	−25	−26	−13
30	2,4	2,3	2,3	2,4	−51	−71	−92	−83	−32	−14	36	11
35	2,2	2,0	2,0	2,2	−85	−100	−113	−99	−45	−115	164	88
40	2,0	1,8	1,8	2,0	−103	−111	−106	−122	124	199	215	204
45	1,9	1,8	2,7	2,8	−116	−110	−108	−121	181	227	244	234
50	1,7	1,7	1,7	1,7	−107	−113	−112	−118	220	252	263	253
55	1,6	1,6	1,7	1,7	−98	−115	−112	−125	229	272	267	302
60	1,7	1,5	1,9	1,8	−103	−111	−64	−112	243	330	238	291
65	2,2	2,0	2,3	2,4	−108	−124	−100	−121	113	222	88	117
70	3,1	3,4	3,2	3,4	−30	−34	−27	−19	−119	−144	−135	−145
75	3,2	3,3	3,2	3,3	61	45	64	69	−110	−144	−122	−116
80	2,9	2,8	2,7	2,8	136	106	129	136	45	−55	10	21
85°S	2,3	2,3	2,2	2,2	240	187	198	220	461	196	244	343

Table III: Frequency (%) of different gradations of cloud cover, 1966-1975, Northern Hemisphere

Latitude	Winter (XII-II)					Spring (III-V)				
	0	1-3	4-6	7-8	9-10	0	1-3	4-6	7-8	9-10
85°	46,1	13,2	17,1	10,4	13,2	43,4	11,5	19,1	11,6	14,4
80	43,7	12,2	17,0	11,3	15,8	39,0	12,7	18,6	14,0	15,7
75	39,0	11,6	17,6	13,6	18,2	34,8	11,6	18,7	18,2	16,7
70	35,5	11,3	15,7	16,2	21,3	30,4	11,0	18,8	21,3	18,5
65	29,7	9,7	15,7	19,1	25,8	24,3	9,9	18,7	26,6	20,5
60	22,1	7,6	16,3	25,8	28,2	17,8	8,2	19,8	30,4	23,8
55	16,1	6,3	16,6	31,7	29,3	13,7	7,3	20,2	32,0	26,8
50	14,0	5,8	16,7	33,0	30,5	13,3	7,3	21,4	31,8	26,2
45	13,2	5,7	18,9	33,1	29,1	14,7	7,9	23,3	29,4	24,7
40	14,3	6,1	21,6	31,9	26,1	15,5	9,2	26,0	27,8	21,5
35	16,3	7,4	25,0	30,1	21,2	18,2	10,0	27,3	26,7	17,8
30	23,8	7,1	26,3	26,4	16,4	23,0	10,2	29,4	22,8	14,6
25	27,4	7,6	30,5	22,5	12,0	25,4	10,5	33,8	19,9	10,4
20	27,6	8,4	35,9	20,1	8,0	26,3	10,5	37,8	18,2	7,2
15	24,7	9,2	31,2	18,8	6,1	23,0	12,3	41,2	17,4	6,1
10	19,1	9,0	34,8	20,1	7,0	12,9	11,0	44,9	22,2	9,0
5	8,6	6,3	43,5	29,0	12,6	5,8	8,1	42,8	29,4	13,9
0	7,6	4,9	46,6	29,9	11,0	6,7	7,6	45,8	28,3	11,6

Latitude	Winter (VI-VIII)					Spring (IX-XI)				
	0	1-3	4-6	7-8	9-10	0	1-3	4-6	7-8	9-10
85°	17,7	6,5	25,9	24,2	25,7	25,6	9,1	19,5	24,7	21,1
80	17,6	7,0	25,5	24,1	25,8	23,4	9,6	19,1	25,6	22,3
75	18,0	6,9	23,2	25,3	26,6	20,9	8,3	18,7	27,9	24,2
70	16,5	6,4	22,5	29,6	25,0	17,2	6,6	18,1	30,7	27,4
65	12,9	5,8	22,7	34,8	23,8	13,9	5,6	16,4	34,0	29,9
60	10,6	5,7	23,2	34,5	26,0	10,5	4,8	17,3	36,5	30,9
55	10,0	5,6	22,9	34,3	27,2	10,7	4,8	17,8	37,9	29,8
50	10,8	6,4	23,7	32,1	27,0	12,8	5,6	20,6	34,8	26,2
45	15,4	7,9	25,5	28,2	23,0	17,4	8,0	22,1	30,8	21,7
40	19,4	9,4	29,7	24,1	17,4	20,1	8,0	26,6	27,8	17,5
35	24,1	8,8	30,9	22,9	13,3	23,0	8,1	29,1	25,8	14,0
30	24,5	7,7	33,3	23,0	11,5	26,4	7,0	32,2	22,9	11,5
25	22,0	8,5	36,9	22,7	9,9	26,4	8,1	35,8	20,7	9,0
20	17,4	8,6	38,1	24,9	11,0	23,7	7,9	39,9	21,2	8,0
15	10,8	7,6	42,1	28,6	10,9	17,9	7,7	41,6	23,8	9,0
10	6,0	4,4	39,7	35,5	14,4	10,0	5,5	40,1	31,2	13,2
5	5,0	3,5	38,6	38,4	14,5	5,7	3,5	40,7	35,8	14,3
0	6,3	4,6	45,9	32,2	11,0	7,4	3,3	45,0	32,8	11,5

Table IV: Frequency (%) of different gradations of cloud cover, 1966-1975, Southern Hemisphere

Latitude	Summer (XII-II)					Autumn (III-V)				
	0	1-3	4-6	7-8	9-10	0	1-3	4-6	7-8	9-10
5°	3,6	14,0	36,8	39,2	6,4	5,6	14,4	42,9	33,7	3,4
10	2,3	12,5	40,7	38,3	6,2	5,5	15,5	44,1	32,3	2,6
15	2,1	13,0	40,9	38,1	5,9	5,9	15,0	44,4	31,5	3,2
20	2,7	14,0	40,3	36,9	6,1	6,8	14,8	43,3	31,1	4,0
25	3,8	16,5	38,4	35,3	6,0	6,5	13,6	41,6	33,5	4,8
30	5,4	16,4	41,2	32,1	4,8	4,5	11,8	41,8	37,5	4,4
35	3,4	10,6	39,6	41,8	4,6	2,4	6,3	35,6	49,9	5,8
40	1,9	7,5	32,4	50,4	7,6	1,4	4,0	28,8	56,7	9,1
45	1,2	5,2	27,8	55,1	10,7	0,9	2,9	23,9	59,2	13,1
50	0,8	3,6	24,8	57,9	12,9	0,6	2,6	21,6	60,5	14,8
55	0,5	2,7	22,3	59,2	15,3	0,4	1,7	19,3	62,4	16,2
60	0,7	2,6	21,1	57,9	17,7	0,5	1,4	17,4	62,2	18,5
65	2,3	6,5	25,0	50,1	16,1	1,9	4,2	21,6	54,9	17,4
70	17,5	16,4	25,7	31,7	8,7	22,1	11,9	20,1	36,5	9,4
75	42,9	17,5	17,5	17,5	4,6	48,5	9,2	14,2	25,3	2,8
80	61,9	14,2	10,4	9,7	3,8	64,4	7,8	10,4	16,7	0,7
85	77,7	9,0	4,6	5,4	3,3	77,2	6,0	7,0	9,4	0,2

Latitude	Winter (VI-VIII)					Spring (IX-XI)				
	0	1-3	4-6	7-8	9-10	0	1-3	4-6	7-8	9-10
5°	5,3	15,8	47,0	29,2	2,7	5,0	11,5	42,1	36,4	5,0
10	8,8	15,5	45,0	27,6	3,1	5,2	12,9	42,4	34,0	5,5
15	10,4	14,6	42,1	29,8	3,1	6,8	12,8	39,6	34,9	5,9
20	10,0	15,5	40,4	33,1	3,6	7,1	13,0	36,3	37,3	6,3
25	8,0	9,4	38,7	37,5	4,4	6,6	12,0	33,7	40,9	6,8
30	5,7	9,6	35,8	43,5	5,3	4,9	11,1	32,0	44,9	7,1
35	2,8	5,6	32,9	50,6	8,1	3,3	7,5	28,8	51,2	9,2
40	1,3	3,8	29,9	54,8	10,2	1,7	5,0	24,5	56,8	11,9
45	1,0	3,1	27,1	56,8	12,0	1,1	4,0	22,0	58,1	14,8
50	1,0	3,1	26,4	57,0	12,5	0,6	3,1	19,2	59,5	17,6
55	1,0	3,1	25,7	57,2	13,1	0,6	2,4	16,9	59,3	20,9
60	1,4	4,0	25,4	55,3	13,9	1,1	2,5	16,3	56,8	23,3
65	4,1	7,0	28,1	49,2	11,6	4,7	6,2	20,5	49,2	19,4
70	23,5	13,2	24,0	33,5	5,8	25,9	13,3	19,1	31,5	10,2
75	49,1	13,1	14,5	21,0	2,3	54,0	9,6	11,8	21,4	3,2
80	65,7	11,2	8,5	13,8	0,8	69,9	7,3	7,5	14,3	1,0
85	76,4	9,0	5,7	8,7	0,2	80,8	5,8	5,2	7,8	0,4

Latitude	Northern Hemisphere $10^2 a$ Wi	Sp	Su	Au	$10^2 b$ Wi	Sp	Su	Au
85	57	50	–12	8	37	38	45	40
80	48	41	–12	3	35	39	44	41
75	35	30	–15	–8	36	41	43	42
70	24	19	–20	–21	36	42	45	43
65	8	4	–29	–34	37	46	50	45
60	–12	–15	–38	–45	42	49	50	48
55	–27	–29	–40	–47	45	51	52	50
50	–36	–28	–36	–33	49	52	50	48
45	–36	–22	–17	–14	51	52	48	46
40	–27	–13	2	0	52	53	48	48
35	–14	–1	17	12	53	54	53	43
30	8	15	21	30	47	53	51	51
25	24	29	21	32	53	57	54	53
20	36	41	11	30	59	61	57	56
15	39	41	–5	12	65	66	65	62
10	28	12	–33	–14	69	72	69	61
5	–15	–22	–40	–33	74	77	71	70
0	–14	–14	–20	–18	78	79	72	71

Latitude	Southern Hemisphere $10^2 a$ Wi	Sp	Su	Au	$10^2 b$ Wi	Sp	Su	Au
85	168	157	134	154	54	45	27	24
80	118	123	104	118	46	46	37	53
75	79	79	69	71	50	44	46	48
70	22	19	10	10	55	48	56	51
65	–49	–59	–65	–78	72	63	74	74
60	–79	–52	–99	–112	81	87	84	85
55	–86	–55	–99	–108	85	73	88	88
50	–84	–99	–79	–98	83	82	94	86
45	–82	–84	–70	–86	86	85	84	85
40	–53	–71	–51	–69	87	79	84	86
35	–51	–48	–25	–45	80	76	84	86
30	–21	–27	–6	–15	75	73	79	81
25	–73	–17	–12	–5	75	72	79	77
20	50	–12	–24	1	74	72	82	78
15	12	–10	–28	1	88	74	85	92
10	12	–12	–29	–10	78	78	84	79
5	27	–14	–23	–4	92	79	79	82

Table V: Parameters of lognormal cloud-cover distribution, 1966-1975

REFERENCES*)

1. Abramovich, K. G. Formation conditions and prediction of low clouds. *Trudy Gidromettsentra SSSR*, 1973, No. 78.

2. Aeroclimatological atlas-handbook of the USSR, No. 2, Climatic characteristics of weather phenomena hazardous to aviation. Edited by A. P. Pen'kov. Gidrometeoizdat, Moscow, 1975.

3. Aeroclimatological atlas-handbook of the USSR, No.3, Statistical characteristics of the spatial and microphysical structure of clouds. Edited by L. S. Dubrovina. Gidrometeoizdat, Moscow, 1975, Vols. 1 and 2.

4. Aeroclimatological handbook of non-Soviet countries, Part 1, Western Europe. Edited by T. N. Deryugina, Z. M. Makhover, L. A. Nudel'man, and A. P. Pen'kov, Gidrometeoizdat, Moscow, 1980.

5. Akvilonova, A. B., and Kutuza, B. G. Radiothermal radiation of clouds. *Radiotekhnika i Elektronika,* 1978, 23 (9): 1972.

6. Akimov, N. M., Volkov, A. D., and Leskov, B. N. Concentration of ice crystals in Ns. *Trudy UkrNIGMI,* 1974, No. 133, p. 102.

7. Alekseeva, I. A., and Mitnik, L. M. Structure of intertropical convergence zone according to radiometric measurements made with *Cosmos 243*. In: The tropical region of the World Oceam and global processes associated with it. Nauka, Moscow, 1973.

8. Amirov, A. D. A method for calculating humidity and temperature in the cumulus-cloud problem. Izv. Akad. Nauk SSSR, *FAO,* 1971, 7 (7): 723.

9. Andreev, V., and Panchev, S. Dynamics of atmospheric thermals. Edited by L. T. Metveev. Gidrometeoizdat, Leningrad, 1975.

10. Antonov, V. S. The application of calculations of cloud-droplet crystallization to problems of precipitation prediction. *Trudy UkrNIGMI,* 1979, No. 170, p. 73.

---

*) With only a few obvious exceptions, references 1 through 330, given in the Cyrillic-alphabetical order, are in Russian (Translator).

11. Antonov, V. S., and Egorovskaya, G. D. Statistical-physical relationships between the complex of thermodynamic parameters of the atmosphere and the intensity of steady precipitation. *Trudy UkrNIGMI*, 1976, No. 134, p. 38.

12. Atlas of climatic characteristics of cloudiness over the Southern Hemisphere. Edited by Z. M. Makhover and L. A. Nudel'man. Gidrometeoizdat, Moscow, 1975.

13. Bagrov, A. N. An operational numerical method for predicting convective phenomena and steady precipitation. *Trudy Gidromettsentra SSSR*, 1971, No. 91, p. 25.

14. Baronov, A. M. Frontal clouds and flying conditions in them. Gidrometeoizdat, Leningrad, 1964.

15. Basharinov, A. E., Gurvich, A. S., and Egorov, S. T. Radio--frequency radiation of the Earth as a planet. Nauka, Moscow, 1974.

16. Bekryaev, V. I., Dovgladyuk, Yu. A., and Zinchenko, A. V. Application of the jet model to calculate cloud convection in the tropics. *Trudy GGO*, 1977, No. 389, p. 32.

17. Bekryaev, V. I., Dovgladyuk, Yu. A., and Zinchenko, A. V. Determination of some properties of an ensemble of convective clouds according to aerological sounding data. *Trudy GGO*, 1979, No. 405, p. 3.

18. Belov, P. N. Numerical methods of weather forecasting. Gidrometeoizdat, Leningrad, 1975.

19. Belyaev, V. I. The Lagrange method in the kinetics of cloud processes. Gidrometeoizdat, Leningrad, 1964.

20. Berlyand, M. E. Contemporary problems in atmospheric diffusion and pollution of the atmosphere. Gidrometeoizdat, Leningrad, 1975.

21. Berlyand, M. E., and Kanchan, Ya. S. Toward a theory of the formation of radiation fogs and their effect on the spread of pollutants. *Trudy GGO*, 1973, No. 293, p. 3.

22. Berlyand, M. E., and Onikul, R. I. Toward a theory of air--mass transformation and river-fog formation. *Trudy GGO*, 1968, No. 207, p. 14.

23. Berlyand, T. G., and Strokina, L. A. The global cloudiness regime. *Trudy GGO*, 1974, No. 388, p. 3.

24. Berlyand, T. G., and Strokina, L. A. The global distribution of the total cloud cover. Gidrometeoizdat, Leningrad, 1980.

25. Bespomestnykh, A. V., and Mal'tsev, V. A. Disturbance of the wind-velocity field in the vicinity of a convective cloud. *Meteorologiya i Gidrologiya,* 1971, No. 10, p. 22.

26. Bibikova, T. N. Some features of orographic wave cloudiness in the Central and Northern Urals. *Meteorologiya i Gidrologiya,* 1978, No. 12, p. 15.

27. Bibikova, T. N. Helical clouds in the Elbrus region. *Meteorologiya i Gidrologiya,* 1979, No. 8, p. 99.

28. Blinova, E. N. On the hydrodynamic theory of long-range weather forecasting. In: Dynamics of large-scale atmospheric processes. Nauka, Moscow, 1965, p. 15.

29. Blinova, E. N. The general circulation of the atmosphere and hydrodynamic long-range weather forecasting. In: Meteorology and hydrology during 50 years of Soviet rule. Edited by E. K. Federov. Gidrometeoizdat, Leningrad, 1967, p. 51.

30. Blinova, E. N. Development of a hydrodynamic theory of long--range weather forecasting. In: Fifty years of the hydrometeorological forecasting center. Gidrometeoizdat, Leningrad, 1979, p. 43.

31. Bobylev, L. N., Vasishcheva, M. A., and Shchukin, G. G. Determination of the integral parameters of the moisture content of a cloudy atmosphere directly from the radio-brightness temperature. *Trudy GGO,* 1977, No. 395, p. 59.

32. Bogorodskii, V. V., Kozlov, A. I., and Tuchkov, L. T. Radiothermal emission of the Earth's covering layers. Gidrometeoizdat, Leningrad, 1977.

33. Boldyrev, V. G. The use of measurements of the outgoing radiation to calculate the temperature of the Earth's surface and the height of an upper cloud boundary. *Trudy Gidromettsentra SSSR,* 1965, No. 8, p. 68.

34. Borisenkov, E. P., and Kuznetsov, I. N. Parametrization of the atmosphere-ocean interaction during stormy weather conditions, as applied to models of the general circulation of the atmosphere. Izv. Akad. Nauk SSSR, *FAO,* 1978, 14 (5): 510.

35. Borovikov, A. M., Gaivoronskii, I. I., Zak, E. G., et al. Cloud physics. Edited by A. Kh. Khrgian. Gidrometeoizdat, Leningrad, 1961 (English translation published by Israel Program for Scientific Translations (Keter), Jerusalem, 1963).

36. Borovikov, A. M., and Mazin, I. P. On the orientation of cloud banks. *Trudy TsAO,* 1967, No. 79, p. 90.

37. Borovikov, A. M., and Mazin, I. P. The microphysical characteristics of clouds. In: Aeroclimatological atlas-handbook of the USSR, No. 3, Vol. 1, Part 2, Gidrometeoizdat, Moscow, 1975, p. 129.

38. Borovikov, A. M., Mazin, I.P., and Nevzorov, A. N. Large particles in clouds. *Trudy VNMS,* 1963, 5: 73.

39. Bortkovskii, R. S. Obtaining more accurate estimates of the heat and moisture exchange of the atmosphere and ocean during a storm. *Trudy GGO,* 1975, No. 326, p. 58.

40. Bugaeva, G. V., and Vel'tishchev, N. F. Numerical modeling of convective cloud vortices. *Trudy GosNITsIPR,* 1976, No. 1, p. 80.

41. Bugaeva, G. V., and Ryazanova, L. A. Seasonal transformations of fields of radiation temperature and cloudiness according to MISZ material. *Trudy TsAO,* 1970, No. 86, p. 135.

42. Buz, A. I. The role of atmospheric fronts and some circulation factors in the development of convection. *Meteorologiya i Gidrologiya,* 1975, No. 9, p. 21.

43. Buikov, M. V. The kinetic-equation method in the theory of clouds. *Trudy VNMS,* 1963, 5: 122.

44. Buikov, M. V. The description of condensation processes in a turbulent atmosphere. *Trudy UkrNIGMI,* 1971, No. 95, p. 3.

45. Buikov, M. V. Numerical modeling of stratiform clouds. A survey. Obninsk, 1978.

46. Buikov, M. V., and Pirnach, A. M. A numerical model of a two-phase stratiform cloud taking microstructure into account. Izv. Akad. Nauk SSSR, *FAO,* 1973, 9 (5): 486.

47. Buikov, M. V., and Pirnach, A. M. Numerical modeling of three-phase stratiform cloudiness taking microstructure into account. *Trudy UkrNIGMI,* 1973, No. 125, p. 3.

48. Buikov, M. V., and Khvorost'yanov, V. I. The formation and evolution of radiation fog and stratus cloudiness in the boundary layer of the atmosphere. Izv. Akad. Nauk SSSR, *FAO,* 1977, 13(4): 356.

49. Buleev, N. I., and Marchuk, G. I. The dynamics of large-scale atmospheric processes. *Trudy IFA Akad. Nauk SSSR,* 1958, No. 2, p. 66.

50. Burkovskaya, S. N., and Minervin, V. E. Air humidity in

clouds. *Trudy TsAO,* 1972, No. 101, p. 61.

51. Burtsev, A. I., and Popova, T. P. The cloud system of cyclones over Siberia according to data of *Cosmos 122. Meteorologiya i Gidrologiya,* 1967, No. 2, p. 9.

52. Bykova, L. P., and Matveev, L. T. The evolution of fields of cloudiness and temperature in a moving cyclone (a quantitative experiment). Izv. Akad. Nauk SSSR, *FAO,* 1966, 2 (9): 905.

53. Van Mieghem, J. Atmospheric energetics. Oxford Univ. Press, 1973.

54. Vasil'ev, V. A., and Alekseeva, V. V. Some estimates of time anomalies of large-scale fields of total cloudiness in extratropical latitudes of the Southern Hemisphere. *Trudy VNIIGMI-MTsD,* 1980, No. 73, p. 98.

55. Vasishcheva, M. A., and Shchukin, G. G. Experimental studies of cloud water contents: Statistical models of the atmosphere. Obninsk, 1977.

56. Vel'tishchev, N. F., and Geokhlanyan, T. Kh. Cellular convection: Laboratory experiments and observations in the atmosphere. *Trudy Gidromettsentra SSSR,* 1974, No. 132, p. 71.

57. Vel'tishchev, N. F., and Zhelnin, A. A. The effect of the unsteady nature of the mean-temperature field on the structure of convective motions. *Trudy Gidromettsentra SSSR,* 1974, No. 148, p. 3.

58. Vel'tishchev, N. F., and Zhelnin, A. A. Convective motions in an airflow that varies its direction with height. *Meteorologiya i Gidrologiya,* 1976, No. 4, p. 18.

59. Vetlov, I. P. A study of cloudiness and outgoing radiation with the aid of *Cosmos 122. Meteorlogiya i Gidrologiya,* 1967, No. 1, p. 3.

60. Vetlov, I. P., and Petrenko, N. V. Cold-front precipitation in the central part of the European USSR during the warm season. Gidrometeoizdat, Leningrad, 1955.

61. Voit, F. Ya., and Mazin, I. P. The water content of cumulus clouds. Izv. Akad. Nauk SSSR, *FAO,* 1972, 8 (11): 1166.

62. Voloshchuk, V. M. The kinetic equation of stochastic coalescence in clouds. *Meteorologiya i Gidrologiya,* 1977, No. 5, p. 3.

63. Voloshchuk, V. M., and Sedunov, Yu. S. Coalescence processes in dispersed systems. Gidrometeoizdat, Leningrad, 1975.

64. Voloshchuk, V. M., and Sedunov, Yu. S. The kinetic equation of evolution of a droplet spectrum in a turbulent medium during the condensation stage of cloud development. *Meteorologiya i Gidrologiya,* 1977, No. 3, p. 3.

65. Vorob'ev, B. M. A numerical model of a steady-state hail process under naturally and artificially regulated conditions. Izv. Akad. Nauk SSSR, *FAO,* 1977, 13 (8): 838.

66. Vorontsov, P. A. Aerological characteristics of sea fogs. *Trudy AANII,* 1959, 228 (1): 38.

67. Voskresenskii, A. I. Water contents of arctic clouds. *Trudy AANII,* 1962, 239: 11.

68. Voskresenskii, A. I., and Matveev, L. T. Water content and turbulent regime of arctic stratiform clouds. *Meteorologiya i Gidrologiya,* 1966, No. 11, p. 14.

69. Vul'fson, N. I. Convective motions in a free atmosphere. Izdatel'stvo Akad. Nauk SSSR, 1961 (English translation published by Israel Program for Scientific Translations (Keter), Jerusalem, 1964).

70. Vul'fson, N. I., and Levin, L. M. A possible effect of vertical heated jets on the development of internal clouds. *Trudy IPG,* 1970, No. 12, p. 52.

71. Gavrilin, B. L. Quasi-solenoidal approximations for the equations of the dynamics of a moist turbulent atmosphere. *Trudy LGMI,* 1968, No. 27, p. 81.

72. Galakhova, T. A. The distribution of ordered vertical motions in frontal-zone regions and the use of these to forecast precipitation and temperature. *Trudy Gidromettsentra SSSR,* 1970, No. 68, p. 54.

73. Gal'perin, B. M., and Kuchumova, L. S. The effect of cloudiness on the radiation of the atmosphere. *Meteorologiya i Gidrologiya,* 1959, No. 8, p. 19.

74. Gal'perin, B. M., and Seryakova, L. P. The fundamental regularities of short-wave radiation and the diurnal radiation balance for different cloudiness. *Trudy LGMI,* 1966, No. 22, p. 11.

75. Glushkova, N. I. A study of the formation mechanism of heavy showers and hail. *Trudy Gidromettsentra SSSR,* 1978, No. 201, p. 57.

76. Godev, N. G., Penenko, V. V., and Obraztsov, N. N. A quantitative experiment on taking into account orography and friction

in models of the dynamics of the atmosphere. *Meteorologiya i Gidrologiya,* 1979, No. 10, p. 5.

77. Godizov, A. G., and Stepanov, A. S. Equations of condensation in a turbulent cloudy medium. Izv. Akad. Nauk SSSR, *FAO,* 1979, 15 (12): 1318.

78. Goisa, N. I., Gorb, A. S., and Dmitrieva, L. R. Albedo of fields of cumuli. *Trudy UkrNIGMI,* 1976, No. 146, p. 58.

79. Goisa, N. I., and Shoshin, V. M. An experimental model of the radiation regime of an "average" stratiform cloud. *Trudy UkrNIGMI,* 1969, No. 82.

80. Golitsyn, G. S. Evaluations of efficiency and rates of convection in laboratory experiments on modeling the genetal circulation. Izv. Akad. Nauk SSSR, *FAO,* 1977, 13 (8): 891.

81. Golitsyn, G. S., and Mokhov, I. I. An evaluation of the sensitivity and role of cloudiness in simple models of the climate. Izv. Akad. Nauk SSSR, *FAO,* 1978, 14 (8): 803.

82. Gorelik, A. G., Kalashnikov, V. V., Kn'yazev, A. V., et al. Determination of the integral water content of rain clouds by means of concurrent radiothermal and radar measurements. Izv. Akad. Nauk SSSR, *FAO,* 1973, 9 (5): 556.

83. Gossard, E. E. Waves in the atmosphere. Elsevier, 1975.

84. Davydova, N. G., and Suzyumova, G. N. Interpretation of combined cloud for all types of digital computers for 1967. In: Materials of meteorological studies. The circulation of the atmosphere. Moscow, 1974, p. 64.

85. Devyatova, V. A. Some characteristics of stratiform clouds. *Meteorologiya i Gidrologiya,* 1973, No. 8, p. 26.

86. Devyatova, V. A. Statistical characteristics of the vertical structure of single-layer stratiform cloudiness and fields of temperature and humidity. *Trudy Gidromettsentra SSSR,* 1975, No. 165, p. 107.

87. Devyatova, V. A. Statistical characteristics of the vertical structure of two-layer stratiform cloudiness and fields of temperature and humidity. *Trudy GosNITsIPR,* 1976, No. 1, p. 48.

88. Dynamics of cumulus clouds. A collection of articles edited by C. Anderson. Russian translation edited by N. I. Vul'fson. Mir, Moscow, 1964.

89. Dmitriev, A. A. Orthogonal exponential functions in hydrometeorology. Gidrometeoizdat, Leningrad, 1973.

90. Dmitriev, A. A., Dremina, V. P., Tronkina, N. N., et al. Parametrization of the vertical humidity distribution using D_M (z, α_1, ... α_M) functions. *Trudy TsVGMO*, 1975, No. 6, p. 136.

91. Dmitriev, A. A., and Mertsalova, O. B. Approximation of an inversion distribution of humidity using D_M (z, α_1, ... α_M) functions for low mechanization. *Trudy TsVGMO*, 1977, No. 9, p. 54.

92. Dmitrieva-Arrago, L. R., Koloskova, L. F., and Orlova, L. S. A check of Smagorinky's graph for determining the cloud cover. *Trudy GGO*, 1969, No. 236, p. 31.

93. Dmitrieva-Arrago, L. R., Parshina, G. V., and Samoilova, L. V. Calculation of fluxes of short-wave radiation under cloudy conditions. *Trudy GGO*, 1972, No. 272, p. 70.

94. Dombkovskaya, E. P., and Mitnik, L. M. Complex analysis of results of satellite measurements of electromagnetic radiation in the optical, infrared, and radio ranges. *Meteorologiya i Gidrologiya*, 1972, No. 6, p. 23.

95. Dorodnitsyn, A. A. The effect of the relief of the earth's surface on air currents. *Trudy TsIP*, 1950, No. 21 (48), p. 3.

96. Drozdov, O. A., and Grigor'eva, A. S. The hydrologic cycle in the atmosphere. Gidrometeoizdat, Leningrad, 1963 (English translation published by Israel Program for Scientific Translations (Keter), Jerusalem, 1965).

97. Dubrovina, L. S., and Verzunova, V. D. The present state of cloud studies using data of ground-based, aircraft, and radar observations. Obninsk, 1970.

98. Dubrovina, L. S., and Verzunova, V. D. Study of the dew-point deficit to determine the probability of clouds at different heights. *Trudy VNIIGMI-MTsD*, 1979, No. 65, p. 79.

99. Dushkin, P. K., Lomonosov, E. G., and Lunin, Yu. N. An attempt at numerical prediction of humidity, cloudiness, and precipitation with a computer. *Meteorologiya i Gidrologiya*, 1960, No. 12, p. 3.

100. Dymnikov, V. P. Parametrization of the amount of nonconvective cloud cover in the problem of the background forecasting of weather. *Trudy ZSRNIGMI*, 1974, No. 11, p. 62.

101. Dymnikov, V. P., and Guseva, N. V. A numerical method of

predicting fields of humidity, stratiform cloudiness, and precipitation. Izv. Akad. Nauk SSSR, *FAO*, 1975, 11 (6): 547.

102. Dymnikov, V. P., and Ishimova, A. V. A nonadiabatic model of short-term weather forecasting. *Meteorologiya i Gidrologiya*, 1979, No. 6, p. 5.

103. Dyubyuk, A. F. The calculation of precipitation. *Doklady TsIP*, 1947, 1 (3): 31.

104. Dyubyuk, A. F., Bibikova, T. N., and Trubnikov, B. N. The conditions of formation of altocumulus lenticular clouds in the Crimea. *Trudy TsAO*, 1963, No. 47, p. 85.

105. Zav'yalova, I. N. The cloudiness regime in Antarctica. *Trudy AANII*, 1976, 328: 156.

106. Zaitsev, V. A., and Ledokhovich, A. A. Instruments for studying fogs and clouds and measuring humidity. Gidrometeoizdat, Leningrad, 1970.

107. Zak, E. G. The cloudiness of anticyclonic inversions. *Trudy NIU GUGMS*, Ser. 2, 1946, No. 14, p. 109.

108. Zakharova, I. M. A mathematical model of the evolution of radiation fog. *Trudy IEM*, 1971, No. 23, p. 157.

109. Zakharova, I. M. Numerical modeling of the formation and development of radiation fog. *Trudy IEM*, 1975, No. 9 (52), p. 124.

110. Zverev, A. S. Fogs and their prediction. Gidrometeoizdat, Leningrad, 1954.

111. Zeitunyan, Kh. N. The nonlinear problem of cloud formation behind a barrier. Izv. Akad. Nauk SSSR, *Ser. Geofiz.*, 1963, No. 9, p. 1427.

112. Ivanov, V. N., and Ordanovich, A. E. Some inverse relations arising during turbulent cellular convection in the atmosphere. Izv. Akad. Nauk SSSR, *FAO*, 1972, 8 (11): 1216.

113. Ivanov, V. N., and Khain, A. P. Dry and moist cellular convection in the atmosphere. Izav. Akad. Nauk SSSR, *FAO*, 1975, 11 (12): 1211.

114. Istomina, L. G. Some evaluations of the spatial scales of cloud fields using aerophotography. Izv. Akad. Nauk SSSR, *FAO*, 1973, 9 (7): 708.

115. Kabanov, A. S., and Klykov, A. E. A model of convection in

the atmosphere with a cloud and numerical experiments on the modification of a cumulus cloud. *Meteorologiya i Gidrologiya*, 1978, No. 3, p. 13.

116. Kadyshnikov, V. M. A pseudo-adiabatic model of steady precipitation for short-term weather forecasting. *Meteorologiya i Gidrologiya*, 1968, No. 2, p. 33.

117. Kasatkin, I. I. Vertical motions in the atmosphere. Byull. Mosk. Obshchestva Ispyt. Prirody, 1914.

118. Kachurin, L. G., Bekryaev, V. I., and Vorob'ev, B. M. Application of the method of turbulent jets of the calculation of convective motions in the atmosphere. In: Trudy VIII Vsesoyuznoi Konf. po Fizike Oblakov i Aktivnym Vozdeistviyam. Gidrometeoizdat, Leningrad, 1970, p. 213.

119. Kibel', I. A. An introduction to hydrodynamic methods of short-term weather forecasting. Gostekhizdat, Moscow, 1957.

120. Kibel', I. A. Determination of cloud boundaries. *Dokl. Akad. Nauk SSSR*, 1965, 160 (4): 815.

121. Kibel', I. A. The hydrodynamics of cloud systems. In: Dynamics of large-scale atmospheric processes. Nauka, Moscow, 1965, p. 278.

122. Kibel', I. A. The hydrodynamic (numerical) short-term forecasting of weather. In: Meteorology and hydrology during 50 years of Soviet rule. Edited by E. K. Federov. Gidrometeoizdat, Leningrad, 1967, p. 67.

123. Kislov, A. V., and Semenov, E. K. Spectral analysis of cloudiness over the Indian Ocean. *Meteorologiya i Gidrologiya*, 1979, No. 5, p. 49.

124. Klinov, F. Ya. The lower layer of the atmosphere in the presence of low-lying clouds and fog. Gidrometeoizdat, Moscow, 1974.

125. Klubovich, K. V. Toward a theory of cloud formation. *Uch. Zap. Minskogo Gos. Ped. Inst., Ser. Fiz. -Mat.*, 1957, No. 7, p. 129.

126. Klyuchnikova, L. A. The formation of advection fogs. *Trudy GGO*, 1955, No. 60, p. 40.

127. Kobysheva, N. V. Indirect calculations of climatic characteristics. Gidrometeoizdat, Leningrad, 1971.

128. Kobysheva, N. V., and Narovlyanskii, G. Ya. The climatological processing of meteorological information. Gidrometeoizdat, Leningrad,

1978.

129. Koviza, A. N. Prediction of an advection fog of cooling and low cloudiness. *Meteorologiya i Gidrologiya,* 1962, No. 1, p. 2.

130. Kogan, E. L. The spatial and temporal evolution of a droplet cumulus cloud in a three-dimensional numerical model. Izv. Akad. Nauk SSSR, *FAO,* 1979, 15 (9): 929.

131. Kozharin, V. S. Regularities in the fluctuations of the height of a lower cloud boundary. *Meteorologiya i Gidrologiya,* 1967, No. 6, p. 29.

132. Kozhevnikov, V. N. A nonlinear problem of orographic perturbation of an airflow. *Izv. Akad. Nauk SSSR, Ser. Geofiz.,* 1963, No. 7, p. 1108.

133. Kozhevnikov, V. N. Orographic perturbations in a two-dimensional steady-state problem. Izv. Akad. Nauk SSSR, *FAO,* 1968, 4 (1): 33.

134. Kozhevnikov, V. N., Bibikova, T. N., and Zhurba, E. V. Orographic perturbations of the atmosphere over the Northern Urals. Izv. Akad Nauk SSSR, *FAO,* 1977, 13 (5): 451.

135. Kozlov, V. N., Lushev, Yu. G., and Matveev, L. T. The use of meteorological-satellite data to diagnose and predict the principal meteorological elements. In: Trudy Vsesoyuz. Konf. po Voprosam Meteorol. Obespecheniya Sverkhzvyk. Aviatsii, 1971, p.60.

136. Kozlov, V. N., and Matveev, L. T. Moisture balance in a cloud. *Meteorologiya i Gidrologiya,* 1970, No. 3, p. 34.

137. Kozlov, V. N., and Matveev, L. T. The use of satellite information to analyze fields of the principal meteorological elements in a cloudy atmosphere. *Meteorologiya i Gidrologiya,* 1971, No. 11, p. 19.

138. Kozlov, V. N., and Matveev, L. T. Numerical modeling of fields of atmospheric moisture content and cloud water reserves over seas and oceans. In: Doklady Vsesoyuz. Simp. po PIGAP. Gidrometeoizdat, Moscow, 1973, p. 10.

139. Kolezhuk, V. T., Levin, A. V., and Tkachenko, A. V. A numerical experimental study of the evolution of the lower boundary of internal stratiform clouds. *Trudy UkrNIGMI,* 1972, No. 118, p. 25.

140. Komarov, V. S. The statistical structure of the humidity field in the free atmosphere over the USSR. *Trudy NIIAK,* 1971, No. 70.

141. Komarov, V. S. The statistical structure of vertical profiles of temperature and humidity in the atmosphere of the Northern Hemisphere. *Trudy VNIIGMI-MTsD*, 1978, No. 57, p. 3.

142. Kondrat'ev, K. Ya., Matveev, L. T., and Shvets, M. E. Fluxes of heat into the atmosphere. In: Global studies of the atmosphere and weather forecasting. Izdatel'stvo Akad. Nauk SSSR, Moscow, 1969, p. 137.

143. Koprova, L. I., and Boldyrev, V. G. The statistical characteristics of the vertical structure of the temperature and humidity fields up to great heights. Izv. Akad. Nauk SSSR, *FAO*, 1970, 6 (2): 154.

144. Kostinskaya, B. I. A method for calculating combinations of low-lying and overall cloudiness. *Trudy NIIAK*, 1966, No. 37, p. 93.

145. Koshelenko, I. V. Fogs. *Trudy UkrNIGMI*, 1977, No. 155.

146. Koshenko, A. M. The phase structure of frontal cloud systems at different stages of development. *Trudy UkrNIGMI*, 1970, No. 96, p. 106.

147. Krystanov, L. K. Selected works on atmospheric physics. Edited by E. K. Federov. Gidrometeoizdat, Leningrad, 1968.

148. Kuznetsov, L. A. A method of predicting fields of humidity, cloudiness, and precipitation. *Meteorologiya i Gidrologiya*, 1962, No. 2, p. 9.

149. Lapcheva, V. P., and Litvin, N. N. The use of data of the MRL network to complete the information about the maximum precipitation in a 12-hour period. *Trudy Gidromettsentra SSSR*, 1978, No. 201, p. 29.

150. Lebedev, S. L. The effect of the gravity fall of cloud droplets on cloud water content. *Izv. Akad. Nauk SSSR, Ser. Geofiz.*, 1963, No. 11, p. 1741.

151. Lebedev, S. L. A three-dimensional unsteady-state model of atmospheric convection cell with a cloud. Izv. Akad. Nauk SSSR, *FAO*, 1966, 2 (1): 14.

152. Lebedev, S. L. The calculation of stratus cloudiness and snowfalls under natural conditions with an artificial modification. In: Studies in cloud physics and active weather modification. Gidrometeoizdat, Leningrad, 1967, p. 112.

153. Lebedeva, N. V. Construction of a convection model and calculation of showery precipitation. *Trudy TsIP*, 1954, No. 31 (58):

p. 3.

154. Levin, L. M., and Sedunov, Yu. S. The turbulent-gravitational coalescence of cloud droplets. *Dokl. Akad. Nauk SSSR*, 1965, 164 (3): 552.

155. Levin, L. M., and Sedunov, Yu. S. A kinetic equation describing microphysical processes in clouds. *Dokl. Akad. Nauk SSSR*, 1966, 170 (2): 323.

156. Lenshin, V. T., Budilova, E. P. Estimating the intensity of atmospheric convection during the diagnosis and prediction of cloud-formation conditions. *Trudy GGO*, 1976, No. 372, p. 13.

157. Leonov, M. P., and Perelet, G. I. Active cloud modification during the cool half-year. Edited by V. G. Morachevskii. Gidrometeoizdat, Leningrad, 1967.

158. Litvinov, I. V. The structure of atmospheric precipitation. Gidrometeoizdat, Leningrad, 1974.

159. Lobanova, V. Ya. Features of the geographical distribution of cloudiness over the Northern Hemisphere. *Trudy NIIAK*, 1967, No. 44, p. 55.

160. Lushev, Yu. G., and Matveev, L. T. A numerical method of short-term prediction of cloudiness. *Dokl. Akad. Nauk SSSR*, 1966, 167 (5): 1042.

161. Lushev, Yu. G., and Matveev, L. T. A dynamic model for predicting stratiform cloudiness. Izv. Akad. Nauk SSSR, *FAO*, 1966, 2 (1): 3.

162. Lushev, Yu. G., and Matveev, L. T. The formation and structure of radiation fogs. Izv. Akad. Nauk SSSR, *FAO*, 1967, 3 (5): 459.

163. Mazin, I. P. Toward a theory of the formation of particle-size spectra of clouds and precipitation. *Trudy TsAO*, 1965, No. 64, p. 57.

164. Mazin, I. P. The relation between supersaturation fluctuations in clouds and fluctuations of temperature and vertical flows. *Trudy TsAO*, 1967, No. 79, p. 3.

165. Mazin, I.P., and Smirnov, V. I. Toward a theory of the formation of a cloud-droplet-size spectrum for stochastic condensation. *Trudy TsAO*, 1969, No. 89, p. 92.

166. Mazurin, N. I. Fields of nonconvective clouds and precipitation zones connected with them. *Trudy VNMS*, 1963, 5: 29.

167. Malkevich, M. S., and Polovina, I. P. The statistical characteristics of the vertical structure of stratus cloudiness and the meteorological elements. Izv. Akad. Nauk SSSR, *FAO,* 1967, 3 (4): 375.

168. Marchuk, G. I. A theoretical model of weather forecasting with a short forecasting period. *Izv. Akad. Nauk SSSR, Ser. Geofiz.,* 1964, No. 5, p. 754.

169. Marchuk, G. I., Dymnikov, V. P., Galin, V. Ya., et al. A hydrodynamic model of the general circulation of the atmosphere and ocean (implementation methods). Izdatel'stvo VTs SO Akad. Nauk SSSR, Novosibirsk, 1975.

170. Marchuk, G. I., Dynmikov, V. P., Lykosov, V. N., et al. A global model of the general circulation of the atmosphere. Izv. Akad. Nauk SSSR ,*FAO,* 1979, 15 (5): 467.

171. Marchuk, G. I., Kontarev, G. R., and Rivin, G. S. Short--term weather forecasting for a limited territory using general equations. Izv. Akad. Nauk SSSR, *FAO,* 1967, 3 (11): 1166.

172. Marchuk, G. I., Kurbatkin, G. P. But, I. V., et al. An operational quasi-geostrophic method of short-term weather forecasting for five levels of the atmosphere. Izv. Akad. Nauk SSSR, *FAO,* 1965, 1 (2): 129.

173. Marchuk, G. I., and Musaelyan, Sh. A. Calculating variations of the total radiant-energy flux for purposes of long-term prediction of large-scale meteorological fields. *Meteorologiya i Gidrologiya,* 1974, No. 8, p. 10.

174. Marchuk, G. I., Penenko, V. V., Aloyan, A. E., and Lazriev, G. L. The numerical modeling of an urban microclimate. *Meteorologiya i Gidrologiya,* 1979, No. 8, p. 5.

175. Matveev, L. T. The wind-velocity distribution in the boundary layer of the atmosphere and determination of the turbulent-mixing parameters. *Meteorologiya i Gidrologiya,* 1949, No. 3, p. 20.

176. Matveev, L. T. Vertical currents in the boundary layer of the atmosphere. *Izv. Akad. Nauk SSSR, Ser. Geofiz.,* 1955, No. 5, p. 453.

177. Matveev, L. T. Vertical currents in the boundary layer of the atmosphere and their relation to the heights of upper and lower cloud boundaries. *Trudy VGMF SA,* 1955, No. 11, p. 3.

178. Matveev, L. T. The qualitative analysis of eddy-formation conditions in the atmosphere and some associated results.

Meteorologiya i Gidrologiya, 1956, No. 4, p. 28.

179. Matveev, L. T. The role of vertical currents and turbulent mixing in the formation and evolution of stratiform cloudiness. In: Studies of clouds, precipitation, and thunderstorm electricity. Gidrometeoizdat, Leningrad, 1957, p. 37.

180. Matveev, L. T. The quantitative characteristics of turbulent mixing in the upper troposphere and lower stratosphere. *Izv. Akad. Nauk SSSR, Ser. Geofiz.,* 1958, No. 7, p. 927.

181. Matveev, L. T. Some problems in the theory of the formation and evolution of stratiform cloudiness. *Trudy AANII,* 1959, 228 (1): 14.

182. Matveev, L. T. Toward a theory of the formation of clouds and aircraft vapor trails under the influence of mixing. *Meteorologiya i Gidrologiya,* 1959, No. 3, p. 3.

183. Matveev, L. T. Problems of formation and prediction of clouds and aircraft vapor trails under the influence of mixing. *Trudy LVVIA im. A. F. Mozhaiskogo,* 1959, No. 273, p. 3.

184. Matveev, L. T. The use of empirical action functions to predict cloud boundaries. *Meteorologiya i Gidrologiya,* 1960, No.9, p. 11.

185. Matveev, L. T. The dependence of the turbulence coefficient on height in the ground layer of the atmosphere. *Izv. Akad. Nauk SSSR, Ser. Geofiz.,* 1960, No. 1, p. 83.

186. Matveev, L. T. The conditions of formation and evolution of clouds under the influence of vertical currents and turbulent mixing. *Izv. Akad. Nauk SSSR, Ser. Geofiz.,* 1961, No. 1, p. 130.

187. Matveev, L. T. The conditions of formation and evolution of clouds. In: Studies of clouds, precipitation, and thunderstorm electricity. Izdatel'stvo Akad. Nauk SSSR, Moscow, 1961, p. 161.

188. Matveev, L. T. Calculating the upper-boundary height and vertical depth of stratiform clouds. *Meteorologiya i Gidrologiya,* 1962, No. 8, p. 3.

189. Matveev, L. T. A study of stratiform clouds as applied to flight conditions in them. In: Material of the scientific conference on aviation meteorology. Gidrometeoizdat, Leningrad, 1963, p. 82.

190. Matveev, L. T. Cloud inversions of dynamic origin and lifted inversions in fog. *Trudy VNMS,* 1963, 5: 10.

191. Matveev, L. T. A method for calculating the height of an upper cloud boundary and the required accuracy in determining fluxes of infrared radiation with the aid of satellites. *Trudy GGO*, 1964, No. 166, p. 214.

192. Matveev, L. T. Fundamentals of general meteorology: Physics of the atmosphere. Gidrometeoizdat, Leningrad, 1965 (English translation published by Israel Program for Scientific Translations (Keter), Jerusalem, 1967).

193. Matveev, L. T. The formation dynamics and the prediction of stratiform cloud systems. In: Dynamics of large-scale atmospheric processes, Nauka, Moscow, 1967, p. 279.

194. Matveev, L. T. The accuracy of air temperature and humidity measurements in the free atmosphere. *Meteorologiya i Gidrologiya*, 1967, No. 6, p. 21.

195. Matveev, L. T. The causes of cloud formation. *Meteorologiya i Gidrologiya*, 1978, No. 8, p. 25.

196. Matveev, L. T. Features of the meteorological regime of a large city. *Meteorologiya i Gidrologiya*, 1979, No. 5, p. 22.

197. Matveev, L. T., and Kozharin, V. S. The role of turbulent mixing in forming the structure of stratiform clouds. *Izv. Akad. Nauk SSSR, Ser. Geofiz.*, 1956, No. 11, p. 1338.

198. Matveev, L. T., and Kozlov, V. N. The formation of fields of humidity and cloudiness in the boundary layer of the atmosphere. *Trudy LGMI*, 1974, No. 51, p. 50.

199. Matveev, L. T., and Ponomareva, O. V. The conditions of formation of low cloudiness and fog close to fronts. *Meteorologiya i Gidrologiya*, 1978, No. 5, p. 33.

200. Matveev, L. T., Sokolov, E. M., and Budovyi, V. D. The statistical relationship between the vertical transparency of the atmosphere and the meteorological parameters. *Meteorologiya i Gidrologiya*, 1975, No. 1, p. 28.

201. Matveev, L. T., and Soldatenko, S. A. Toward a theory of the formation and prediction of evaporation fogs. *Meteorologiya i Gidrologiya*, 1977, No. 2, p. 24.

202. Matveev, L. T., and Soldatenko, S. A. Toward a theory of advection fogs. Izv. Akad. Nauk SSSR, *FAO*, 1978, 14 (2): 149.

203. Matveev, L. T., Tsitsenko, G. V., and Vasil'ev, V. A. The statistical-physical analysis of cloudiness. In: The application

of statistical methods in meteorology. Edited by G. V. Gruz and N. N. Aksarin. Gidrometeoizdat, Moscow, 1978, p. 139.

204. Matveev, Yu. L. On cloud-water-content distribution functions. *Trudy VNIIGMI - MTsD,* 1979, No. 65, p. 73.

205. Matveev, Yu. L. The zonal cloudiness field of the Southern Hemisphere according to satellite data. *Trudy VNIIGMI - MTsD,* 1980, No. 73, p. 89.

206. Matveev, Yu. L., and Soldatenko, S. A. A model of stationary-front cloudiness. *Meteorologiya i Gidrologiya,* 1981, No. 2, p. 40.

207. Matveev, Yu. L., and Soldatenko, S. A. The moisture content of a vertical air column. *Trudy VNIIGMI - MTsD,* 1981, No. 94, p. 39.

208. Mason, B. J. The physics of clouds, Second Edition, Clarendon Press, Oxford, 1971.

209. Meleshko, V. P., and Shvets, M. E. Prediction of the dew-point deficit. *Trudy GGO,* 1960, No. 114, p. 3.

210. Meleshko, V. P., Shneerov, B. E., Dmitrieva-Arrago, L. R., and Shvets, M. E. A hydrodynamic three-level model of the general circulation of the atmosphere. *Meteorologiya i Gidrologiya,* 1979, No. 6, p. 21.

211. Melkaya, I. Yu. A steady-state model of radiation fog. *Izv. Akad. Nauk SSSR, FAO,* 1968, 4 (2): 220.

212. Mel'nichuk, Yu. V., and Chernikov, A. A. An operational method of detecting turbulence in clouds and precipitation. *Trudy TsAO,* 1973, No. 110, p. 3.

213. Merkulovich, V. M., and Stepanov, A. S. Effects of hygroscopicity and surface-tension forces in the condensation growth of cloud droplets with turbulence. Izv. Akad. Nauk SSSR, *FAO,* 1977, 13 (2): 163.

214. Minervin, V. E. The water content of clouds over the European USSR. *Trudy VNMS,* 1963, 5: 64.

215. Minervin, V. E. Water reserves of clouds. *Trudy TsAO,* 1965, No. 64, p. 44.

216. Minervin, V. E., Mazin, I. P., and Burkovskaya, S. N. Some new data on cloud water contents. *Trudy TsAO,* 1958, No. 19, p. 3.

217. Mitnik, L. M. Determination of the effective temperature of

droplet cloud formations according to the thermal radiation of the atmosphere in the shf range. *Trudy Gidromettsentra SSSR*, 1974, No. 148, p. 115.

218. Mogila, L. N., and Polovina, I. P. Some results of a study of frontal cloud layers not yielding precipitation. *Trudy UkrNIGMI*, 1970, No. 92, p. 103.

219. Molchanov, P. A. An aerological portrayal of the structure of the atmosphere for passage of warm and cold air masses. *Meteorologiya i Gidrologiya*, 1937, No. 4-5, p. 83.

220. Molchanov, P. A. Aerology. Gidrometeoizdat, Leningrad-Moscow, 1938.

221. Monin, A. S. Weather forecasting as a physics problem. Nauka, Moscow, 1969.

222. Monin, A. S., and Obukhov, A. M. The fundamental regularities of turbulent mixing of the atmosphere. *Trudy Geofiz. Inst. Akad. Nauk SSSR*, 1954, No. 24 (151), p. 163.

223. Morachevskii, V. G. Thermodynamics of phase transitions of water in a polluted atmosphere. *Mezhvozovskii Sbornik Trudov*, 1981, No. 73, p. 53 (LPI).

224. Morachevskii, V. G., Nikandrova, G. T., and Chikirova, G. A. The variability of the characteristics of ground radiation fogs. *Trudy GGO*, 1971, No. 278, p. 106.

225. Mullamaa, Yu-A. R., Kuusk, A. D., Okhvril', Kh. A., et al. Cloudiness and radiation. Izdatel'stvo Akad. Nauk ESSR, Tartu, 1975.

226. Musaelyan, Sh. A. Climatic data on the earth's cloud cover and the possibility of using them to solve problems in synoptic meteorology. *Trudy Gidromettsentra SSSR*, 1968, No. 30, p. 68.

227. Musaelyan, Sh. A. The nature of some superlong-term atmospheric processes. Gidrometeoizdat, Leningrad, 1978.

227a. Nadezhina, E. D., and Simonov, V. V. The formation and transformation of advection fogs. *Trudy GGO*, 1965, No. 167, p. 59.

228. Neizvestnyi, A. I., and Onishchenko, L. I. The experimental determination of the condensation coefficient for distilled water. Izv. Akad. Nauk SSSR, *FAO*, 1979, 15 (10): 1052.

229. Nigmatulin, R. I. Use of the methods of continuum mechanics to describe multiphase media. *Prikl. Matem. Mekh.*, 1970, 34 (6):

1097.

230. Nikandrov, V. Ya. The artificial modification of clouds and fogs. Gidrometeoizdat, Leningrad, 1959.

231. Obukhov, A. M., and Basharinov, A. E. A study of the atmosphere on the basis of *Cosmos 243* data on radiothermal self-radiation. *Kosmich. Issled.*, 1971, 9 (1): 66.

232. Obukhov, A. M., and Golitsyn, G. S. Estimates of the lower boundary and depth of the cloud layer on Venus. *Kosmich. Issled.*, 1968, 6 (5): 759.

233. Ovsyanikov, V. V. A hydrodynamic method for predicting humidity, cloudiness, and precipitation. *Trudy sIP*, 1961, No. 81, p. 92.

234. Onikul, R. I. The formation of Irkuk fogs. *Trudy GGO*, 1962, No. 127, p. 69.

235. Optical densities of clouds. A. L. Kosarev, I. P. Mazin, A. N. Nevzorov, and V. F. Shugaev. *Trudy TsAO*, 1976, No. 124.

236. Orlova, E. M. The short-term prediction of precipitation. Edited by B. D. Uspenskii. Gidrometeoizdat, Leningrad, 1979.

237. Osipova, G. I. Some characteristics of fogs and the northwestern part of the European USSR. *Trudy GGO*, 1959, No. 72, p. 66.

238. Osipova, G. I. The development of cloudiness in connection with the moisture content of the atmosphere and moisture transport in it. *Trudy GGO*, 1971, No. 262, p. 163.

239. Panchev, S., and Atanasov, D. The structure of the baroclinic Ekman planetary boundary layer. *Meteorologiya i Gidrologiya*, 1979, No. 5, p. 28.

240. Pastushkov, R. S. The development of cumulus clouds in an atmosphere with vertical wind shear. *Meteorologiya i Gidrologiya*, 1969, No. 4, p. 26.

241. Penenko, V. V., Aloyan, E. A., and Lazriev, G. L. A numerical model of local atmospheric processes. *Meteorologiya i Gidrologiya*, 1979, No. 4, p. 24.

242. Peskov, B. E. The role of ordered vertical motions in the friction layer during the development of dense convective clouds and thunderstorms. *Trudy TsIP*, 1964, No. 136, p. 61.

243, Peskov, B.E. A study of the meteorological conditions of thunderstorm formation by means of a systematic discrimant analysis.

Trudy Gidromettsentra SSSR, 1977, No. 185, p. 24.

244. Petrenko, N. V. Some improvements in the method of predicting radiation fog and visibility. *Trudy TsIP*, 1966, No. 157, p. 12.

245. Petrenchuk, O. P. Experimental studies of the atmospheric aerosol. Gidrometeoizdat, Leningrad, 1979.

246. Petrichenko, I. A. The prediction of precipitation with the aid of data on cloud water content. *Meteorologiya i Gidrologiya*, 1971, No. 10, p. 93.

247. Petrichenko, I. A., and Galakhova, T. A. Quantitative methods of predicting steady and showery precipitation. In: Fifty years of the center for hydrometeorological forecasts. Gidrometeoizdat, Leningrad, 1979, p. 94.

248. Petrova, L. V., and Feigel'son, E. M. The role of radiation in cloud development. *Izv. Akad. Nauk SSSR, Ser. Geofiz.*, 1964, No. 8, p. 1247.

249. Petrova, L. I., and Ivanov, V. N. Some parameters of cellular convection according to meteorological-satellite and high-mast observations. *Meteorologiya i Gidrologiya*, 1973, No. 6, p. 26.

250. Pirnach, A. M. The effect of ice-particle aggregation on the development of precipitation in mixed stratiform clouds. *Trudy UkrNIGMI*, 1979, No. 170, p. 32.

251. Podurskaya, V. N. The water-content distribution in stratus and stratocumulus clouds during the cold season over the Ukraine. *Trudy UkrNIGMI*, 1971, No. 106, p. 59.

252. Podgurskaya, V. N. The conditions preceding the appearance of St-Sc clouds during the cold season over the Ukraine. *Trudy UkrNIGMI*, 1976, No. 144, p. 84.

253. Polovina, I. P. Modification of internal stratiform clouds. Gidrometeoizdat, Leningrad, 1971.

254. Ponomarenko, I. N., and Zabolotskaya, T. N. The water content of frontal clouds over the Ukrainian steppes according to experimental data. *Trudy UkrNIGMI*, 1965, No. 48, p. 52.

255. Ponomarenko, I. N., and Koshenko, A. M. Cloud systems of fronts as related to the distribution of vertical motions. Izv. Akad. Nauk SSSR, *FAO*, 1968, 4 (9): 930.

256. Ponomarenko, I. N., Koshenko, A. M., and Gaidai, L. I. Phase states of clouds in frontal zones over the Ukraine. *Trudy UkrNIGMI*,

1966, No. 56, p. 67.

257. Popov, S. M. Some statistical characteristics of the vertical structure of temperature and humidity fields. Izv. Akad. Nauk SSSR, *FAO*, 1965, 1 (1): 18.

258. Prikhod'ko, G. F. Artificial precipitation from convective clouds. Gidrometeoizdat, Leningrad, 1968.

259. The prediction of heavy precipitation with forecasting periods of up to 36 hr. *Trudy Gidromettsentra SSSR*, 1976, No. 190.

260. Prokh, L. Z. A characterization of fogs of the Ukraine. *Trudy UkrNIGMI*, 1966, No. 55, p. 43.

261. Prokh, L. Z. The microstructure of fogs at Kiev. *Trudy UkrNIGMI*, 1968, No. 70, p. 136.

262. Pchelko, I. G. Aviation meteorology. Gidrometeoizdat, Leningrad, 1963.

263. Reshetov, G. D. Convective cloudiness at fronts. *Trudy Gidromettsentra SSSR*, 1967, No. 7, p. 41.

264. Reshetov, G. D. A method of predicting hail so as to promote flight safety. *Trudy Gidromettsentra SSSR*, 1978, No. 201, p. 3.

265. Rozenberg, G. V. The covering phase of an aerosol and its role in the atmosphere. Izv. Akad. Nauk SSSR, *FAO*, 1979, 15 (10): 1094.

266. Romanov, N. N. Cloud characteristics according to large-volume observations with a Tu-104 aircraft. *Trudy NVMS*, 1963, 5: 37.

267. Romov, A. I. Vertical motions at atmospheric fronts. Izv. Akad. Nauk SSSR, *FAO*, 1966, 2 (6): 561.

268. Romov, A. I. A numerical study of the dynamics of atmospheric fronts. *Trudy UkrNIGMI*, 1968, No. 75, p. 3.

269. Sadokov, V. P. The dynamics of vertical motions in a stratified cloudy atmosphere. *Meteorologiya i Gidrologiya*, 1971, No. 10, p. 12.

270. Sadokov, V. P. A hydrodynamic numerical model for describing large-scale atmospheric processes. *Trudy Gidromettsentra SSSR*, 1977, No. 192, p. 22.

271. Sedunov, Yu. S. The fine structure of clouds and its role in forming the cloud-particle spectrum. Izv. Akad. Nauk SSSR, *FAO*, 1965, 1 (7): 722.

272. Sedunov, Yu. S. The physics of the formation of the liquid-droplet phase in the atmosphere. Gidrometeoizdat, Leningrad, 1972.

273. Selezneva, E. S. Atmospheric aerosols (condensation nuclei). Gidrometeoizdat, Leningrad, 1966.

274. Selezneva, E. S. Some results of a study of internal clouds. In: Studies of clouds, precipitation, and thunderstorm electricity. Gidrometeoizdat, Leningrad, 1957, p. 7.

275. Smirnov, V. I. The rate of aerosol-particle growth by coalescence and condensation. *Trudy TsAO*, 1969, No. 92.

276. Smith, K. Principles of applied climatology. Halsted Press, 1975.

277. Snopkov, V. G. The total water-vapor content in the trade-wind zone of the Atlantic and its variability in time. *Meteorologiya i Gidrologiya*, 1971, No. 10, p. 103.

278. Solomatina, I. I. Condensation during the mixing of two air masses. *Trudy GGO*, 1957, No. 69, p. 45.

279. Sonechkin, D. M. Objective methods of cloud diagnosis. *Trudy Gidromettsentra SSSR*, 1968, No. 30, p. 39.

280. Spiridonova, Yu. V. Scales of cloud formations and the classification of cloudiness fields on the basis of satellite data. *Trudy VNIIGMI - MTsD*, 1976, No. 2, p. 63.

281. Handbook of statistical characteristics of the temperature and humidity in a free atmosphere. Edited by V. S. Komarov. Gidrometeoizdat, Moscow, 1972, 1975 (Part 1, 1972; Part 4, 1975).

282. Mediterranean cyclones in a cloudiness field. T. P. Popova, R. Runkanu, T. Tentser, and V. Sharov. Edited by T. P. Popova. Gidrometeoizdat, Leningrad, 1975.

283. Stepanenko, V. D. Radar in meteorology. Gidrometeoizdat, Leningrad, 1973.

284. Stepanov, A. S. The condensation growth of cloud droplets in a turbulent atmosphere. Izv. Akad. Nauk SSSR , *FAO*, 1975, 11 (1): 27.

285. Stepanov, A. S. The effect of turbulence on the cloud-droplet-size spectrum with condensation. Izv. Akad. Nauk SSSR, *FAO*, 1976, 12 (3): 281.

286. The stochastic structure of cloudiness and radiation fields.

Edited by Yu-R. R. Mullamaa. Izdatel'stvo Akad. Nauk ESSR, 1972.

287. Sulakvelidze, G. K. Rainstorms and hail. Gidrometeoizdat, Leningrad, 1967 (English translation published by Israel Program for Scientific Translations (Keter), Jerusalem, 1968).

288. Sulakvelidze, G. K., Glushkova, N. I., and Fedchenko, L. M. The prediction of hail, thunderstorms, and showery precipitation. Gidrometeoizdat, Leningrad, 1970.

289. Surmava, A. A. A quantitative study of the motion of atmospheric fronts with the aid of a two-layer model. Izv. Akad. Nauk SSSR, *FAO*, 1977, 13 (4): 347.

290. Tverskaya, N. P. The statistical characteristics of the vertical humidity profile for cumuli . *Meteorologiya i Gidrologiya*, 1974, No. 8, p. 56.

291. Tverskaya, N. P., and Sokolova, T. P. The statistical characteristics of the temperature and humidity profiles in the atmosphere on a day with convective cloudiness. *Trudy GGO*, 1975, No. 356, p. 45.

292. Timofeev, M. P. Toward a theory of advection fogs. *Izv. Akad. Nauk SSSR, Ser. Geofiz.*, 1955, No. 6, p. 514.

293. Timofeev, N. A. Calculation (according to the cloudiness) of the total solar radiation at the ocean surface over short time intervals. *Morskie Gidrofiz. Issled.*, 1970, No. 4, p. 168.

294. Timofeev, N. A. The vertical distribution of the absolute air humidity and moisture content in the atmosphere over oceans. *Meteorologiya i Gidrologiya*, 1979, No. 8, p. 55.

295. Titov, V. I. Atlas of climatic characteristics of cloudiness according to satellite data. Gidrometeoizdat, Moscow, 1980.

296. Titov, V. I., and Golovleva, V. N. Some quantitative characteristics of cloudiness in the Northern Hemisphere. *Trudy VNIIGMI MTsD*, 1976, No. 2. p. 28.

297. Titov, V. I., and Musaelyan, T. Sh. Features of the cloudiness distribution over the Northern Hemisphere in summer 1975. *Trudy VNIIGMI - MTsD*, 1979, No. 66, p. 32.

298. Tkachenko, A. V. The rate of variation of the characteristics of stratiform cloudiness. *Trudy UkrNIGMI*, 1966, No. 54, p. 19.

299. Tkachenko, A. V., and Kolezhuk, V. T. Fluctuations of the lower boundary of layer clouds (numerical experiment). *Trudy*

UkrNIGMI, 1974, No. 133, p. 43.

300. Tkachenko, A. V., and Levin, A. V. The rate and acceleration of height variations of layer clouds. *Trudy UkrNIGMI*, 1970, No. 92, p. 58.

301. Tkachenko, A. V., and Yukhimenko, E. I. Evaluating the effect of vertical motions and turbulent exchange on the rate of variation of layer-cloud heights. *Trudy UkrNIGMI*, 1969, No. 82, p. 16.

302. TROPEKS-72. Edited by M. A. Petrosyants et al. Gidrometeoizdat, Leningrad, 1974.

303. TROPEKS-74, Vol. 1, The atmosphere. Edited by M. A. Petrosyants et al. Gidrometeoizdat, Leningrad, 1976.

304. Fogs. M. E. Berlyand, P. A. Vorontsov, P. N. Krasikov, et al. Gidrometeoizdat, Leningrad, 1961.

305. Uspenskii, B. D. An operational numerical prediction of steady precipitation. *Meteorologiya i Gidrologiya*, 1965, No. 2, p. 3.

306. Uspenskii, B. D. The quantitative prediction of steady and showery precipitation. *Meteorologiya i Gidrologiya*, 1970, No. 1, p. 11.

307. Uspenskii, B. D., Mertsalov, A. N., Orlova, E. M., et al. A synoptic-hydrodynamic scheme for the quantitative prediction of steady and showery precipitation. *Trudy Gidromettsentra SSSR*, No. 157, p. 3.

308. Ushakova, T. V. The water balance of nimbostratus clouds. *Trudy LGMI*, 1972, No. 48, p. 76.

309. Fal'kovich, A. I. The modeling of a cloud ensemble. *Meteorologiya i Gidrologiya*, 1979, No. 8, p. 24.

310. Federov, E. K., and Mamina, E. F. The water balance of a cloud system. *Izv. Akad. Nauk SSSR, Ser. Geofiz.*, 1957, No. 5, p. 658.

311. Feigel'son, E. M. Radiative heat exchange and clouds. Gidrometeoizdat, Leningrad, 1970.

312. Feigel'son, E. M., and Frolova, N. G. The calculation of cloudiness. Izv. Akad. Nauk SSSR, *FAO*, 1965, 1 (3): 241.

313. Physics of the atmosphere and problems of climate. Edited by G. S. Golitsyn and A. M. Yaglom. Nauka, Moscow, 1980.

314. Khaman, K. The dynamics of stratiform clouds. *Meteorologiya i Gidrologiya,* 1972, No. 7, p. 100.

315. Khvorost'yanov, V. I. Inverse correlations between turbulence, vertical currents, and cloudiness in the boundary layer of the atmosphere. Izv. Akad. Nauk SSSR, *FAO,* 15 (8): 804.

316. Khrgian, A. Kh. The effect of the Ural range on cloudiness and precipitation. *Meteorologiya i Gidrologiya,* 1961, No. 3, p. 11.

317. Khrgian, A. Kh. Some properties of low stratiform cloudiness. *Trudy TsAO,* 1965, No. 64, p. 11.

318. Khrgian, A. Kh. The physical processes of cloud-layer formation. *Vestnik MGU, Ser. 3, Fizika i Astronomiya,* 1967, No. 1, p. 82.

319. Khrgian, A. Kh. The structure of stratiform clouds. Izv. Akad. Nauk SSSR, *FAO,* 1977, 13 (11): 1150.

320. Tsarev, A. M. The effect of a city on the fog regime. *Trudy GGO,* 1977, No. 387, p. 147.

321. Chapurskii, L. I. The spectral brightness characteristics of clouds, the atmosphere, and the underlying surface in the wavelength interval from 0.3 to 2.5 μm. *Trudy GGO,* 1966, No. 196, p. 110.

322. Chapurskii, L. I., and Andreeva, N. I. Teleindication of cloudiness in the spectral ranges 0.35-0.85 μm and 1.2-3.0 μm. *Meteorologiya i Gidrologiya,* 1978, No. 8, p. 41.

323. Shakina, N. P. Theoretical studies of atmospheric fronts and cyclones of the middle latitudes. *Meteorologiya i Gidrologiya,* 1978, No. 7, p. 108.

324. Shakina, N. P. Instability in the atmosphere. *Meteorologiya i Gidrologiya,* 1979, No. 6, p. 106.

325. Shvets, M. E. Water-vapor condensation in the atmosphere. *Izv. Akad. Nauk SSSR, Ser. Geofiz.,* 1955, No. 6, p. 547.

326. Shvets, M. E. Precomputation of the field of relative humidity. *Trudy GGO,* 1959, No. 81, p. 3.

327. Shvets, M. E., and Shneerov, B. E. Calculation of the main characteristics of an ensemble of cumuli in the tropics. *Trudy GGO,* 1976, No. 367, p. 15.

328. Shishkin, N. S. Clouds, precipitation, and thunderstorm

electricity. Gidrometeoizdat, Leningrad, 1964.

329. Shishkin, N. S., and Ponomarenko, S. I. Use of the slice method to predict the amount of showery precipitation. In: Inform. Sbornik No. 2. Gidrometeoizdat, Leningrad, 1972, p. 8.

330. Shmeter, S. M. The physics of convective clouds. Gidrometeoizdat, Leningrad, 1972.

331. Arakawa, A., and Schubert, W. Interaction of a cumulus cloud ensemble with a large-scale environment, Pt. 1. *J. Atm. Sci.*, 1974, 31: 674,

332. Asai, T. Three-dimensional features of thermal convection in a plane couette flow. *J. Met. Soc. Japan, Ser. 2*, 1970, 48 (1): 18.

333. Asai, T. Thermal instability of a sheer flow turning the direction with height. *J. Met. Soc. Japan, Ser. 2*, 1972, 50 (6): 525.

334. Barlett, J. T., and Jonas, P. R. On the dispersion of the size of droplets growing by condensation in turbulent clouds. *Q. J. Roy. Met. Soc.*, 1972, 98 (415): 150.

335. Benwell, G. R. R., and Timpson, M. S. Further work with the Bushby-Timpson 10-level model. *Q. J. Roy. Met. Soc., London*, 1968, 94: 12.

336. Bernhardt, K. Über den Einfluss der planetarischen Grenzschicht über Land und Meer auf die grossraumige atmosphärische Dynamik. *Z. Met.*, 1974, 24 (1-2): 17.

337. Benwell, G. R. R., Gadd, A. J., Keers, J. F., et al. The Bushby-Timpson 10-level model on a fine mesh. *Met. office, Sci. Paper*, 1971, 32: 59.

338. Betts, A. K. A composite cumulonimbus budget. *J. Atm. Sci.*, 1973, 30: 597.

339. Betts, A. K., and Silva Dias, M. F. Unsaturated downdraft thermodynamics in cumulonimbus. *J. Atm. Sci.*, 1979, 36: 1061.

340. Brown, J. M. Mesoscale unsaturated downdrafts driven by rainfall evaporation - a numerical study. *J. Atm. Sci.*, 1979, 36: 313.

341. Bushby, F. H. and Timpson, M. S. A 10-level atmospheric model and frontal rain. *Q. J. Roy. Met. Soc., London*, 1967, 93: 1.

342. Byers, H. R., and Braham, R. R. The thunderstorm. Washington, 1949.

343. Cheng, C. P., and House, R. A. Sensitivity of diagnosed convective fluxes to model assumptions. *J. Atm. Sci.*, 1980, 37: 774.

344. Cho, H. R. Contribution of cumulus cloud life-cycle effects to the large-scale heat and moisture budget equations. *J. Atm. Sci.*, 1977, 34: 87.

345. Cho, H. R., and Cheng, L. Parametrization of horizontal transport of vorticity by cumulus convection. *J. Atm. Sci.*, 1980, 37: 812.

346. Cho, H. R., Cheng, L., and Bloxam, R. M. The representation of cumulus cloud effects in the large-scale vorticity equation. *J. Atm. Sci.*, 1979, 36: 127.

347. Cho, H. R., and Ogura, Y. A relationship between the cloud activity and the low-level convergence as observed in Reed--Recker's composite easterly waves. *J. Atm. Sci.*, 1974, 31: 2058.

348. Clark, T. Numerical modelling of the dynamics and microphysics of warm cumulus convection. *J. Atm. Sci.*, 1973, 30 (5): 857.

349. Clouds and convection. GARP Publ. series, *WMO*, 1972, No. 8. p. 20.

350. Deardorff, J. W. On the entrainment rate of a stratocumulus-topped mixed layer in a strong inversion. *Q. J. Roy. Met. Soc.*, 1976, 102 (433): 563.

351. Deardorff, J. W. Cloud top entrainment instability. *J. Atm. Sci.*, 1980, 37: 131.

352. Döös, B. R. A theoretical analysis of lee wave clouds observed by *Tiros I. Tellus*, 1962, 14 (3): 301.

353. Einaudi, F., and Lalas, D. P. The propagation of acoustic-gravity waves in a moist atmosphere. *J. Atm. Sci.*, 1973, 30 (3): 365.

354. Einaudi, F., and Lalas, D. P. Some new properties of Kelvin-Helmholtz waves in an atmosphere with and without condensation effect. *J. Atm. Sci.*, 1974, 31 (8): 1995.

355. Fisher, E. L., and Caplan, P. An experiment in prediction of fog and stratus. *J. Atm. Sci.*, 1963, 20: 117.

356. Fletcher, N. H. The chemical physics of ice. Cambridge Univ. Press, London, 1970.

357. Foot, G. B., and Frankhauser, J. C. Airflow and moisture budget beneath a northeast Colorado hailstorm. *J. Appl. Met.*, 1973, 12: 1330.

358. Fraedich, K. A mass budget of an ensemble of transient cumulus clouds determined from direct cloud observations. *J. Atm. Sci.*, 1976, 33: 262.

359. Frank, W. M. The life cycles of GATE convective systems. *J. Atm. Sci.*, 1978, 35: 1256.

360. Gerrity, J. P. A physical-numerical model for the prediction of synoptic scale low cloudiness. *Mon. Wea. Rev.*, 1967, 95 (5): 261.

361. Gray, W. M. Cumulus convection and larger scale circulations. I. Broadscale and mesoscale consideration. *Mon. Wea. Rev.*, 1973, 101: 839.

362. Haman, K. On the influence of convective clouds on the large-scale stratification. *Tellus*, 1969, 21: 40.

363. Haman, K. E. On the motion of a three-dimensional quasi-steady convective storm in shear. *Mon. Wea. Rev.*, 1978, 106 (11): 1622.

364. Herman G., and Goody, R. Formation and persistence of summer time Arctic stratus clouds. *J. Atm. Sci.*, 1976, 33 (8): 1537.

365. Hill, G. E. On the orientation of cloud bands. *Tellus*, 1968, 20 (1): 132.

366. House, R. A. Structure and dynamics of a tropical squall-line system observed during GATE. *Mon. Wea. Rev.*, 1977, 105: 1540.

367. House, R. A., Cheng, C. P., Leary, C. A., and Gamache, J. F. Diagnosis of cloud mass and heat fluxes from radar and synoptic data. *J. Atm. Sci.*, 1980, 37: 754.

368. Hudlow, M. D. Mean rainfall patterns for the three phases of GATE. *J. Appl. Met.*, 1979, 18: 1656.

369. Hudson, H. R. On the relationship between horizontal moisture convergence and convective cloud formation. *J. Appl. Met.*, 1971, 10: 755.

370. Johnson, R. B. The role of synoptic scale precipitation downdrafts in cumulus and synoptic scale interactions. *J. Atm. Sci.*, 1976, 33: 1890.

371. Kessler, E. On the distribution and continuity of water substance in atmospheric circulation. *Met. Monogr., Amer. Met. Soc.,* 1969, 10: 755.

372. Konrad, T. G. The alignment of clear air convective cells. In: *Proc. Int. conf. of cloud physics,* Toronto, 1968, p. 539.

373. Krishnamurti, R. On the transition to turbulent convection. Pt. 1: The transition from two- to three-dimensional flow. *J. Fluid Mech.,* 1970, 42 (2): 295.

374. Krishnamurti, R. Some further study of the transition. *J. Fluid Mech.,* 1973, 60 (2): 285.

375. Krishnamurti, T. N., Ramanathes, Y., Pan, H. L., et al. Cumulus parametrization and rainfall rates. 1. *Mon. Wea. Rev.,* 1980, 108: 465.

376. Kuettner, J. P. Cloud bands in the Earth's atmosphere. Observations and theory. *Tellus,* 1971, 23 (4-5): 404.

377. Kuo, H. L. Further studies on the parametrization of the influence of cumulus convection on large-scale flow. *J. Atm. Sci.,* 1974, 31: 1231.

378. Lalas, D. P., and Einaudi, F. On the stability of a moist atmosphere in the presence of background wind. *J. Atm. Sci.,* 1973, 30 (7): 795.

379. Lalas, D. P., and Einaudi, F. On the correct use of the wet adiabatic lapse rate in stability criteria of a saturated atmosphere. *J. Appl. Met.,* 1974, 13 (2): 318.

380. Leary, C. A., and House, R. A. The contribution of mesoscale motions to the mass and heat fluxes of an intense tropical convective system. *J. Atm. Sci.,* 1980, 37: 784.

381. Lewis, J. M., Ogura, Y., and Gidel, L. Large-scale influences upon the generation of mesoscale disturbance. *Mon. Wea. Rev.,* 1974, 102: 545.

382. Lewis, W. Forecasting 700 mb dewpoint depression by a three-dimensional trajectory technique. *Mon. Wea. Rev.,* 1957, 85 (9): 297.

383. Lilly, D. K. The dynamical structure and evolution of thunderstorms and squall lines. *Ann. Rev. Earth Planet. Sci.,* 1979, 7: 117.

384. Lipps, F. B. Two-dimensional numerical experiments in thermal

convection with vertical shear. *J. Atm. Sci.*, 1971, 28 (1): 3.

385. London, J. A. Study of the atmospheric heat balance. New York Univ. final report, 1957, p. 122.

386. Lopez, R. E. Cumulus convection and larger scale circulation. Pt. 2: Cumulus and mesoscale interactions. *Mon. Wea. Rev.*, 1973, 101: 856.

387. Lopez, R. E. The lognormal distribution and cumulus cloud population. *Mon. Wea. Rev.*, 1977, 105: 865.

388. Manabe, S., Smagorinsky, J., Holloway, J. L., and Stone, H. M. Simulated climatology of a general circulation model with a hydrologic cycle. *Mon. Wea. Rev.*, 1970, 98 (3).

389. Manabe, S., Smagorinsky, J., and Strickler, R. F. Simulated climatology of a general circulation model with a hydrological cycle. *Mon. Wea. Rev.*, 1965, 93: 769.

390. Mason, B. J. The Physics of Clouds. 2nd Edition. Oxford Univ. Press, London, 1971.

391. Mason, B. J. Clouds, Rain and Rainmaking. 2nd Edition. London, Oxford Univ. Press, 1975.

392. Martin, D. W., and Suomi V. E. A satellite study of cloud clusters over the tropical North Atlantic Ocean. *Bull. Amer. Met. Soc.*, 1972, 53: 135.

393. Matsumoto, S., Ninomiya, K., and Akiyama, T. Cumulus activities in relation to the mesoscale convergence field. *J. Met. Soc. Japan*, 1967, 45: 292.

394. Matveev, L. T. On the formation and development of layer clouds. *Tellus*, 1964, 16 (2): 139.

395. Moncrieff, M. W., and Miller, M. J. The dynamics and simulation of tropical squall lines. *Q. J. Roy. Met. Soc.*, 1976, 102: 373.

396. Mossop, S. C. Concentrations of ice crystals in clouds. *Bull. Amer. Met. Soc.*, 1970, 51 (6).

397. Nelson, L. D. A numerical study on the initiation of warm rain. *J. Atm. Sci.*, 1971, 28: 752.

398. Ninomiya, K. Heat and water budget over the Japan Sea and the Japan Islands in winter season - with special emphasis on the relation among the supply from sea surface, the convection

transfer and heavy snowfall. *J. Met. Soc. Japan,* 1968, 46: 343.

399. Nitta, T. Observational determination of cloud mass flux distribution. *J. Atm. Sci.,* 1975, 32: 73.

400. Nicholls, J. M. Aircraft measurements in the GARP Atlantic Tropical Experiment. *Met. Mag.,* 1979, 108: 349.

401. Ogura, Y. The evolution of a moist convective element in a shallow conditionally unstable atmosphere: a numerical calculation. *J. Atm. Sci.,* 1963, 20: 610.

402. Ogura, Y., and Cho, H. R. Diagnostic determination of cumulus cloud populations from larger scale variables. *J. Atm. Sci.,* 1973, 30: 1276.

403. Ogura, Y., and Cho, H. R. On the interaction between the subcloud and the cloud layers in the tropical regions. *J. Atm. Sci.,* 1974, 31: 1276.

404. Ogura, Y., and Lion, M. T. The structure of a midlatitude squall line - a case study. *J. Atm. Sci.,* 1980, 37: 553.

405. Ogura, Y., and Phillips, N. A. A scale analysis of deep and shallow convection in the atmosphere. *J. Atm. Sci.,* 1962, 19: 173.

406. Ogura, Y., and Takahashi, T. Numerical simulation of the life cycle of a thunderstorm cell. *Mon. Wea. Rev.,* 1971, 99: 895.

407. Ogura, Y., and Takahashi, T. The development of warm rain in a cumulus model. *J. Atm. Sci.,* 1973, 30: 262.

408. Ocoyama, K. A theory of parametrization of cumulus convection. *J. Met. Soc. Japan,* 1971, 49: 744.

409. Palm, E. On the tendency towards hexagonal cells in steady convection. *J. Fluid Mech.,* 1960, 8: 183.

410. Parametrization on subgrid scale processes. GARP Publ. series, *WMO,* 1972, No. 8

411. Pruppacher, H. R. (ed.). Cloud dynamics. *Pure and Appl. Geophysics,* 1975, 113: 713.

412. Pruppacher, H. R., and Klett, J. D. Microphysics of clouds and precipitation. D. Reidel Publ. Co, Dordrecht-Boston-London, 1978.

413. Randoll, D. A. Entrainment into a stratocumulus layer with distrubuted radiative cooling. *J. Atm. Sci.,* 1980, 37: 148.

414. Reihl, H., and Malkus, J. S. On the heat balance in the equatorial through zone. *Geophysica,* 1958, 6: 503.

415. Report of the JOG study conference on climate models: performance, intercomparison and sensitivity studies. Edited by W. L. Gates. GARP publ. series , 1979, 22 (1):

416. Report of the JOG study conference on parametrization of extended cloudiness and radiation for climate models. Geneva, WMO, 1978.

417. Rogers, R. R., and Vali Gabor. Recent developments in meteorological physics. *Physics Reports, Amsterdam,* 1978, 48 (2)

418. Rosenkranz, P. W., Staelin, D. H., and Grody, N. C. Typhoon June (1975) viewed by scanning microwave spectrometer. *J. Geophys. Res.,* 1978, 83 (4): 1857.

419. Schubert, W. H., Wakefield, J. S., Steiner, E. J., and Cox, S. K. Marine stratocumulus convection. Pt. 1: Governing equations and horizontally homogeneous solutions. *J. Atm. Sci.,* 1979, 36: 1286.

420. Schubert, W. H., Wakefield, J. S., Steiner, E. J., and Cox, S. K. Marine stratocumulus convection. Pt. 2: Horizontally inhomogeneous solutions. *J. Atm. Sci.,* 1979, 36: 1308.

421. Smagorinsky, J. On the dynamical prediction of large-scale condensation bv numerical methods. *Geophys. Monogr., Washington,* 1960, No. 5, p. 71.

422. Smagorinsky, J., and Collins, G. O. On the numerical prediction of precipitation. *Mon. Wea. Rev.,* 1955, 83 (3): 53.

423. Somerville, R. C. J. Numerical simulation of small-scale thermal convection in the atmosphere. *Lect. Notes Phys.,* 1973, 19: 238.

424. Soong, S. T. Numerical simulation of warm rain development in an axisymmetric cloud model. *J. Atm. Sci.,* 1974, 31: 1262.

425. Soong, S. T., and Ogura, Y. A comparison between axisymmetric and slab-symmetric cumulus cloud models. *J. Atm. Sci.,* 1973, 30: 879.

426. Staelin, D. H., Cassel, A. L., Kunzi, F., et al. Microwave atmospheric temperature sounding: effects of clouds on *Nimbus-5* satellite data. *J. Atm. Sci.,* 1975, 32 (10): 1970.

427. Steiner, J. T. A three-dimensional model of cumulus development. *J. Atm. Sci.*, 1973, 30 (2): 414.

428. Takahashi, T. Numerical simulation of tropical showers. *J. Atm. Sci.*, 1974, 31 (1): 2190

429. Takahashi, T. Tropical showers in an axisymmetric cloud model. *J. Atm. Sci.*, 1975, 32 (7).

430. Takahashi, T. Warm rain, giant nuclei, and chemical balance - a numerical model. *J. Atm. Sci.*, 1976, 33 (2): 269.

431. Twomey, S. Influence of the aerosol on optical properties of clouds. In: [416], Appendix E.,

432. Toong, H. D., and Staelin, D. H. Passive microwave spectrum measurements of atmospheric water vapor and clouds. *J. Atm. Sci.*, 1970, 27 (5): 781.

433. Warner, J. The miscrostructure of cumulus cloud. Pt. 1: General feature of the droplet spectrum. *J. Atm. Sci.*, 1969, 26 (5): 1049.

434. Wilhelmson, R. The life cycle of a thunderstorm in three dimensions. *J. Atm. Sci.*, 1974, 31: 1629.

435. Wilhelmson, R., and Ogura, Y. The pressure perturbation and the numerical modelling of a cloud. *J. Atm. Sci.*, 1972, 29: 1295.

436. Williams, K. T., and Gray, W. M. Statistical analysis of satellite observed trade wind cloud clusters in the western North Pacific. *Tellus*, 1973, 25: 313.

437. Yanai, M., Esbensen, S., and Chu, J. Determination of bulk properties of tropical cloud clusters from large-scale heat and moisture budget. *J. Atm. Sci.*, 1973, 30: 611.

438. Zdunkowski, W. G., and Nielsen, B. C. A preliminary prediction analysis of radiation fog. *Pure and Appl. Geophysics*, 1969, 75: 2.

439. Zdunkowski, W. G., and Barr, A. E. A radiactive-conductive model for the prediction of radiation fog. *Boundary Layer Met.*, 1972, 3: 152.

440. Zipser, E. J. The role of organized unsaturated convective downdrafts in the structure and rapid decay of an equatorial disturbance. *J. Appl. Met.*, 1969, 8: 799.

441. Zipser, E. J. Mesoscale and convective-scale downdrafts as distinct components of squall-line circulation. *Mon. Wea. Rev.*, 1977, 105: 1568.

INDEX

ATMOSPHERIC SCIENCES LIBRARY

Already published

Atmospheric Turbulence and Air Pollution Modelling

Edited by F. T. M. Nieuwstadt and H. van Dop ISBN 90-277-1365-6